Aromatic Substitution by the S_{RN}1 Mechanism

Aromatic Substitution by the $S_{RN}1$ Mechanism

Roberto A. Rossi
Rita H. de Rossi

ACS Monograph 178

AMERICAN CHEMICAL SOCIETY
WASHINGTON, D. C. 1983

Library of Congress Cataloging in Publication Data

Rossi, Roberto A.
   Aromatic substitution by the $S_{RN}1$ mechanism.

   (ACS monograph; 178)

   Includes index.

   1. Substitution reactions. 2. Aromatic compounds.
   I. Rossi, Rita H. II. Title. III. Series.
QD281.S67R67   1983   547′.604593   82-22829
ISBN 0-8412-0648-1   ACMOAG   1983

ACS Monographs

Marjorie C. Caserio, *Series Editor*

FOREWORD

ACS Monograph Series was started by arrangement with the interallied Conference of Pure and Applied Chemistry, which met in London and Brussels in July 1919, when the American Chemical Society undertook the production and publication of Scientific and Technologic Monographs on chemical subjects. At the same time it was agreed that the National Research Council, in cooperation with the American Chemical Society and the American Physical Society, should undertake the production and publication of Critical Tables of Chemical and Physical Constants. The American Chemical Society and the National Research Council mutually agreed to care for these two fields of chemical progress.

The Council of the American Chemical Society, acting through its Committee on National Policy, appointed editors and associates to select authors of competent authority in their respective fields and to consider critically the manuscripts submitted. Since 1944 the Scientific and Technologic Monographs have been combined in the Series. The first Monograph appeared in 1921, and up to 1972, 168 treatises have enriched the Series.

These Monographs are intended to serve two principal purposes: first to make available to chemists a thorough treatment of a selected area in form usable by persons working in more or less unrelated fields to the end that they may correlate their own work with a larger area of physical science; secondly, to stimulate further research in the specific field treated. To implement this purpose the authors of Monographs give extended references to the literature.

ABOUT THE AUTHORS

ROBERTO ARTURO ROSSI, currently a professor in the Department of Organic Chemistry at the Facultad de Ciencias Quimicas, Universidad Nacional de Córdoba, graduated from the same in 1966 as Bioquimico and in 1968 as Doctor en Bioquimica. From 1970 to 1972 he did postdoctoral work with Joseph F. Bunnett at the University of California at Santa Cruz. Dr. Rossi is a research member in the Consejo Nacional de Investigaciones Cientificas y Técnicas in Argentina. His current research interests include the reaction of radicals with nucleophiles, organometallic chemistry, chemistry of radical anions, and electron transfer reactions.

RITA HOYOS DE ROSSI works as an associate professor in the Department of Organic Chemistry, Facultad de Ciencias Quimicas, Universidad Nacional de Córdoba. She graduated from the same as Bioquimica in 1966 and as Doctor en Bioquimica in 1969. She then did postdoctoral work with Joseph F. Bunnett and Claude F. Bernasconi from 1970 to 1972 at the University of California at Santa Cruz. She also is a research member in the Consejo Nacional de Investigaciones Cientificas y Técnicas. Her current research work involves the study of the mechanism of general base catalyzed reactions in nucleophilic aromatic substitution, as well as catalysis reactions by electron transfer.

To our daughter and son,
Gabriela and Enrique

CONTENTS

Preface, **xi**

1. Aromatic Nucleophilic Substitutions, **1**

   Mechanisms, **1**
   Steps Participating in the $S_{RN}1$ Mechanism, **7**

2. Nucleophiles Derived from the IVA Group of Elements, **11**

   Carbon Nucleophiles, **11**
   Silicon Nucleophiles, **50**
   Germanium Nucleophiles, **51**
   Tin Nucleophiles, **51**

3. Nucleophiles Derived from the VA Group of Elements, **59**

   Nitrogen Nucleophiles, **59**
   Phosphorus Nucleophiles, **67**
   Arsenic Nucleophiles, **72**
   Antimony Nucleophiles, **76**

4. Nucleophiles Derived from the VIA Group of Elements, **79**

   Oxygen Nucleophiles, **79**
   Sulfur Nucleophiles, **80**
   Selenium Nucleophiles, **92**
   Tellurium Nucleophiles, **95**

5. Participating Substrates, **101**

   Reaction of Benzene Derivatives in Ammonia, **101**
   Reactions of Polycyclic Hydrocarbons in Ammonia, **117**
   Reactions of Heterocyclic Compounds in Ammonia, **117**
   Comparative Reactivity in Ammonia, **123**
   Reaction in Other Solvents, **126**
   Reaction Stimulated by Electrodes, **135**

6. Molecular Orbital Considerations, **143**

   PMO Applied to $S_{RN}1$ Reactions, **143**
   *Coupling of Phenyl Radicals with Hydrocarbon-Derived
       Carbanions,* **149**
   Coupling of Aryl Radicals with Ketone Enolate Anions, **151**
   Systems with Other Low Lying Antibonding MOs, **154**

7. The Initiation Step, **161**

   Spontaneous Reactions, **161**
   Reactions Stimulated by Solvated Electrons, **163**
   Photostimulated Reactions, **170**
   Reactions Stimulated by Electrodes, **176**

8. Chain Propagation Steps, **187**

   Decomposition of a Radical Anion, **188**
   Coupling of an Aryl Radical with a Nucleophile, **192**
   Electron Transfer Reactions, **194**
   Competing Reactions, **205**
   Entrainment, **225**
   Inhibition of $S_{RN}1$ Reactions, **227**
   Comparative Reactivity, **233**

9. The Termination Step, **239**

   *Termination Steps that Depend on the $S_{RN}1$ Initiation
       Reaction,* **239**
   *Termination Steps Independent of the Nature of the
       Initiation Step,* **241**

10. Other Related Mechanisms, **249**

    Photonucleophilic Aromatic Substitution, **249**
    $S_{RN}1$ Reaction in Aliphatic Systems, **258**
    $S_{RN}1$ Reaction in Vinylic Systems, **273**
    Miscellaneous Reactions Involving Electron Transfer, **275**

Appendix: Reactions in Liquid Ammonia, **283**

Index, **289**

PREFACE

THE FIELD OF NUCLEOPHILIC AROMATIC SUBSTITUTION began with the discovery that aryl compounds with an appropriate leaving group and strong electron withdrawing groups in the o- or p- positions could react with nucleophiles under mild conditions by a two-step mechanism that involves the formation of a Meisenheimer complex.

For a long time it was generally believed that unactivated aromatic compounds could not react with nucleophiles unless submitted to drastic conditions such as high temperature and pressure. About 30 years ago it was found that unactivated aryl halides could be substituted through a benzyne intermediate and a great amount of work has been done in this area. Arynes can be generated by several means and their reactions are interesting routes to the synthesis of important compounds.

In 1970 Bunnett and Kim proposed a new mechanism for nucleophilic aromatic substitution of unactivated aryl compounds, and during the last 12 years a number of studies in the field have demonstrated the generality of the mechanism in regard to the variety of substrates and nucleophiles.

This monograph is intended to compile the information available regarding these nucleophilic aromatic substitution reactions. Literature through December 1980 is cited as well as some papers that appeared in 1981 and were made available to us through the courtesy of the authors. We tried to be objective in the presentation of the results. The explanations given are in some cases based on concepts firmly established, but in others the interpretation is speculative and attempts to stimulate more research in the area.

We are greatly indebted to Joseph F. Bunnett: he introduced us to the field, he has had many enlightening discussions with us, and he has had the courtesy to read this monograph and give us many valuable suggestions. In addition, this monograph could not have been finished without the help of Marjorie C. Caserio who took the task of correcting and proofreading the manuscript. We also thank Architect Tómas Vargas for doing the original drawings for the figures.

ROBERTO ROSSI

RITA HOYOS DE ROSSI
Universidad Nacional de Cordoba
Cordoba, Argentina

December 1982

Aromatic Nucleophilic Substitutions

Mechanisms

During the last three decades, and after the landmark review of Bunnett and Zahler (*1*), organic chemists have recognized that aromatic compounds can undergo nucleophilic substitution just as easily as they undergo electrophilic substitution. The available mechanisms vary greatly depending on the aromatic moiety, the nucleophile, and the reaction conditions. Generally these aromatic nucleophilic substitution mechanisms can be summarized as shown in Scheme I.

Intermediates **1–4**, after one or several reaction steps, lead to the substitution product, as in Reaction 5 (*2, 3*).

Although a one-step mechanism is formally possible, present evidence does not indicate a concerted one step mechanism. On the other hand, the literature contains abundant evidence for multistep mechanisms of nucleophilic aromatic substitutions.

Aromatic substrates bearing nitrogen as the leaving group, such as diazonium salts seem to be the only compounds able to form aryl cations **1**. Consequently, only **5** has been found to react by this mechanism (reaction 6) (*4, 5*).

Formation of aryl carbanions **2** as precursors of benzyne intermediates takes place in strongly basic solution. Although benzynes can be formed by several other means, these do not involve a carbanion intermediate. Reactions involving benzyne intermediates have the drawback that more than one product is obtained (Reactions 7, 8). However, their advantage is that they are quite general regarding the substituent in the aromatic moiety. Thus, nucleophilic substitution by this mechanism has been observed with substrates bearing both electron-withdrawing and electron-releasing substituents (*6, 7, 8*).

With substrates bearing an electron-withdrawing group, the most popular mechanism involves the formation of a σ-complex such as **3**.

0065–7719/83/0178–0001$06.00/1
© 1983 American Chemical Society

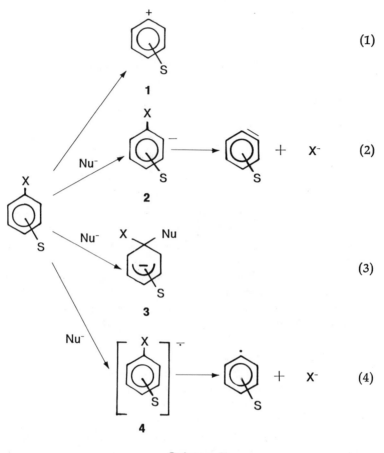

(1)

(2)

(3)

(4)

Scheme I

Historically, this was the first mechanism proposed for nucleophilic aromatic substitution, and the subject has been reviewed frequently (1, 4, 5, 9, 10). A restrictive condition for this mechanism seems to be the presence of an electron-withdrawing group in the aromatic ring. However, an unsubstituted aromatic substrate does react apparently by this mechanism under drastic conditions, such as high temperature, or in dipolar aprotic solvents (11).

Most studies have been carried out with substrates containing at least one activating group; the mechanism amy be different for unactivated substrates. While studying reactions of unactivated aryl halides with amide ion in liquid ammonia Bunnett and Kim (12) observed some unusual results that could not be explained by the proposed benzyne mechanism. The reaction under study was the formation of 5- and 6-

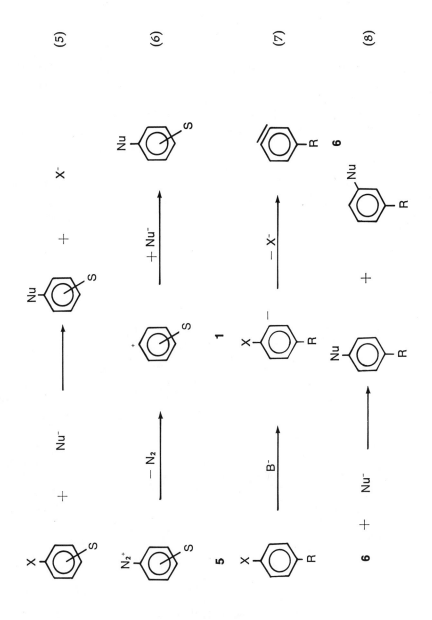

amino-1, 2, 4-trimethylbenzenes from 5- and 6-halo -1, 2, 4-trimethyl-benzenes, (**7** and **8**, Scheme II).

Both isomers **7** and **8** were expected to give the same ratio, **10:11**, of products, and the ratio was expected to be independent of the halogen (X) employed. This expectation was fulfilled with X being a chloro or bromo leaving group, where the product ratio of **10:11** was 1.46. However, with an iodo leaving group, the mixture of aminotrimethyl-benzenes obtained was always enriched in the product in which the amine group assumed the same position as the leaving iodide. That is, **7** gave more of **10** and **8** gave more of **11**.

A mechanism involving σ-complexes (as in Reaction 3) was considered unlikely at the low temperatures at which the reactions were carried out.

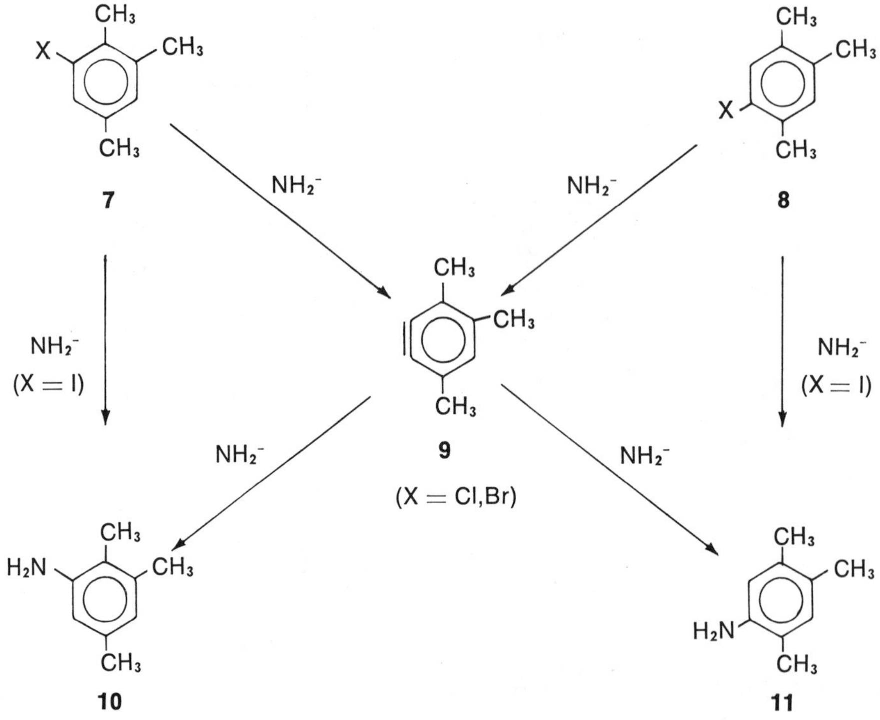

Scheme II

An important observation was that the reactions were catalyzed by solvated electrons from dissolution of alkali metals in liquid ammonia. This fact provided a clue to the mechanism, which was subsequently suggested to involve radicals and radical anions (Scheme III) (13).

$$ArI + e^- \rightarrow (ArI)^{\bar{\cdot}} \tag{9}$$

$$(ArI)^{\bar{\cdot}} \rightarrow Ar^{\cdot} + I^- \tag{10}$$

$$Ar^{\cdot} + NH_2^- \rightarrow (ArNH_2)^{\bar{\cdot}} \tag{11}$$

$$(ArNH_2)^{\bar{\cdot}} + ArI \rightarrow ArNH_2 + (ArI)^{\bar{\cdot}} \tag{12}$$

Scheme III

where ArI = 5- or 6-iodo-1,2,4-trimethylbenzene.

It was postulated that in the initiation step (Reaction 9), the aryl halide captures an electron. The radical anion formed dissociates to form an aryl radical and iodide ion (Reaction 10). This aryl radical reacts with amide ions to form a new radical anion (Reaction 11) which, by electron transfer reaction to the substrate aryl iodide, gives the substitution product and the radical anion of the substrate (Reaction 12). Reactions 10–12 are the propagation steps of the chain mechanism. Adding Reactions 10–12, we have the substitution process (Reaction 13). Although the overall reaction is a nucleophilic aromatic substitution, it involves radicals and radical anions as intermediates.

$$ArX + NH_2^- \rightarrow ArNH_2 + X^- \tag{13}$$

This mechanism was termed *Substitution Radical Nucleophilic, Unimolecular*, or $S_{RN}1$, due to its similarity to the S_N1 mechanism.

$$\frac{S_N1}{ArN_2^+ \rightarrow Ar^+ + N_2} \qquad \frac{S_{RN}1}{(ArX)^{\bar{\cdot}} \rightarrow Ar^{\cdot} + X^-}$$

A radical nucleophilic substitution mechanism, with reaction steps similar to Reactions 9–12, was suggested by Kornblum (14), and Russell (15), independently. For example, it was known that p-nitrobenzyl chloride reacts with the sodium salt of 2-nitropropane to yield 92% of the C-alkylated product and only 6% of the O-alkylated product (isolated as p-nitrobenzaldehyde) (Reaction 14). However, the usual reaction of this nucleophile with alkyl halides leads to O-alkylation (Reaction 15) (16).

Kornblum et al. found that the C-alkylation product not only depends on the p-nitro group, but also on the leaving group. In addition, the C-alkylation reaction is inhibited with very good electron acceptors,

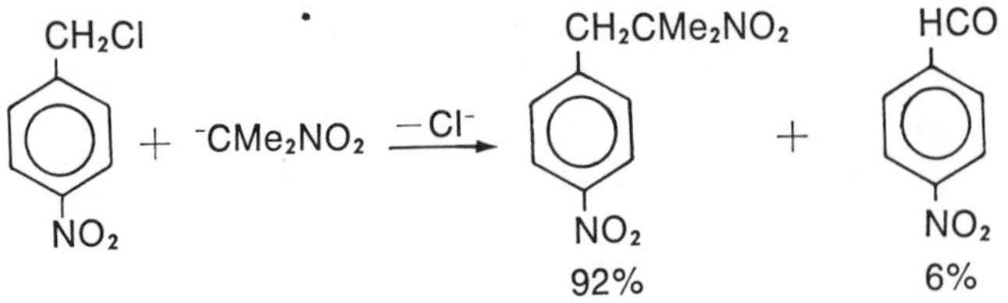

$$\text{(14)}$$

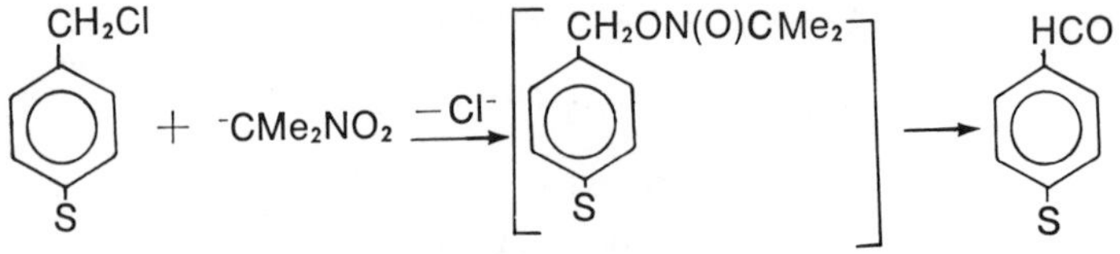

$$\text{(15)}$$

S = CN, CF₃, N⁺Me₃Cl, MeCO, Me, Br

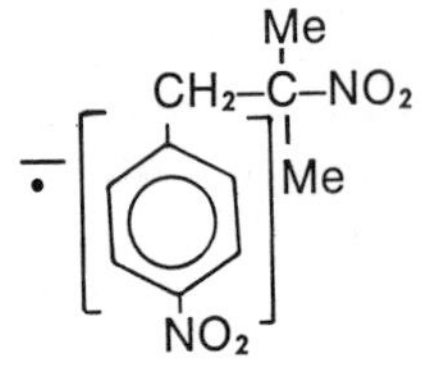

12

whereas the O-alkylation reaction becomes relatively important. Thus, in the presence of 0.2 equivalent of p-dinitrobenzene the yield of C-alkylation decreased to 6%, but the O-alkylation increased to 88% yield (*17*).

This evidence suggested that O-alkylation of the sodium salt of nitroalkanes is actually a direct nucleophilic displacement of chloride from the alkylating agent by the oxygen of the nitroparaffin anion. Also, an alternative mechanism that leads to the C-alkylation product in the p-nitrobenzyl series competes with the direct displacement mechanism. The competing mechanism proposed involves radical and radical anions as intermediates and is depicted in Reactions 16–19.

Based on a different approach, the same mechanism was proposed concurrently by Russell and Danen (*15*). They showed that the coupling reaction between the 2-nitro-2-propyl anion and p-nitrobenzyl chloride, or 2-chloro-2-nitropropane, is catalyzed by light, and the radical anion of the p-nitrobenzyl-2-nitro-2-propyl coupling product (**12**) was detected by electron spin resonance (ESR) spectroscopy in ethanol or dimethylformamide solution.

The subject of $S_{RN}1$ reactions at aliphatic sites has been reviewed by Kornblum (*16*), and others (*18*). Thus we will deal briefly with some

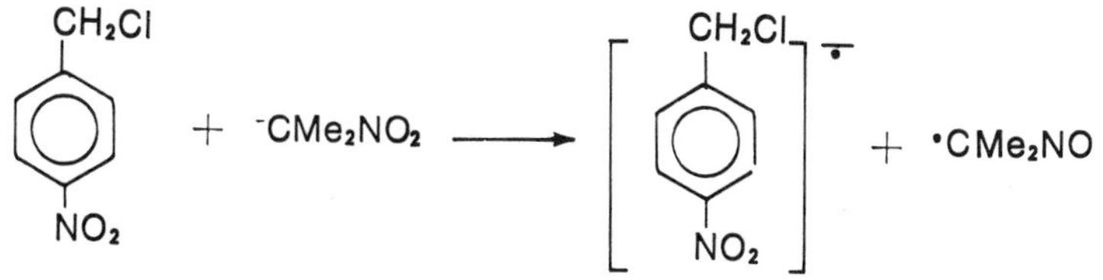

$$\text{(16)}$$

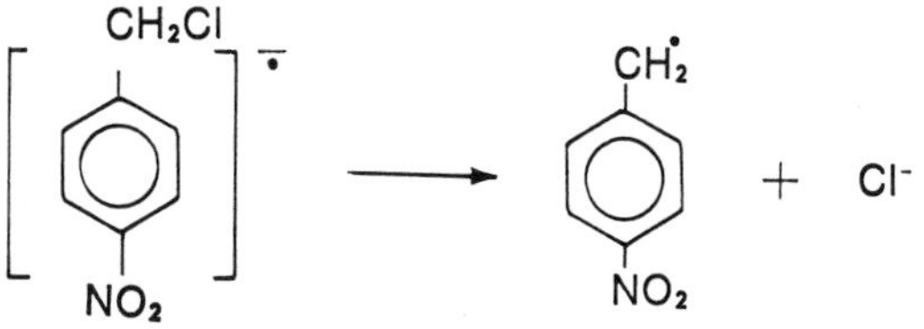

$$\text{(17)}$$

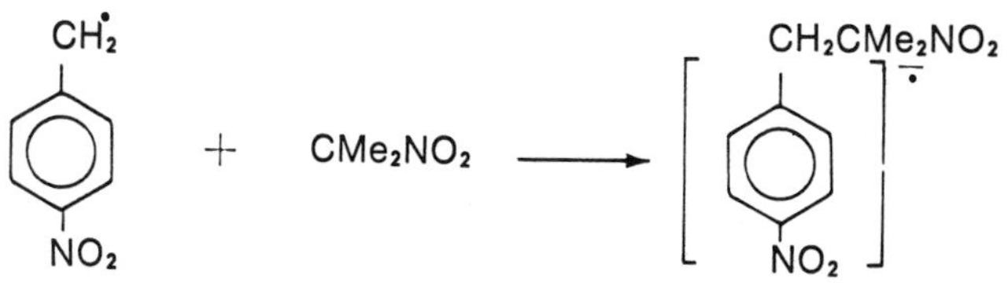

$$\text{(18)}$$

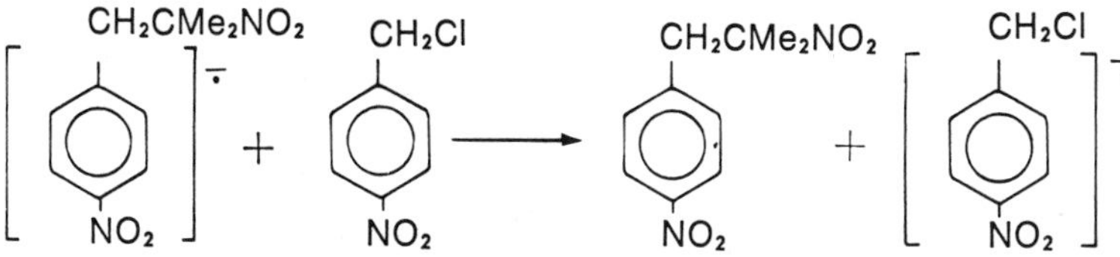

$$\text{(19)}$$

aspects of the mechanism in Chapter 10. The discussion hereafter deals mainly with $S_{RN}1$ reactions at aromatic sites.

Steps in the $S_{RN}1$ Mechanism

The main steps that comprise the $S_{RN}1$ mechanism are the initiation step, the chain propagation steps, and the termination steps (*see* Scheme IV).

The first reactive intermediate of this mechanism, a radical anion, is formed when an aromatic substrate with an appropriate nucleofugal group receives an electron (Reaction 20). This occurs either by reaction with a solvated electron from dissolution of alkali metals in liquid ammonia, or from a cathode, or by electron transfer from another radical anion, or by some other chemical reaction.

In Step 21 of the propagation cycle, the radical anion dissociates to form an aryl radical and the anion of the nucleofugal group. If the substrate ArX is a cation, the nucleofugal group leaves as a neutral species.

In Step 22 the aryl radical reacts with the nucleophile to form a new σ-bond and a new radical anion. Step 23 is an electron transfer of the radical anion to the substrate, which completes the chain propagation sequence. Termination steps also occur, and they depend on the reaction conditions.

Figure 1 is another representation of the propagation steps of Scheme IV. In the inner circle are the starting materials, ArX and the nucleophile Z^-. In the middle are the intermediates that participate in this reaction, aryl radicals, radical anions derived from the substrate, and the substitution product. In the outer circle are the products.

A reaction that produces one of the three reactive intermediates will start the reaction, provided that the sequence of reactions 21–23 occurs. Conversely, a reaction that interrupts the sequence by destroying any one of the intermediates will stop the cycle. If the central circle (where the reactive intermediates are) is subtracted, a substitution reaction occurs (Reaction 24).

initiation step $\quad\quad\quad\quad ArX + e^- \rightarrow (ArX)^-$ $\quad\quad\quad\quad\quad\quad$ (20)

chain propagation $\quad\quad\quad (ArX)^- \rightarrow Ar^{\cdot} + X^-$ $\quad\quad\quad\quad\quad\quad$ (21)

$$Ar^{\cdot} + Nu^- \rightarrow (ArNu)^- \quad\quad\quad\quad (22)$$

$$(ArNu)^- + ArX \rightarrow ArNu + (ArX)^- \quad\quad\quad (23)$$

Scheme IV

$$ArX + Z^- \rightarrow ArZ + X^- \quad\quad\quad\quad (24)$$

The subject of radical nucleophilic aromatic substitution has been reviewed (18–20). However, the continuously growing number of reported results calls for a new review, and this book is intended to report

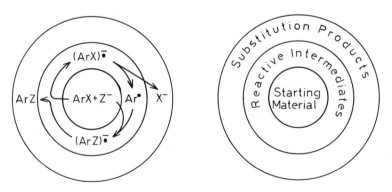

Figure 1. Schematic representation of the propagation steps of the $S_{RN}1$ mechanism.

the body of results in a comprehensive way and also to give some new insight into the mechanism.

Possibly some other mechanism or mechanisms could better explain some of the results summarized in this book. We include here all the examples that react in the way originally proposed by Kim and Bunnett, because, at this time, the experimental data are best interpreted in terms of the Kim–Bunnett mechanism.

Chapters 2–4 deal with the nucleophiles that undergo this reaction, arranged according to the periodic group to which the atom that forms the carbon–nucleophile bond belongs.

Chapter 5 deals with the substrates that participate. Because most of the $S_{RN}1$ reactions reported have been carried out in liquid ammonia, results belonging to this solvent form the main body of this chapter. The reactions of substrates derived from benzene with one leaving group, and those with two, have been treated separately. Polycyclic hydrocarbons and heterocyclic compounds are also dealt with in separate sections. Limited data exist in regard to the comparative reactivity of substrates, and these are also summarized in Chapter 5.

Reactions carried out in other solvents are grouped in a separate section within Chapter 5. Reactions stimulated by electrodes are also included here because they present some particular features not encountered in reactions promoted by chemical means. Some aspects of this subject were reviewed previously (21).

Some repetition about some of the results reported is unavoidable, because of the necessity to discuss the reactions from the standpoint of the reacting nucleophiles, and then again from that of the aromatic substrate. Thus, some reactions are mentioned twice. However, we hope that we have not overlooked any pertinent data.

In Chapter 6 some aspects of the coupling of nucleophiles with aryl radicals, or the reverse reaction of leaving group expulsion, are treated in the framework of molecular orbital theory, mainly the Perturbational Molecular Orbital (PMO) approach.

Chapters 7–9 are devoted to consideration of the steps that participate. Although factors related to these steps are discussed in perhaps a somewhat speculative way, we hope to stimulate further work in the area.

Chapter 10 contains some of the literature data that involve mechanisms that are in some way related to the $S_{RN}1$ mechanism, namely, photonucleophilic aromatic substitution, aliphatic $S_{RN}1$ reactions, vinylic $S_{RN}1$ reactions and other miscellaneous reactions involving electron transfer. This compilation is far from complete.

Finally, the Appendix includes some experimental details related to work in liquid ammonia, which we think may be of interest to the experimentalist not familiar with this particular solvent.

Literature Cited

1. Bunnett, J. F.; Zahler, R. E., *Chem. Rev.* **1951**, *49*, 275.
2. Zoltewicz, J. A. *Topics in Current Chemistry* **1975**, *59*, 33.
3. Bernasconi, C. F. *Chimia*, **1980**, *34*, 1.
4. Pietra, F. *Quart. Rev. Chem. Soc.* **1969**, *23*, 504.
5. Bunnett, J. F., *Quart Rev.* **1958**, *12*, 1.
6. Hoffmann, R. W. "Dehydrobenzene and Cycloalkynes"; Academic Press: New York, 1967.
7. Gilchrist, R. L.; Rees, C. W. "Carbenes, Nitrenes and Arynes"; Nelson: London, 1969.
8. Kauffmann, T.; Wirthwein, R. *Angew. Chem. Int. Ed. Engl.* **1971**, *10*, 20.
9. Bernasconi, C. F. M. T. P. *Int. Rev. Sci.*, Org. Chem. Ser. One, **1973**, *3*, 33.
10. Miller, J. "Aromatic Nucleophilic Substitution"; Elsevier: New York, 1968.
11. Cogolli, P.; Maiolo, F.; Testaferri, L.; Tingoli, M.; Tiecco, M. *J. Org. Chem.* **1979**, *44*, 2642.
12. Kim, J. K.; Bunnett, J. F. *J. Am. Chem. Soc.* **1970**, *92*, 7463.
13. Ibid. 7464.
14. Kornblum, N.; Michel, R. E.; Kerber, R. C. *J. Am. Chem. Soc.* **1966**, *88*, 5662.
15. Russell, G. A.; Danen, W. C. *J. Am. Chem. Soc.* **1966**, *88*, 5663.
16. Kornblum, N. *Angew. Chem. Int. Ed. Engl.* **1975**, *14*, 734.
17. Kerber, R. C.; Urry, G. W.; Kornblum, N. *J. Am. Chem. Soc.* **1965**, *87*, 4520.
18. Beletskaya, I. P.; Drozd, V. N. *Russian Chem. Rev.* **1979**, *48*, 431.
19. Bunnett, J. F. *Acc. Chem. Res.*, **1978**, *11*, 413.
20. Wolfe, J. F.; Carver, D. R. *Org. Prep. Proc. Int.* **1978**, *10*, 225.
21. Saveant, J. M. *Acc. Chem. Res.* **1980**, *13*, 323.

Nucleophiles Derived from the IVA Group of Elements

Carbon Nucleophiles

Conjugated Hydrocarbons. All reactions of hydrocarbon anions with aryl derivatives have been carried out in liquid ammonia. Because the pK_a of ammonia is about 32.5 (*1*), any conjugated hydrocarbon that has a lower pK_a is converted into its conjugate base by reaction with the amide ion in ammonia. The simplest unsaturated hydrocarbon, propene, has an estimated pK_a of 35.5 for the allylic hydrogen. More extensively conjugated systems should have lower pk_a values (*2*).

The carbanion **2** derived from 1,3-pentadiene **1** and amide ion (Reaction 1) did not react with bromobenzene at −78°C. However, when potassium metal was added, until electrons were in excess, reaction occurred to form a complex mixture of products (Reaction 2), including 5-phenyl-1-pentene (20%), 1-phenyl-2-pentene (6%), 5-phenyl-1,3-pentadiene (18%), and 1-phenyl-1,3-pentadiene (13%). Diphenylated products were also observed (*2*).

$$CH_3-CH=CH-CH=CH_2 \quad \xrightarrow[(-NH_3)]{NH_2^-} \quad (CH_2\text{⋯}CH\text{⋯}CH\text{⋯}CH\text{⋯}CH_2)^- \quad (1)$$
$$\textbf{1} \qquad\qquad\qquad\qquad\qquad\qquad\qquad\qquad\qquad \textbf{2}$$

$$\textbf{2} + PhBr \quad \xrightarrow{K} \quad Ph-CH_2CH_2CH_2CH=CH_2 \quad (20\%) \quad (2)$$
$$+ Ph-CH_2CH=CHCH_2CH_3 \quad (\ 6\%)$$
$$+ Ph-CH_2CH=CHCH=CH_2 \quad (18\%)$$
$$+ Ph-CH=CHCH=CHCH_3 \quad (13\%)$$

In the products obtained, it is interesting that phenylation occurred only in position 1 of **2**. Thus, by catalytic hydrogenation of the product mixture a 74% yield of 1-phenylpentane was obtained (*2*).

0065-7719/83/0178-0011$13.50/1

These results suggest that phenyl radical reacts with **2** to give a radical anion **3** (Reaction 3) which can react with another electron and be reduced to phenylpentenes (Reaction 4), or it can transfer the odd electron to bromobenzene to give 5-phenyl-1,3-pentadiene (Reaction 5). The reduction products of Reaction 4 probably do not come from reduction of **4**, because **4** is quickly converted to the 1-phenylpentadienide ion (Reaction 6) in the strongly basic environment. Upon acidification, protonation of **5** occurs at both the 1 and 5 positions, forming two isomeric phenylpentadienes.

$$Ph\cdot + 2 \longrightarrow Ph-CH_2 - (CH = CHCH = CH_2)^{\overline{\cdot}} \tag{3}$$
$$\mathbf{3}$$

$$3 + e^- + \xrightarrow{NH_3} PhCH_2CH_2CH_2CH = CH_2$$
$$+ PhCH_2CH = CHCH_2CH_3 \tag{4}$$

$$3 + PhBr \longrightarrow PhCH_2CH = CHCH = CH_2 + (PhBr)^{\overline{\cdot}} \tag{5}$$
$$\mathbf{4}$$

$$4 + 2 \longrightarrow (Ph-CHCHCH(HCH_2)^- + 1 \tag{6}$$
$$\mathbf{5}$$

Another carbanion phenylated by bromobenzene in potassium metal solutions in ammonia was (p-anisyl)propenide (**6**). Phenylation occurred at positions 1 and 3 of **6** to almost the same extent. After catalytic hydrogenation, propane derivatives **7** and **8** were obtained (Reaction 7).

The carbanion **9** derived from indene afforded 3-phenylindene **10** in good yield (Reaction 8). The yield of 1-phenylindane (the reduction product of the radical anion) was low and suggests that electron transfer to bromobenzene, as in Reaction 5, occurred to a significant extent.

The potassium cyclopentadienide ion is insoluble in ammonia, and no phenylation products were obtained when it was treated with bromobenzene under potassium metal stimulation (3). The monoanion and dianion of acetylene were also insoluble and no reaction was observed (4). Other examples are reported in Table I (2).

The carbanions mentioned were used only in reactions stimulated by potassium metal, and it was not determined whether they are reactive in photostimulated reactions. However, pentadienide ion **2** reportedly does not react with bromobenzene under photostimulation, and aniline is the only product found when the arylation of **2** is attempted (5).

Irradiation of 2-lithio-1,3-dithiane in the presence of bromobenzene in liquid ammonia was reported to give complete conversion of the halide, but the yield of 2-phenyl-1,3-dithiane was less than 5% (6). Under similar conditions, 1-hexyne anion failed to interact with bromobenzene (6).

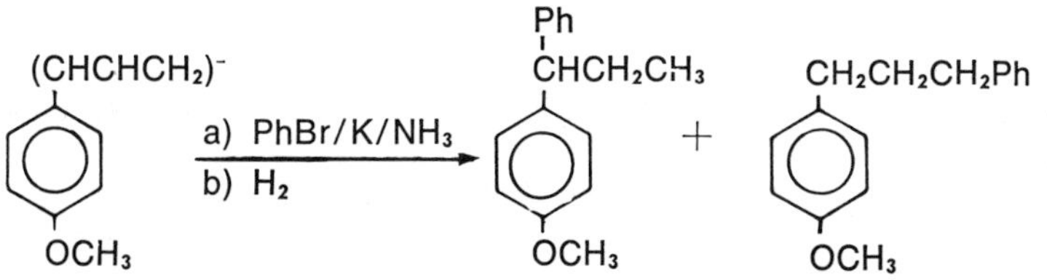

$$PhX + {}^-CH_2COCH_3 \xrightarrow{K} PhCH_2COCH_3 + PhCH_2CHCH_3 \quad (9)$$

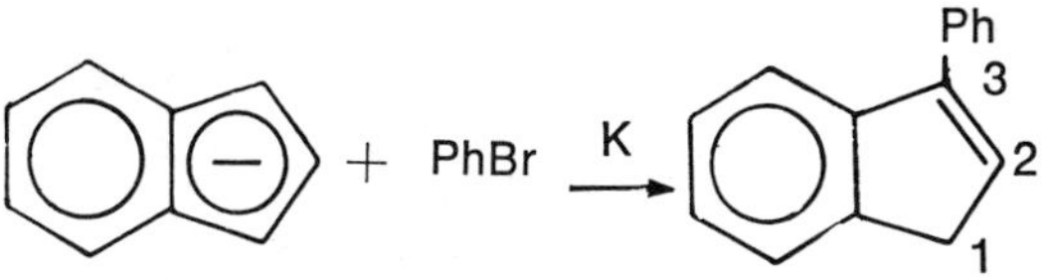

$$\begin{array}{ccc} & & | \\ 11 & 12 & OH \\ & & 13 \end{array}$$

Ketone Enolate Anions. ALKALI METAL STIMULATION. Ketone enolates are the most studied nucleophiles in arylation reactions by the $S_{RN}1$ mechanism. The first report that acetone enolate anion **11** can be arylated by the $S_{RN}1$ mechanism dealt with the reaction of halobenzenes in liquid ammonia stimulated by solvated electrons (7). The main products obtained were phenylacetone **12** and 1-phenyl-2-propanol **13** (Reaction 9).

The reaction was investigated with substrates bearing several leaving groups, halogens, ${}^+NMe_3$, OPh and SPh, the yield of **12** and **13** being strongly dependent on the nature of X. The ratio of **13:12** increases as the reduction potential of PhX becomes more negative. Under strongly basic conditions, **12** is quickly deprotonated, and protected toward reduction; **13** cannot be formed then by the reduction of **12**. However, the $S_{RN}1$ mechanism can adequately account for its formation. When the phenyl radical reacts with the acetone enolate anion, a ketyl type radical anion **14** is formed (Reaction 10) which can transfer its extra electron to the substrate PhX in a chain propagation step (Reaction 11), or it can react with another electron to give the alkoxide in a termination step (Reaction 12).

An earlier interpretation was that the product of Reaction 12 is independent of the identity of PhX, contrary to Reaction 11 which pre-

TABLE I

Reaction of Carbanions with Bromobenzene and Potassium Metal in Liquid Ammonia

Source of Carbanion	Principal Products, Yield
1,3-Pentadiene	1-phenylpentane, 74%[a]
	diphenylpentanes, 7%[a]
1-(p-Anisyl)propene	1-phenyl-1-(p-anisyl)propane, 13%[a]
	3-phenyl-1-(p-anisyl)propane, 36%[a]
	diphenyl-1-(p-anisyl)propanes, 33%[a]
Indene	3-phenylindene, 32%
	diphenylindenes, 12%
Fluorene	9-phenylfluorene, 57%
	9,9-diphenylfluorene, 23%

[a]After catalytic hydrogenation.
Note: Selected data are taken from Ref. 2. Copyright 1973, American Chemical Society.

$$Ph^{\cdot} + {}^{-}CH_2COCH_3 \longrightarrow Ph-CH_2-\overset{\cdot}{\underset{\underset{\textbf{14}}{O_-}}{C}}-CH_3 \qquad (10)$$

$$\underset{\textbf{11}}{}$$

$$\underset{\textbf{14}}{Ph-CH_2-\overset{\cdot}{\underset{O_-}{C}}-CH_3} + PhX \longrightarrow PhCH_2COCH_3 + (PhX)^{\bar{\cdot}} \qquad (11)$$

$$\underset{\textbf{12}}{}$$

$$\textbf{14} + e^- \xrightarrow[(-NH_2^-)]{NH_3} Ph-CH_2-\underset{O_-}{CH}-CH_3 \qquad (12)$$

dicts that the rate of electron transfer to PhX—and hence the rate at which reduction occurs—should depend on the substrate PhX (7). However, results obtained in competition experiments, when a mixture of aryl iodide and aryl chloride was allowed to react with solvated electrons in the presence of acetone enolate anions, indicate that in fact the formation of both products depends on the ArX. The rationale is described in Chapter 7.

In the reaction of m-methoxyiodobenzene **15a** with **11**, the ratio of ketone **16** to alcohol **17** was 6.9 (Reaction 13), which is similar to the ratio

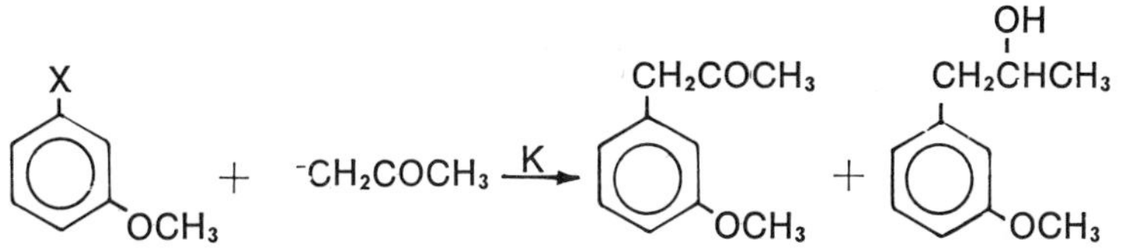

15a, X = I

15b, X = Cl

of 6.6 for **12:13**, with iodobenzene as substrate. With m-methoxychlorobenzene **15b** the ratio of **16:17** was 0.51, which is similar to the 0.55 obtained for the ratio of **12:13**, with chlorobenzene as substrate (8). But, when a mixture of iodobenzene and m-methoxychlorobenzene (**15b**) was allowed to react with **11**, a value of 23 was found for the ratio of **12:13** (from iodobenzene), and a value of 0.64 was found for the ratio of **16:17** (from **15b**), as if each substrate were reacting separately. Moreover, when a mixture of chlorobenzene and m-methoxyiodobenzene (**15a**) was allowed to react with solvated electrons in the presence of **11**, the ratio of **12:13** (from chlorobenzene) was 0.47 and the ratio of **16:17** (from **15a**) was 28. These results will be discussed further in Chapter 7.

In the reaction of 1-chloronaphthalene and acetone enolate, stimulated by potassium metal in liquid ammonia, the products obtained were the ketones **18, 19,** and **20** (9). No traces of the alcohol **21** were found. This behavior contradicts the results obtained from the reaction of **11** with halobenzenes, in which both the ketone **12** and the alcohol **13** were found. This behavior was attributed to the different nature of the radical anion intermediate when the aryl radical couples with **11**. When the aryl is a phenyl group, the radical anion formed is the ketyl type **14,** but, when the aryl is 1-naphthyl, the intermediate radical anion is **22** in which the odd electron is part of the pi system of the aromatic ring. Reaction of **22** with another solvated electron leads to reduction of the naphthalene ring and formation of **19** and **20**.

Nevertheless, when 1-chloronaphthalene was allowed to react with the anion derived from acetophenone and potassium metal, no reduction of the naphthalene ring was observed, suggesting that the radical anion intermediate is of the ketyl type **23**.

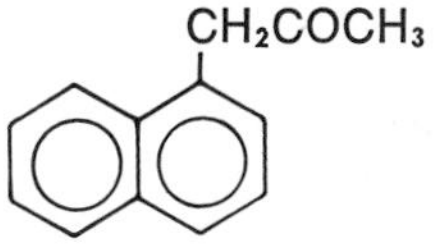

18 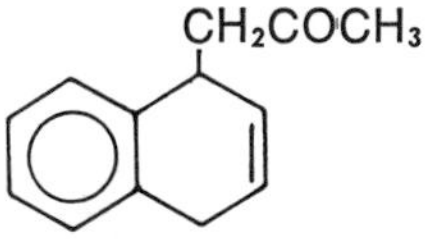   **19** 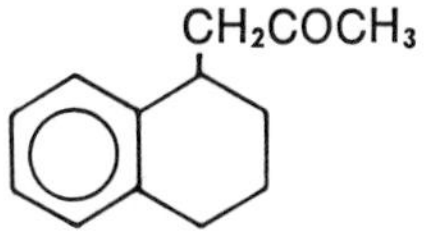   **20**

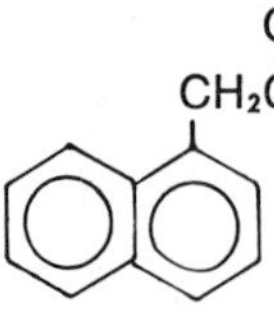

From molecular orbital considerations, the structure of the radical anion intermediate should depend on the energy of the lowest unoccupied molecular orbital (LUMO), and the order benzene > $RCOCH_3$ > naphthalene > RCOPh is in good agreement with the postulated structures **14, 22,** and **23** (9, 10).

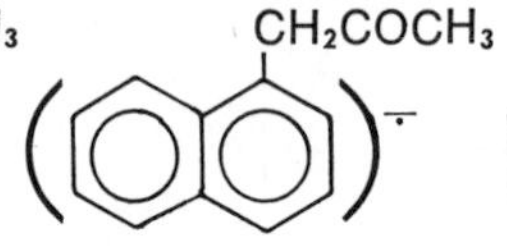

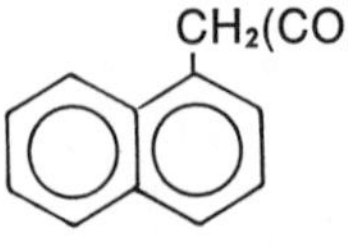

3-Bromothiophene **24a** reacts with **11** under sodium or potassium metal stimulation to produce 20–30% of ketone **25a,** representing monothienylation, and the secondary alcohol **27a.** Smaller yields (in the range of 2–10%) of dithienylation products **26a** and **28a** were also formed (Reaction 14) (11).

Although Reaction 14 resembles the corresponding reaction of bromobenzene with **11;** there are some differences in detail. First, dithienylation to form ketone **26a** and the alcohol **28a** is more prominent than with bromobenzene. Furthermore, the proportion of secondary alcohol **27a** to the ketone **25a** is higher than with the bromobenzene reaction.

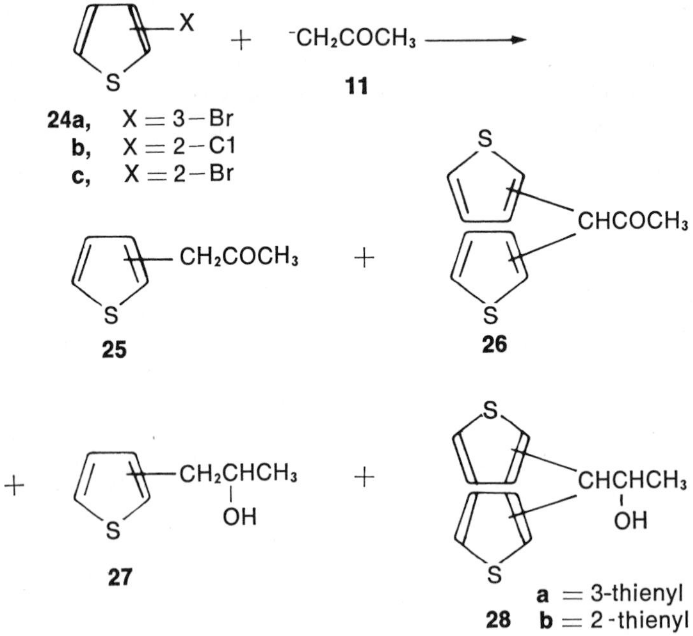

In comparison with 3-bromothiophene, the 2-halothiophenes **24b** and **24c** behaved quite differently (under solvated electron stimulation) and only small amounts of ketone **25b** were formed together with 8–20% of **27b**. The same low yield of substitution products were found with halopyridines and under alkali metal stimulation **11** (*12*).

The reactions of 4-phenyl- and 4-*tert*-butyl-5-halopyrimidines **29** and the enolate ions of acetone **12,** acetophenone and pinacolone, stimulated by potassium metal in liquid ammonia, afforded good yields of substitution products (Reaction 15) (Table II) (*13*).

PHOTOSTIMULATED REACTIONS. The reactions of halobenzenes with acetone enolate ions occur not only when stimulated with solvated electrons, but also when irradiated with near UV light (290–350 nm). A simple tungsten bulb is enough to produce the reaction. The initiation is probably the transfer of one electron to the substrate halobenzene forming its radical anion, and starting the reaction (Reactions 16–17) (*14*).

$$R^1 = Ph, Bu^t \qquad R^2 = Me$$
$$X = Cl, Br \qquad\qquad Ph$$
$$Bu^t$$

$$PhX + {}^-CH_2COCH_3 \xrightarrow{h\nu} (PhX)^{\overline{\cdot}} \tag{16}$$

$$(PhX)^{\overline{\cdot}} \longrightarrow Ph^{\cdot} + X^- \tag{17}$$

With this procedure, a high yield of phenylacetone is obtained uncontaminated with the alcohol or other reduced species that are formed under potassium metal stimulation. One by-product usually obtained is the α, α-diarylketone, which represents double arylation. Introduction of a third aryl ring was attempted without success, even in the presence of a large excess of bromobenzene (*15*).

Another by-product formed under potassium metal stimulation is benzene, which presumably is formed by reduction of the phenyl radical intermediate by reaction with a solvated electron followed by protonation of the resulting phenyl anion (Reaction 18).

$$Ph^{\cdot} + e^- \longrightarrow Ph^- \xrightarrow{NH_3} PhH + NH_2^- \tag{18}$$

TABLE II

Alkali Metal Stimulation with Ketone Enolate Ions in Liquid Ammonia

ArX	Ketone Enolate Ion	α-Arylketone Yield %	α-Arylalcohol Yield %	Ref.
Fluorobenzene	acetone	3	46	7
Chlorobenzene	"	68[a]		7
Bromobenzene	"	67	10	7
Iodobenzene	"	71[a]		7
Diphenyl ether[b]	"	—	4.5	7
Phenyl sulfide	"	18	71	7
Phenyltrimethyl- ammonium iodide	"	46	18	7
p-Bromotoluene	"	57	17	7
p-Tolyltrimethyl- ammonium iodide	"	30	42	7
p-Anisyltrimethyl- ammonium iodide	"	20	39	7
p-Bromoanisole	"	36	10	7
Diethylphenyl phosphate	"	5	56	15
2-Isopropylphenyl diethyl phosphate	"	6	24	15
2-t-Butylphenyl diethyl phosphate	"	1	3.6	15
1-Chloronaphthalene	"	23	[c]	9
1-Iodonaphthalene	"	6	[d]	9
3-Bromothiophene	"	18–29	22–32	11
2-Bromothiophene	"	3	3	11
2-Chlorothiophene	"	2	22	11
5-Bromo-4- phenylpyrimidine	"	42		13
5-Chloro-4- phenylpyrimidine	"	47		13
5-Bromo-4-t- butylpyrimidine	"	30		13
5-Chloro-4-t- butylpyrimidine	"	42		13
2-Chloroquinoline	"	43		20
1-Chloro- naphthalene	acetophenone	57		9
5-Bromo-4- phenylpyrimidine	"	60–65		13
5-Chloro-4- phenylpyrimidine	"	60		13
5-Bromo-4-t- butylpyrimidine	"	35		13

TABLE II (continued)

Alkali Metal Stimulation with Ketone Enolate Ions in Liquid Ammonia[a]

ArX	Ketone Enolate Ion	α-Arylketone Yield %	α-Arylalcohol Yield %	Ref.
5-Chloro-4-t-butylpyrimidine	"	67		13
5-Bromo-4-phenylpyrimidine	pinacolone	45–50		13
5-Chloro-4-phenylpyrimidine	"	50		13
5-Bromo-4-t-butylpyrimidine	"	65		13
5-Chloro-4-t-butylpyrimidine	"	65		13

Note: The yield varied with the experimental conditions.
[a]After the oxidation of the alcohol.
[b]93% of starting material.
[c]69% of dihydro and tetrahydronaphthylacetones.
[d]84% of dihydro and tetrahydronaphthylacetones.

When **11** is prepared by reaction of acetone with potassium metal in ammonia, a byproduct, isopropoxide ion, is formed. In the reaction of halobenzenes with **11** similarly prepared, a substantial amount of benzene is produced.

On the other hand, in the photostimulated reaction of **11** prepared by the acid–base reaction with amide or t-butoxide ions (which do not lead to isopropoxide ions) the yield of benzene was very low. These results imply that the formation of isopropoxide and benzene byproducts is interrelated in some way. In fact, the benzene formed in the former reaction is attributed to hydrogen atom abstraction of phenyl radical from isopropoxide ion, to give benzene and the ketyl **31** (Reaction 19).

$$Ph^{\cdot} + CH_3-\underset{\underset{O_-}{|}}{CH}-CH_3 \longrightarrow PhH + CH_3-\underset{\underset{O_-}{|}}{\overset{\cdot}{C}}-CH_3 \qquad (19)$$

31

Hydrogen atom abstraction from alkoxides by aryl radicals to form the reduction product and the radical anion of the ketone is a well known reaction (*16*).

Ketones bearing β-hydrogens, such as the 2,4-dimethyl-3-pentanone enolate ion **32**, react sluggishly with iodobenzene to form the phenylated ketone (32%) accompanied by benzene (20%) and a dimeric product **34**. The formation of **34** is thought to follow Reaction 20, where a phenyl radical abstracts a β-hydrogen atom from **32** forming benzene and the radical anion intermediate **33**. Because **33** seems to be unable to transfer the odd electron to iodobenzene, Reaction 20 becomes a pre-termination step (*17*). The termination step is the disproportionation of **33** as in Reaction 21.

$$Ph^{\cdot} + (CH_3)_2\overline{C}COCH(CH_3)_2 \longrightarrow PhH + (CH_2{=}C{-}COCH(CH_3)_2)^{\doteq} \quad (20)$$

$$\underset{\textbf{32}}{} \qquad\qquad\qquad \underset{\textstyle CH_3}{}$$

33

$$33 \longrightarrow CH_2{=}\underset{\underset{CH_3}{|}}{\overset{\overset{O}{\|}}{C}}{-}C{-}CH(CH_3)_2 \; + \; \underset{H_3C}{\overset{{}^{-}CH_2}{\diagdown}}C{=}C\underset{CH(CH_3)_2}{\overset{O^{-}}{\diagup}} \qquad (21)$$

32

$$\longrightarrow (CH_3)_2CHCO{-}\underset{\underset{CH_3}{|}}{\overset{\overset{CH_3}{|}}{C}}{-}CH_2{-}\underset{\underset{H}{|}}{\overset{\overset{CH_3}{|}}{C}}{-}CO{-}CH(CH_3)_2$$

34

The abstraction of a β-hydrogen atom was also shown to be an important competing reaction during intramolecular photostimulated $S_{RN}1$ reactions (*18*). However, it does not seem to be very important with 3-pentanone enolate ion as good yields of arylation products were found (*19*). Even **32** can be arylated in good yield by 2-bromopyridine (*12*) or 2-chloroquinoline (*20*).

Enolates from arylketones, such as acetophenone enolate ion **35,** react poorly with aryl halides under photostimulation. No reaction of **35** occurred with iodobenzene (*19*), or with 2-bromopyridine (*12*), but with 1-chloronaphthalene about an 8% yield of substitution product was obtained (*9*). However, with more intense and longer irradiation, iodobenzene reportedly reacts with **35** to give a 67% yield of the substituted product (*6*).

The lithium salt of **35** reacted with 2-chloroquinoline to give yields as high as 82% of substitution product (*21*), but the sodium salt was much less effective (*20*). With 4-phenyl or 4-*t*-butyl-5-halopyrimidines, good yields of substitution products were obtained (Reaction 22) (*13*)

The steric bulk of the enolate ions seems to play only a minor part in the overall facility of the reaction. However, bulky groups ortho to the nucleofugal substituent will block substitution with hindered enolate anions.

Although monoanions of β-diketones fail to undergo $S_{RN}1$ reaction with aryl and hetaryl halides, 1,3-dianions of β-diketones can be arylated at the terminal carbanion site as shown in Reactions 23–24 (*19, 22*).

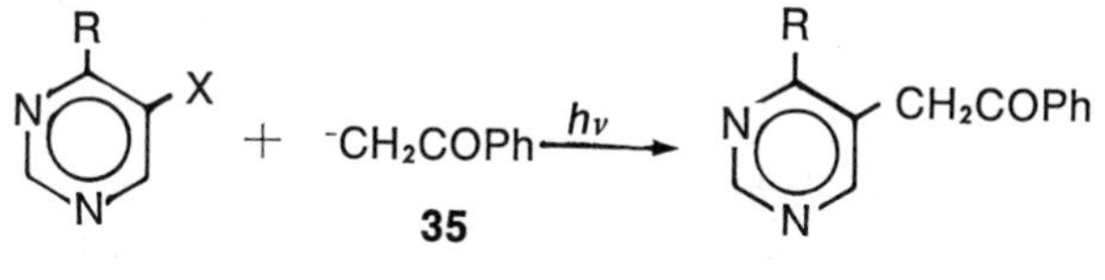

$$(22)$$

R = Ph, t-butyl

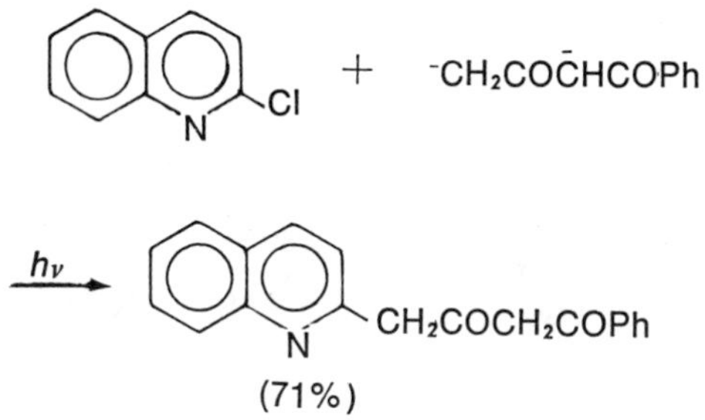

$$(23)$$

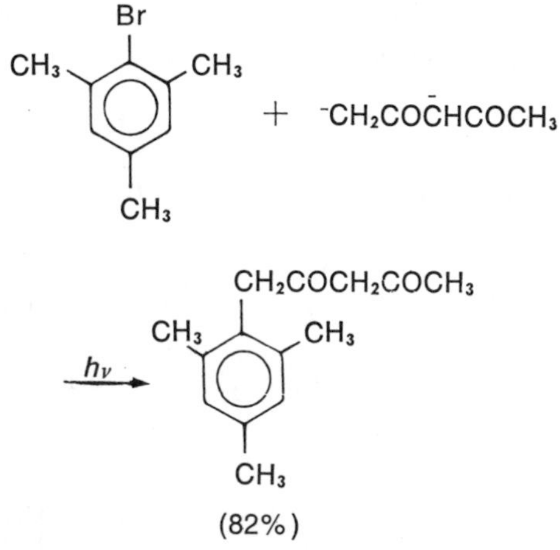

(71%)

$$(24)$$

(82%)

Enolate anions of α-dicarbonyl compounds, with one carbonyl group protected as an acetal, such as 2-oxopropanal dimethyl acetal, have been reported to react under photostimulation with 2-chloroquinoline to give 80% yields of the substitution product (Reaction 25). Under the same conditions iodobenzene does not react (*23*).

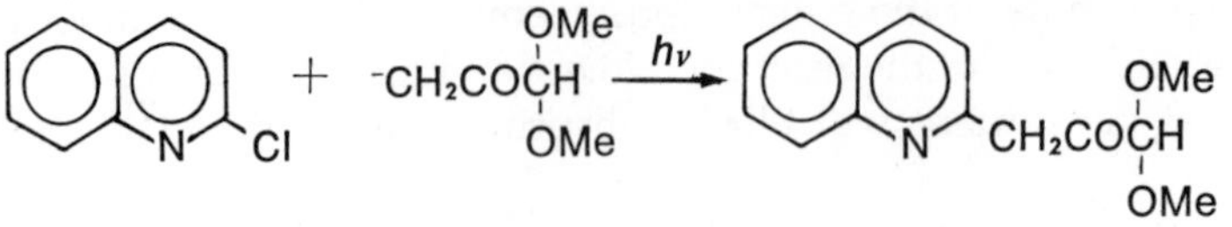

(25)

Enolate ions from the dimethyl acetals of 2-oxopropanol and 3-oxobutanone were successfully used in the synthesis of indoles (*see* Table III).

Unsymmetrical Dialkyl Ketones. Unsymmetrical ketones can form more than one enolate ion, and each enolate can be arylated, although not necessarily at the same rate. For instance, butanone **36** can form three enolate ions **37, 38a,** and **38b** (Reaction 26) (*24*).

$CH_3COCH_2CH_3 \xrightarrow{B^-}$
36

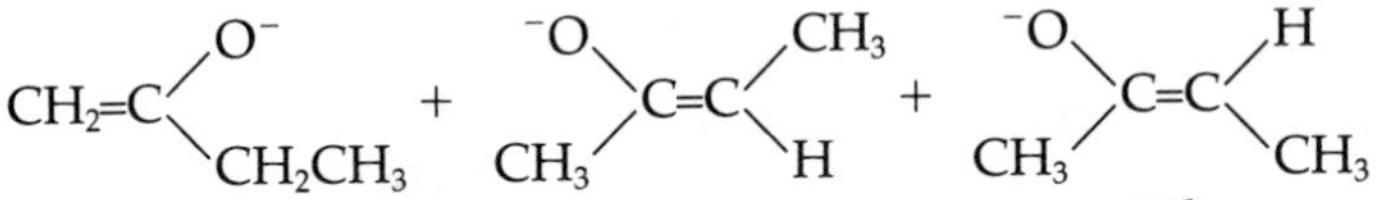

(26)

37 **38a** **38b**

Reacting these enolate ions with bromo- or iodobenzene gave a ratio of roughly 1.6 of the ketones **39:40** (note that the enolates **38a** and **38b** lead to the same 3-phenyl-2-butanone **40**) (*2*).

$PhCH_2COCH_2CH_3$ **39**

$PhCH(CH_3)COCH_3$ **40**

The photostimulated reaction of *a*-iodoanisole with **36** gave both ketones **39a** and **40a** (*25*).

The reaction of 3-methyl-2-butanone enolate ions **41** and **42** with

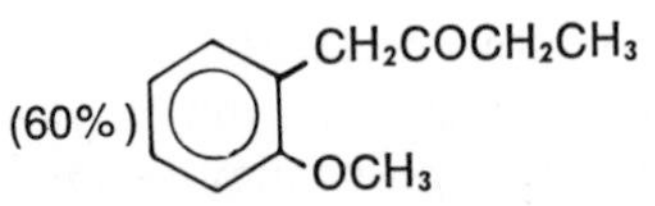

39a

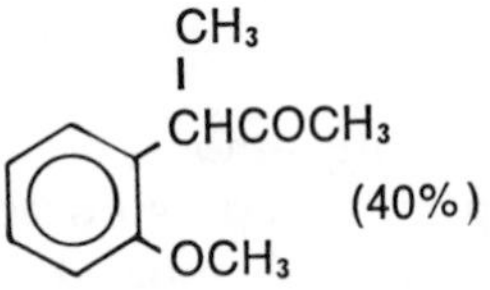

40a

Table III

Photostimulated Reactions of Aryl and Hetaryl Substrates with Ketone Enolate Ions in Liquid Ammonia

ArX	Source of Enolate	Irradiation, Time (min)	α-Aryl-ketone %	Ref.
PhI	Acetone	5	67	14
PhBr	"	11	85	14
PhCl	"	180	61	14
PhF	"	200	60	14
PhSPh	"	30	66	14
$Ph_3S^+Cl^-$	"	70	75	14
PhSePh	"	30	95	14
PhOPh	"	250	14	14
$PhN^+Me_3I^-$	"	60	57	14
$PhOP(O)(OEt)_2$	"	250	13	14
$2,4(MeO)_2C_6H_3Br$	"	120	76	15
$3,5(MeO)_2C_6H_3I$	"	60	68	15
$2,4,6(MeO)_3C_6H_2I$	"	90	92	15
$2-^-O_2CC_6H_4Br$	"	90	85	15
$E-^-O_2CC_6H_4Br$	"	140	80	15
$4-^-O_2CC_6H_4Br$	"	90	70	15
$2,5(Isopropyl)_2C_6H_3I$	"	130	78	15
$2,5(t\text{-Butyl})_2C_6H_3I$	"	130	26	15
$2,4,6(Me)_3C_6H_2I$	"	32	82	15
$2,4,6(Et)_3C_6H_2Br$	"	120	70	15
$2,4,6(Isopropyl)_3C_6H_2Br$	"	150	2	15
$2,4,6(Isopropyl)_3C_6H_2I$	"	180	16	15
4-Bromobiphenyl	"	150	69	15
$m\text{-}FC_6H_4I$	"	70	56	15
$m\text{-}CF_3C_6H_4I$	"	90	35	15
$p-^-CH_2COC_6H_4Br$	"	100	—	15
$p\text{-}NMe_2C_6H_4I$	"	—[a]	90	64
$p\text{-}NH_2C_6H_4I$	"	—[a]	33	23
$m\text{-}NMe_2C_6H_4I$	"	—[a]	82	64
$m\text{-}NH_2C_6H_4I$	"	—[a]	66	23
$p-^-OC_6H_4Br$	"	85	—	15
$m\text{-}NO_2C_6H_4Cl$	"	150	—	15
$m\text{-}NO_2C_6H_4Br$	"	150	—	15
1-Chloronaphthalene	"	90	88	9
2-Chloroquinoline	"	60	90	20
2-Bromopyridine	"	15	100	12
2-Chloropyridine	"	60	85	12
2-Fluoropyridine	"	120	40	12
3-Bromopyridine	"	15	65	12
4-Bromopyridine	"	15	28	12

Continued on next page.

Table III (continued)

Photostimulated Reactions of Aryl and Hetaryl Substrates with Ketone Enolate Ions in Liquid Ammonia

ArX	Source of Enolate	Irradiation, Time (min)	α-Aryl-ketone %	Ref.
3-Bromothiophene	"	60	51	11
2-Bromothiophene	"	60	31	11
2-Chlorothiophene	"	60	17	11
o-CH$_3$OC$_6$H$_4$I	"	15	67	25
4-Ph-5-bromopyrimidine	"	75	25–30	13
4-Ph-5-chloropyrimidine	"	75	20–25	13
4-t-Butyl-5-bromopyrimidine	"	75	70–75	13
4-t-Butyl-5-chloropyrimidine	"	75	60–65	13
2-Chloropyrimidine	"	15	15	33
PhBr	pinacolone	90	90	19
p-Br$_2$C$_6$H$_4$	"	120	65^c	28
o-CH$_3$OC$_6$H$_4$I	"	15	100	25
2-Bromopyridine	"	90	94	12
2,6-Dibromopyridine	"	60	89^c	12
2,6-Dichloropyridine	"	60	86^c	12
4-Phenyl-5-bromopyrimidine	"	75	60–65	13
4-Phenyl-5-chloropyrimidine	"	75	45	13
4-t-Butyl-5-bromopyrimidine	"	75	95	13
4-t-Butyl-5-chloropyrimidine	"	75	90–95	13
2-Chloropyrimidine	"	15	32	33
2,6-Dimethoxy-4-chloropyrimidine	"	15	98	33
3-Chloro-6-methoxypyridazine	"	15	72	33
PhBr	3-pentanone	70	80	19
Bromomesitylene	"	130	14	19
Iodomesitylene	"	130	24	19
2-Chloroquinoline	"	60	68	20
PhBr	2,4-dimethyl-3-pentanone	90	6	19
PhI	"	180	32	19
2-Chloroquinoline	"	15	98	20

Table III (continued)

Photostimulated Reactions of Aryl and Hetaryl Substrates with Ketone Enolate Ions in Liquid Ammonia

ArX	Source of Enolate	Irradiation, Time (min)	α-Aryl-ketone %	Ref.
2-Bromopyridine	"	60	97	12
2-Chloropyrimidine	"	15	88	33
3-Chloro-6-methoxypyridazine	"	60	20	33
PhBr	4-heptanone	120	80	19
PhBr	$CH_3OCH_2COCH_3$	130	1	19
PhBr	Cyclobutanone	150	90	19
PhBr	Cyclopentanone	150	64	19
2-Chloroquinoline	"	60	63	20
PhBr	Cyclohexanone	60	72	19
2-Bromopyridine	Cyclohexanone	60	47	12
PhBr	Cyclooctanone	210	95	19
PhBr	2-indanone	150	90	19
PhBr	Cyclohexa-2-en-1-one	100	—	19
PhI	acetophenone	188	—	19
2-Chloroquinoline	"	180	14	20
1-Chloronaphthalene	"	180	8	9
4-Phenyl-5-bromopyrimidine	"	75	63	13
4-Phenyl-5-chloropyrimidine	"	75	60–65	13
4-t-Butyl-5-bromopyrimidine	"	75	85–90	13
4-t-Butyl-5-chloropyrimidine	"	75	90	13
2-Chloroquinoline	propiophenone	60	50	20
PhBr	2-butenone	50	80[d]	14
PhBr	3-methyl-2-butanone	120	90[e]	14
$o(MeO)C_6H_4I$	"	15	66[f]	14
2-Chloroquinoline	"	60	75[g]	20[a]

[a] Not reported.
[b] Disubstituted product.
[c] 3-Phenyl-2-butanone, 61%; 1-Pehnyl-2-butanone, 19%.
[d] 1-Phenyl-3-methyl-2-butanone, 81%; 3-phenyl-3-methyl-2-butanone, 9%.
[e] Only 1(o-anisyl)-3-methyl-2-butanone was reported.
[f] 1(2-quinolyl)-3-methyl-2-butanone, 62%; 3(2-quinolyl)-3-methyl-2-butanone, 13%.

PhBr yields mostly the arylation product of **42**. The same result has been observed with other aryl substrates (Reaction 27); *see also* Table II.

$$(CH_3)_2\overline{C}COCH_3 \qquad\qquad (CH_3)_2CHCOCH_2^-$$
$$\textbf{41} \qquad\qquad\qquad \textbf{42}$$

$$ArX + (CH_3)_2CHCOCH_3 \xrightarrow{B^-, h\nu} ArCH_2COCH(CH_3)_2 + Ar\overset{\overset{\textstyle CH_3}{|}}{\underset{\underset{\textstyle CH_3}{|}}{C}}COCH_3 \quad (27)$$

$$\textbf{43} \qquad\qquad \textbf{44}$$

ArX	43:44	Ref.
Bromobenzene	16	6
Bromobenzene	9	2
2-Bromopyridine	7	12
2-Chloroquinoline	4.8	20

The photostimulated reaction of *o*-iodoanisole with 3-methyl-2-butanone enolate ion gave 66% (isolated yield) of ketone **45**. Formation of ketone **46** was not observed which may be due to steric hindrance in the formation and reaction of enolate **41**. It has been suggested that in these reactions the product distribution is governed by the relative populations of the isomeric enolate ions, either kinetically or thermodynamically determined. This suggestion is supported (qualitatively) by studies indicating that an equilibrium mixture of **41** and **42** contains 98% of **42** (*26*).

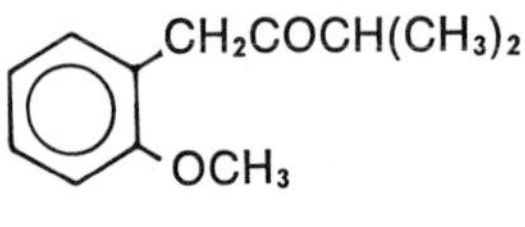

45 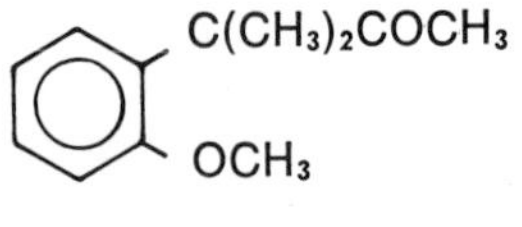   **46**

Reaction with Disubstitution Substrates. In the photostimulated reaction of 2,6-dibromopyridine **47a** or 2,6-dichloropyridine **47b** with pinacolone enolate ion **48** in liquid ammonia, the disubstituted product **49** was obtained in high yield (Reaction 28) (*12*). Formation of **49** appears to arise directly from the starting dihalopyridines without build-up of the monosubstitution compounds. Evidence for the direct formation of **49** was obtained from an experiment in which **47a** and **48** were irradiated for

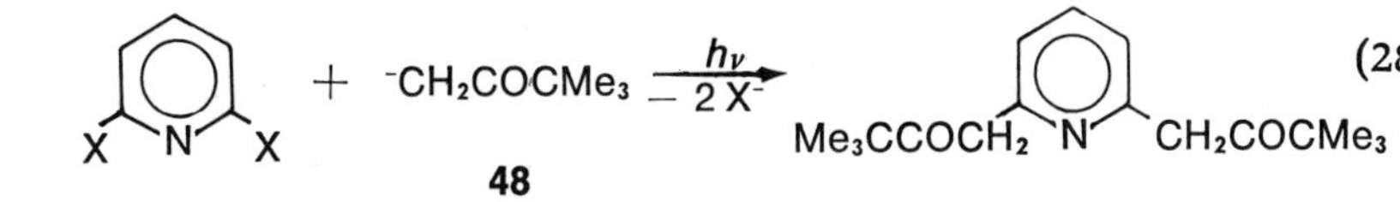

(28)

47a, X = Br
47b, X = Cl

45 seconds. Under these conditions, none of the monosubstitution pyridine was detected, but starting materials and **49** were found in a ratio of 1:3 (*12*).

These results, together with the behavior of other dihaloaryl compounds and nucleophiles, are characteristic of the $S_{RN}1$ mechanism. When a haloaryl radical such as **50** couples with a nucleophile, a new radical anion is formed. For the radical anion **51** to give the disubstitution product without intermediate formation of the monosubstitution product, it must expel the halide ion faster than it can transfer an electron to the starting substrate. The aryl radical formed (**52**) then couples with another enolate anion **48** to give **49** (Reactions 29–31).

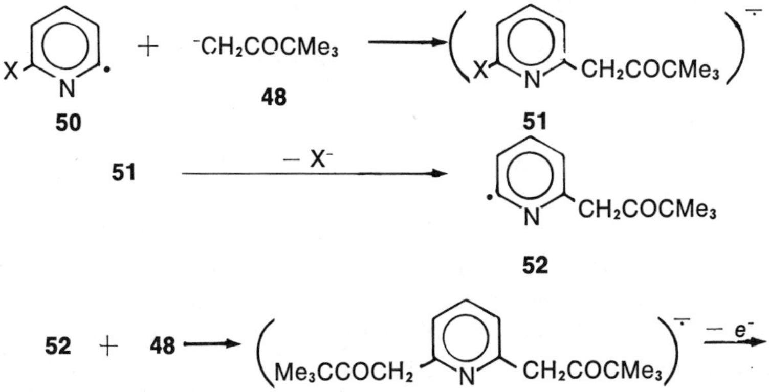

(29)

(30)

(31)

The photostimulated reaction of *o*-dibromobenzene with excess pinacolone enolate ion in ammonia afforded the disubstitution product, 1,2-bis(3,3-dimethyl-2-oxobutyl)benzene, in 62% yield together with 35% of the unreacted *o*-dibromobenzene. Again, there was no evidence of the monosubstitution product (Reaction 32) (*27*). The photostimulated

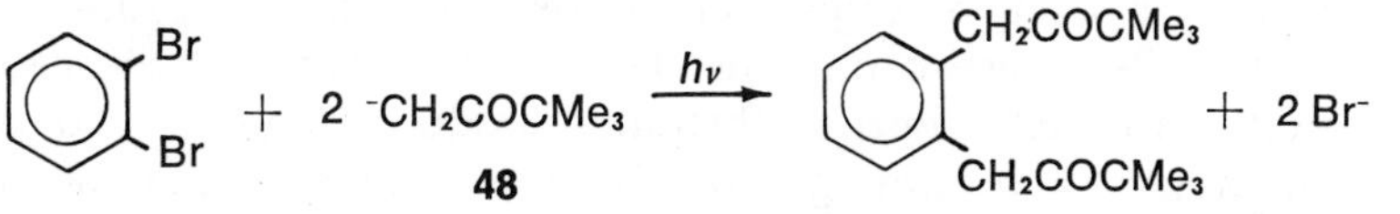

(32)

reaction of o-dibromobenzene with excess acetone enolate ion afforded acetylindenes, probably by aldol condensation of the o-diacetonylbenzene formed.

p-Dihalobenzenes **53** were allowed to react with ketone enolate ions **54** under photostimulation with excess potassium t-butoxide to allow the disubstitution products formed to ionize (Reaction 33–34). The anion (**56**)

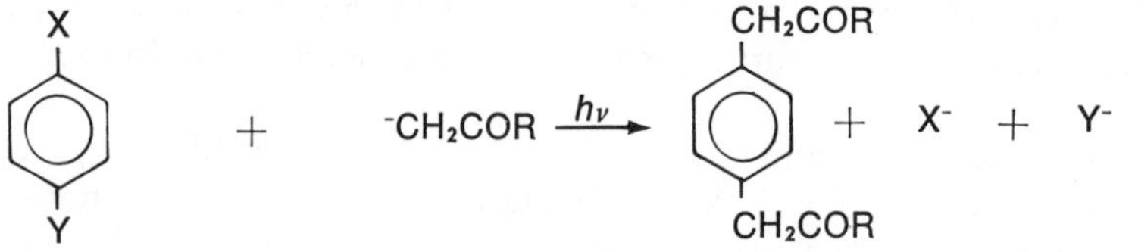

53a, X = Y = Cl   **54a,** R = CH₃   **55a,** R = CH₃
53b, X = Y = Br   **54b,** R = C(CH₃)₃   **55b,** R = C(CH₃)₃
53c, X = Br, Y = I

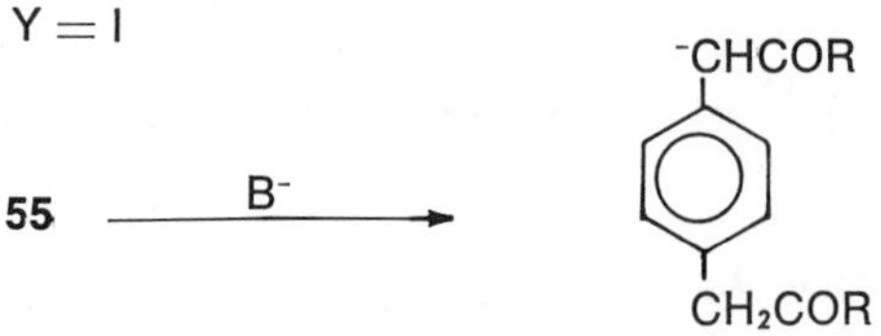

formed was expected to act as a nucleophile that could react with aryl radicals to produce oligomeric and/or polymeric products (*28*). However, in the photostimulated reaction of **53** with **54a** in excess t-butoxide base (ratio 1:1:4), the amount of halide ion eliminated was only 25–60% of the theoretical value (considering two halide atoms per molecule of **53**). Yet, in the photostimulated reaction of **53a** with a large excess of the starting enolate **54a** or **54b,** chloride ion was eliminated to the extent of 83% and 93% respectively. The disubstitution products were formed in high yields indicating that Reaction 33 probably proceeds to completion (**55b** was isolated in 65% yield) (*28*).

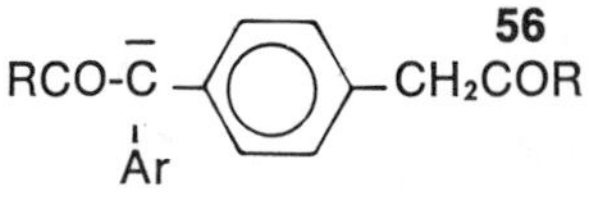

The amount of halide ion eliminated in the reaction when the ratio of reagents was 1:1:4 indicates that the reaction may not proceed to completion under these conditions. It was suggested that the cause could be the insolubility of the substituted products, together with decomposition. (At the beginning of reaction the solutions were clear, but under irradiation they became dark green and a solid was formed.)

Another reason could be the formation of structures like **57,** which are unreactive toward aryl radicals.

Exceptions to the general observation that dihaloaryl compounds react to give disubstitution products are known. As an example, in the photostimulated reaction of *m*-fluoroiodobenzene and acetone enolate ion **11,** *m*-fluorophenylacetone **59** was formed in 56% yield. This result indicates that the radical anion **58,** formed in the first $S_{RN}1$ substitution, transfers its extra electron to the starting material, and no C–F bond fragmentation occurs (*15*).

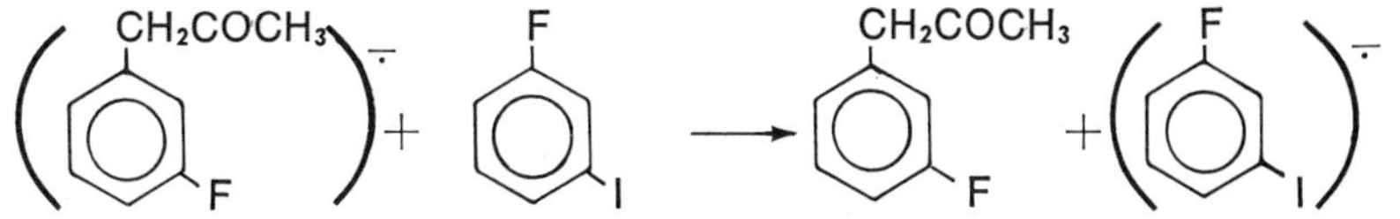

$$(35)$$

58   **59**

Ring Closure Reactions. An interesting type of enolate arylation involves intramolecular $S_{RN}1$ reactions. The first recognition of this reaction was the synthesis of cephalotaxinone **61** from the iodoketone **60,** with potassium *t*-butoxide in liquid ammonia under photostimulation (Reaction 36) (*29*). This approach led to product **61** in 94% yield. Compared with the corresponding aryne type of reaction (15% yield), Ni(0) catalyzed reaction (30–35% yield), and the $S_{RN}1$ reaction stimulated with potassium metal (45% yield), Reaction 36, appears to be the best method (*29*).

Intramolecular arylation of enolates from iodo compounds, such as **62,** demonstrated that cyclization can be achieved efficiently (Reaction 37)

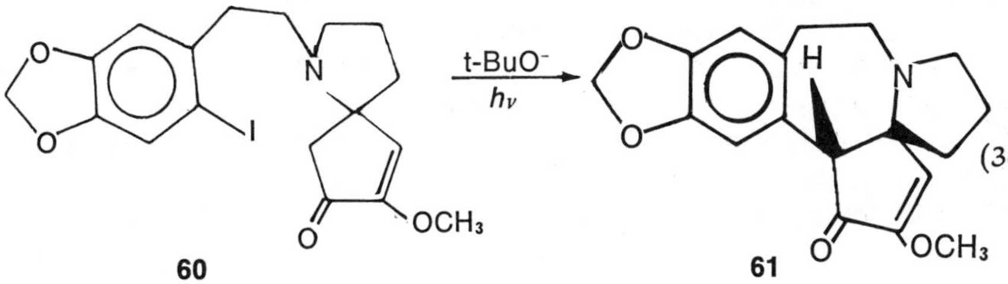

$$(36)$$

60   **61**

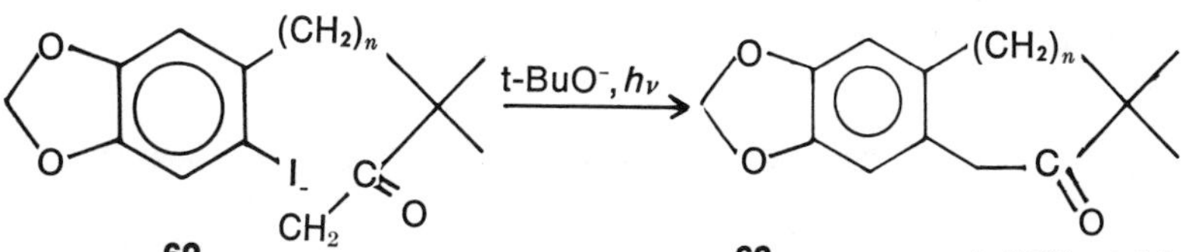

$$(37)$$

62   **63**

n = 1, 99% yield
n = 3, 73% yield
n = 5, 25-35% yield

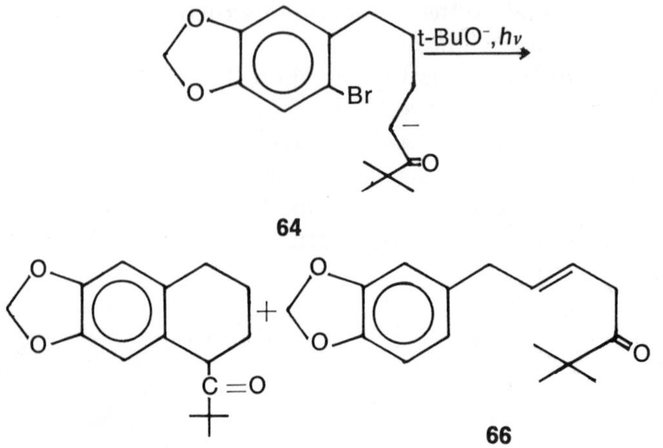

64

65

66

(38)

(*18*). When n = 1 a 99% yield of **63** is obtained, and with n = 5 the ten membered ring is obtained in reasonable yield (35%). However, when a β-hydrogen is available, the hydrogen atom abstraction competes with cyclization. For instance, with irradiation, bromoketone **64** with potassium *t*-butoxide gave only 13% yield of the cyclization product **65,** and a 65% yield of the β, γ unsaturated ketone **66** (*18*).

Formation of **66** was proposed to arise by way of the α, β-unsaturated isomer formed by intramolecular hydrogen atom transfer from the β-position of the carbonyl group to the intermediate aryl radical. Direct evidence for this mechanism was obtained from a deuterium labeling experiment. The labeled ketone **67** was subjected to the usual photostimulated $S_{RN}1$ conditions giving the products indicated in Reaction 39. The ratio of 69:68, which measures the competition between hydrogen transfer and cyclization, is much larger for ·the labeled compound (~ 5:1) than for the unlabeled (1:2). The difference was attributed to an isotope effect on the partitioning of the aryl radical intermediate (Reaction 40) (*6*).

The cyclic ketones **69a** and **69b** arise from the two possible enolates of ketone **67.** The relative yield of the two cyclization products varied with the number of methylene groups in the carbon chain (Reaction 41) (*6, 18*).

ELECTROCHEMICALLY GENERATED REACTIONS.   It was reported that the reaction of 2-chloroquinoline **70** with the enolate ions of acetone **11** and acetophenone **35** in liquid ammonia can be promoted electrochemically (*30*). By cyclic voltammetry, **70** presents two waves, the first (wave H, Figure 1) corresponds to the irreversible two-electron hydrogenolysis of the C–Cl bond (Reactions 42–45), followed by a one-electron reversible

(39)

(40)

(41)

wave (wave D, Figure 1) corresponding to the reduction of quinoline, formed at the first wave, into its anion radical (Reaction 46).

$$2\text{–ClQ} + e^- \longrightarrow (2\text{–ClQ})^{\overline{\cdot}} \tag{42}$$
$$\mathbf{70}$$

$$(2\text{–ClQ})^{\overline{\cdot}} \longrightarrow 2\text{–Q}^{\cdot} + Cl^- \tag{43}$$

$$2\text{–Q}^{\cdot} + e^- \longrightarrow 2\text{–Q}^- \tag{44}$$

$$2\text{–Q}^- \xrightarrow{NH_3} Q + NH_2^- \tag{45}$$

$$Q + e^- \rightleftharpoons (Q)^{\overline{\cdot}} \tag{46}$$

$$Q = \text{quinoline}$$

In the presence of acetone enolate ions, there is a change in the cyclic voltammogram. A decrease in the first and second waves is evident, and a new one-electron reversible wave appears (wave S, Figure 1) corresponding to the formation of the substitution product.

The wave attributed to the substitution product is located at potentials markedly more negative than the wave of quinoline. This indicates that the reduction of the enolate anion is involved rather than the reduction of the corresponding ketone. These facts were interpreted as follows: the 2-quinolyl radical formed in Reactions 42–43 reacts with **11** to give the radical anion **71** which, by homogeneous electron transfer to **70**, gives the substitution product. In the basic medium the product is converted into the enolate anion (Reactions 47–49).

$$2\text{–Q}^{\cdot} + \mathbf{11} \longrightarrow (2\text{–QCH}_2\text{COCH}_3)^{\overline{\cdot}} \tag{47}$$
$$\mathbf{71}$$

$$\mathbf{71} + \mathbf{70} \longrightarrow (2\text{–ClQ})^{\overline{\cdot}} + 2\text{–QCH}_2\text{COCH}_3 \tag{48}$$

$$2\text{–QCH}_2\text{COCH}_3 + \mathbf{11} \longrightarrow CH_3COCH_3 + 2\text{–Q}\overline{C}HCOCH_3 \tag{49}$$

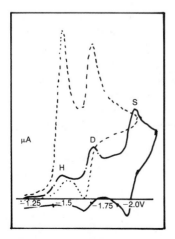

Figure 1: Cyclic voltammetry of 2-chloroquinoline 1.75×10^{-3} M (- - -) with out nucleophile added (—) in the presence of $^-CH_2COCH_3$ 2.5×10^{-2} M. Scan rate of 0.2 Vs^{-1}. (Reproduced with permission from Ref. 30. Copyright 1979, American Chemical Society.)

Similar results were obtained with acetophenone enolate **35** and other enolate anions (*30*).

Spontaneous Reactions.   In an investigation of photostimulated reactions of bromo- and iodobenzene with acetone enolate ions in dimethyl sulfoxide solution, Bunnett et al. observed that significant reaction occurred without light at room temperature. For example, iodobenzene reacted with **11** at about 23°C to release 31% of iodide ion in only 12 min in the dark. In ammonia the dark reaction seems to be much slower because there was only 2–8% of reaction after 180 min in the dark (*14*).

The dark reaction in dimethyl sulfoxide was thoroughly investigated with iodobenzene and pinacolone enolate ions **48**. The reaction was

$$PhI + {}^-CH_2COC(CH_3)_3 \xrightarrow[25°C]{DMSO} PhCH_2COC(CH_3)_3 + I^- \qquad (50)$$

48

shown to follow the integrated rate law of Equation 51, where *a* repre-

$$-x + a \ln\frac{a}{a-x} = k_c t \qquad (51)$$

sents the initial concentration of **48,** and *x* the concentration of iodide ion at any time. Although the authors were not able to ascribe any chemical significance to Equation 51, it serves for comparing reactivity.

Potassium iodide has no effect in reaction 50 indicating that *x* is not iodide acting as an inhibitor. Addition of 10% di-*t*-butyl nitroxide (based on iodobenzene concentration) decreases the rate by a factor of about 20. *p*-Dinitrobenzene has a dramatic effect on the reaction because it lowers the rate by a factor of 10 at a concentration of only 0.03 mole percent. On the other hand, oxygen strongly accelerates the reaction, but the product is destroyed under these conditions. The effect of added substances strongly suggests that the $S_{RN}1$ mechanism is operative.

Furthermore, competition experiments between bromo- and iodobenzene indicate that reactivity of the halides in the dark and under illumination is the same in both cases ($k_I/k_{Br} \sim 6$), which is strong evidence that the same mechanism prevails under both sets of conditions.

An aryne mechanism may be rejected because the dark reaction of *p*-iodo and *m*-iodotoluenes gave substitution products uncontaminated with the rearranged products. Furthermore, 2-iodomesitylene **72**—a substance incapable of being converted to a benzyne derivative—reacts with **48** to give the ketone **73** (Reaction 52).

However, in the reaction of *m*-iodoanisole with **48,** along with the substituted product **74,** the methylthio-substituted phenol **75** was

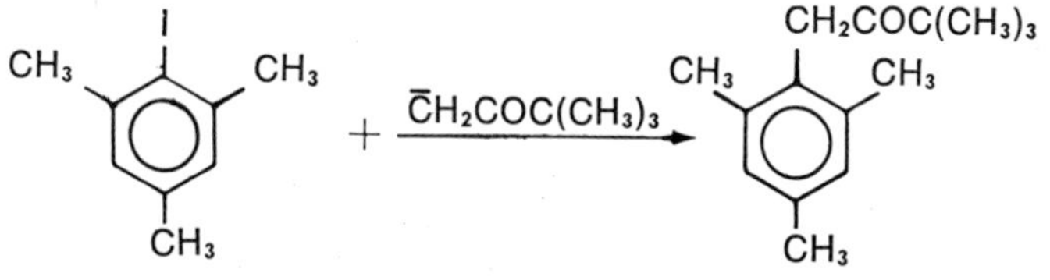

(52)

(50-58%)
74

(11-15%)
75

formed. Formation of **75** was thought to come from the addition of dimethyl sulfoxide to 3-methoxybenzyne **76**, as was suggested for the action of potassium t-butoxide in dimethyl sulfoxide on an aryl bromide (Reactions 53–55) (*32*).

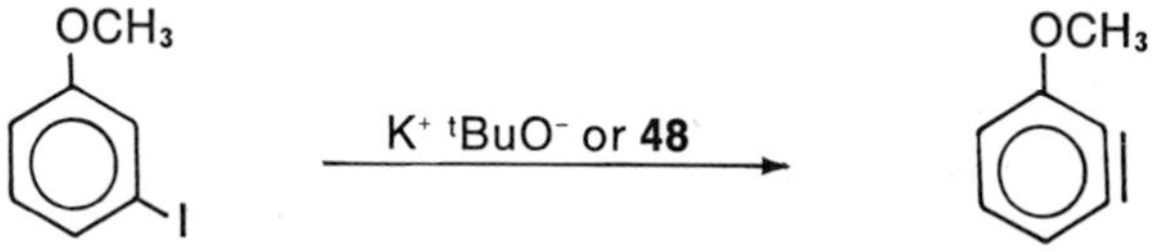

(53)

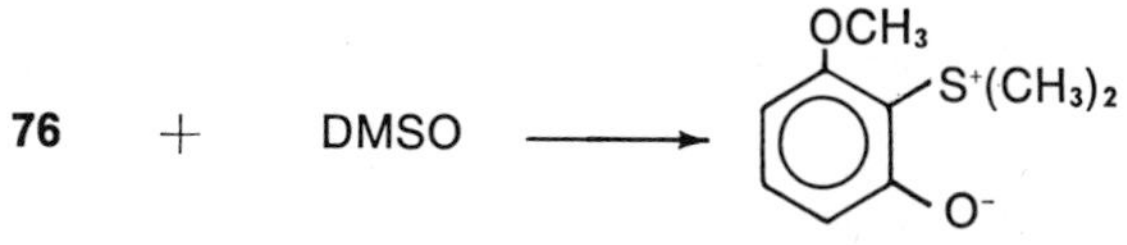

76

(54)

77

77 $\longrightarrow$ **75** (55)

Some of the product **74** may come from addition of **48** to the aryne **76**, although experiments in the presence of p-dinitrobenzene (which almost completely inhibits the $S_{RN}1$ mechanism) indicate that most of **74** comes from the $S_{RN}1$ mechanism.

Pyrimidines, pyridazines and pyrazines (appropriately substituted) were found to react in the dark with enolate ions of acetone, pinacolone, diisopropyl ketone and acetophenone in liquid ammonia. However, the only substrate that gave satisfactory displacement of chloride (82–98%)

was 2-chloropyrazine (Reaction 56) (*33*). Inhibition by di-*t*-butyl nitroxide proves the radical nature of these reactions.

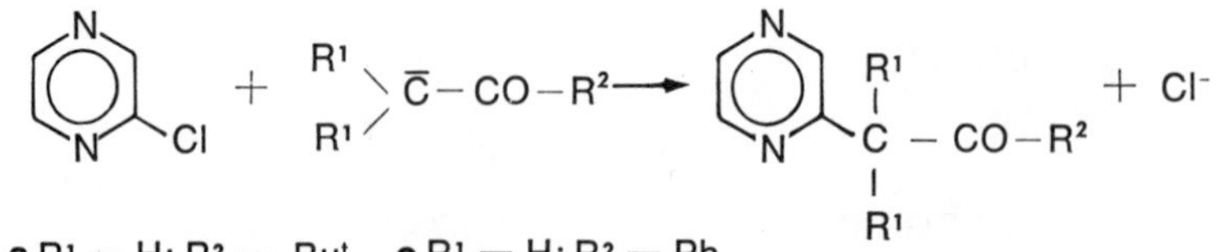

a $R^1 = H$; $R^2 = Bu^t$ **c** $R^1 = H$; $R^2 = Ph$

b $R^1 = Me$; $R^2 = Pr^i$ **d** $R^1 = H$; $R^2 = Me$

Esters, Aldehyde, and *N,N*-Disubstituted Amide Enolate Ions. ESTER ENOLATE IONS.   Although there is no detailed study of the enolate ions derived from esters, some experiments have been done that reveal the behavior of these anions in photostimulated reactions in liquid ammonia.

The lithium or potassium enolate of *t*-butyl acetate **78** reacted under photostimulation in liquid ammonia to give the mono- and disubstituted products **79** and **80** (Reaction 57, Table IV) (*18*).

$$ArX + {}^-CH_2CO_2t\text{-}Bu \xrightarrow{hv} ArCH_2CO_2t\text{-}Bu + Ar_2CHCO_2t\text{-}Bu \quad (57)$$
$$\textbf{78}\textbf{79}\textbf{80}$$

Table IV

Photostimulated Reactions of Ester Enolate Ions in Ammonia

ArX	$^-CRR'CO_2$-*t-butyl*	*Products, Yield*	*Ref.*
PhBr^a	R=R'=H	—	4
PhBr	R=R'=H	61% (monosubstituted)	4
		37% (disubstituted)	—
PhBr	R=H; R'=Me	60% (monosubstituted)	6
PhBr	R=R'=Me	11% (monosubstituted)	6
p-(MeO)C$_6$H$_4$Br	R=R'=H	67% (monosubstituted)	18
		29% (disubstituted)	—
p-(MeO)C$_6$H$_4$Br	R=R'=Me	5% (monosubstituted)	18
		50% (reduced)	—

^{a}Dark reaction

As found previously with ketone enolate ions, when the ester enolate has a β-hydrogen available, hydrogen atom abstraction competes with the coupling reaction. Under the same experimental conditions as for arylation of the enolate **78** (which gave more than 95% of mono- and disubstituted products), the photostimulated reaction of p-bromoanisole with the enolate of t-butyl-2-methylpropionate gave only 5% of the arylation product, and about 50% of the reduction product, anisole (18).

The photostimulated reaction of p-bromoanisole with the labeled enolate ion **81**, produced 56% of anisole, 35% labeled with deuterium, confirming the major role of hydrogen atom abstraction from the enolate (Reaction 58).

ALDEHYDE ENOLATE IONS.   There are only a few reports of photostimulated reactions in liquid ammonia of aldehyde enolate ions. These ions are known to be formed in liquid ammonia (34), but in the photostimulated reaction with bromobenzene or 1-chloronaphthalene there was little reaction that led to dehalogenation and substitution products (4).

The photostimulated reaction of p-iodoaniline with acetaldehyde enolate ion **82** gave only the reduction product, aniline (23). Aniline was also a prominent product from the reaction of **82** with o-iodoaniline. But in this case the nucleophilic reaction leading ultimately to indoles also occurred (Reaction 59) (23). It is not clear why the o-isomer gave the substitution product, while the para isomer did not. On the other hand the reaction of o-iodoaniline with ketone enolate ions gave predominantly indole derivatives.

If we compare the reactivity of aldehyde enolate ions with ketone enolate ions of similar structure, we see that the latter are more reactive in the photostimulated $S_{RN}1$ reactions, whereas more reduction products are formed with the former.

N,N-DISUBSTITUTED AMIDE ENOLATE IONS.   The ion of acetamide **84** in liquid ammonia failed to undergo reaction with aryl halides under photostimulation (36), or by electron stimulation from a cathode (30). However, the enolate ions of N,N-disubstituted amides **85** gave good yields of substitution products under photostimulation in liquid ammonia (36).

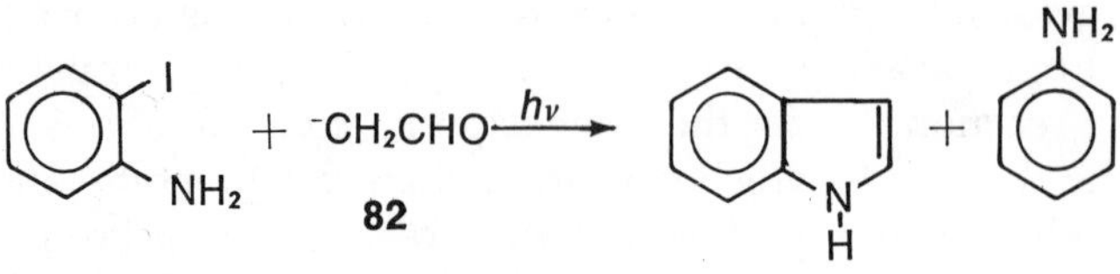

(59)

$$CH_3CONH^- \qquad {}^-CH_2CONRR'$$

$$84 \qquad\qquad 85$$

a, R = CH₃, R′ = Ph

b, R = R′ = CH₃

c, R = R′ = −(CH₂)₅−

d, R = R′ = −(CH₂)₂O(CH₂)₂−

With iodobenzene and **85** without light there was 34% of reaction, but a higher yield was obtained under photostimulation. Reactions of various aryl halides with **85a** or **85b** gave good product yields. The enolate ion of **85c** was insoluble in ammonia, and there was little reaction with bromobenzene. However, **85d,** which is soluble in ammonia, gave high yields of products. Note that the change of a −CH₂− of piperidine by an −O− in morpholine changes the solubility in ammonia drastically (Table V).

When the solution of N-(p-chlorobenzyl)-N-methylacetamide enolate ion **86** in liquid ammonia was irradiated, 80% yield of chloride ion was found. The solid formed in these reactions was almost insoluble in all common solvents, except dimethyl sulfoxide. In the dark, there was no reaction. The solid product could not be fully identified. However, its high melting point, low solubility, and the high yield of chloride ion suggest that there was a coupling of several molecules of substrate to give oligomeric and/or polymeric products (28).

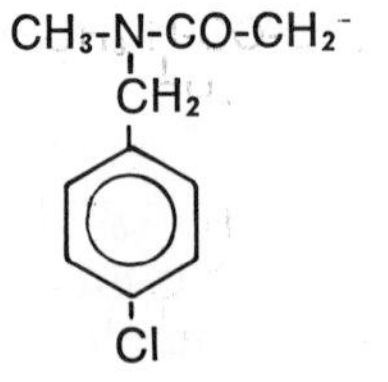

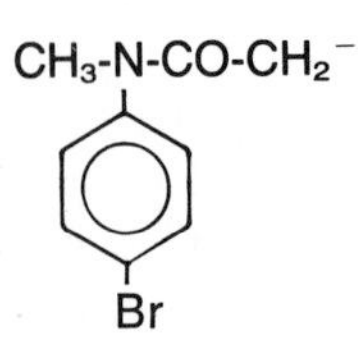

On the other hand, the photostimulated reaction of the enolate **87** gave only 11% yield of bromide ion, despite the fact that bromide normally is a better leaving group than chloride. It was suggested that the difference in reactivity of **86** and **87** is due to the fact that, in **87**, the nucleophilic center interacts with the leaving group through the pi sys-tem thereby decreasing its reactivity. In substrate **86**, both the nucleo-philic center and the leaving group are separated by an sp^3 carbon atom (*28*).

Further evidence pertaining to the reactivity of **87** was obtained from the reaction of **87** with chlorobenzene (1:5 molar ratio of chlorobenzene and **87**). It was found that all the chlorobenzene reacted, together with 53% of **87**. These results show that the pi system interaction of the anion with the aromatic ring decreases the reactivity of **87** towards bromide loss, probably due to a slow initiation step. The addition of a more reactive substrate can entrain the reaction. In this case, a neutral molecule with a poor leaving group is a better substrate than one with a better leaving group when the latter is conjugated with negative charge (28).

The photostimulated reaction of *N*-alkyl-*N*-acyl-*o*-chloroaniline en-olate ions **89** undergo cyclization to afford oxindoles **90** in tetrahydro-furan–hexane solution. The enolate ions were prepared by acid–base

Table V

Photostimulated Reaction of Aryl Halides with *N,N*-Disubstituted Amide Enolate Ions in Ammonia

ArX	R'RNCOCH$_2$K R	R'	Irradiation Time (min)	Substitution Product, Yield (%)
PhCl	CH$_3$	Ph	120	72
PhCl	CH$_3$	Ph	120[a]	—
PhBr	CH$_3$	Ph	120	60
PhI	CH$_3$	Ph	30	80[b]
PhI	CH$_3$	Ph	30[a]	34
1-Chloronaphthalene	CH$_3$	Ph	120	50
9-Bromophenanthrene	CH$_3$	Ph	90	80
PhBr	–(CH$_2$)$_5$–		120	9[c]
PhBr	CH$_3$	CH$_3$	120	80[d]
PhBr	–(CH$_2$)$_2$O(CH$_2$)$_2$–		180	56[e]
PhCl	–(CH$_2$)$_2$O(CH$_2$)$_2$–		180	75[f]

[a]Dark reaction.
[b]10% Disubstitution product.
[c]As bromide ion.
[d]18% disubstitution product.
[e]12% disubstitution product.
[f]18% disubstitution product.
Source: Reproduced from Ref. 36. Copyright 1980, American Chemical Society.

reaction of **88** with lithium diisopropyl amide (LDPA) (37) (Reaction 60). The dianion of N-acyl-o-chloroanilines (**89**, R = Li) also gave good yields of oxindoles **90**.

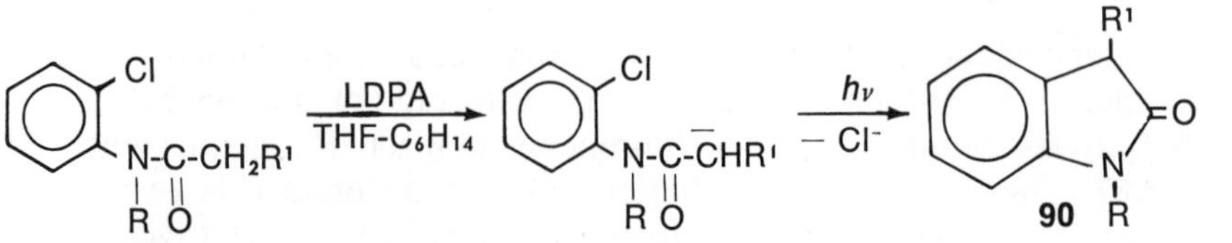

	R	R¹	
a	CH₃	Ph	64%
b	CH₃	H	82%
c	CH₃	m-C₄H₉	73%
d	PhCH₂	H	32%
e	H	Ph	63%
f	H	H	74%
g	H	CH₃	73%

In addition, pyridine derivatives such as **91** gave good yields of azaoxindoles **92** (Reaction 61) (37). However, this reaction must be carried out at −78°C to prevent decomposition of the enolate ion. The dianion derived from N-acetyl-2-chloro-3-aminopyridine failed to give the photostimulated ring closure reaction (37).

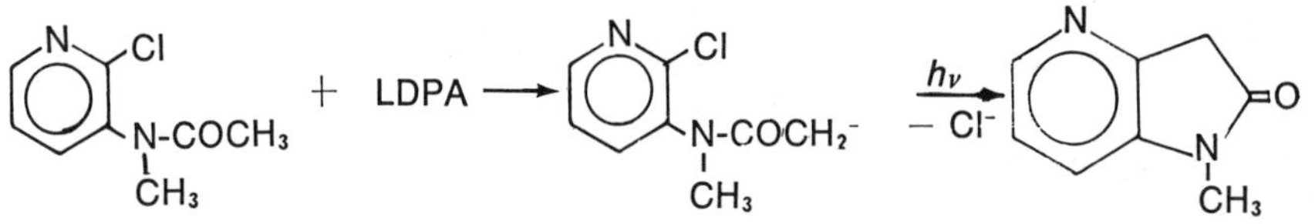

These reactions apparently proceed by the $S_{RN}1$ mechanism. The benzyne mechanism was rejected because the reactions did not occur in the dark. Quenching the dark reaction with D_2O resulted in quantitative recovery of starting material containing 0.95 deuterium atom in the respective acetyl methyl group. Also, these reactions were inhibited by di-t-butyl nitroxide (37).

In particular, isolation of oxindoles from 3-substituted-2-chloro-anilides **93** rules out an aryne mechanism (Reaction 62). These reactions

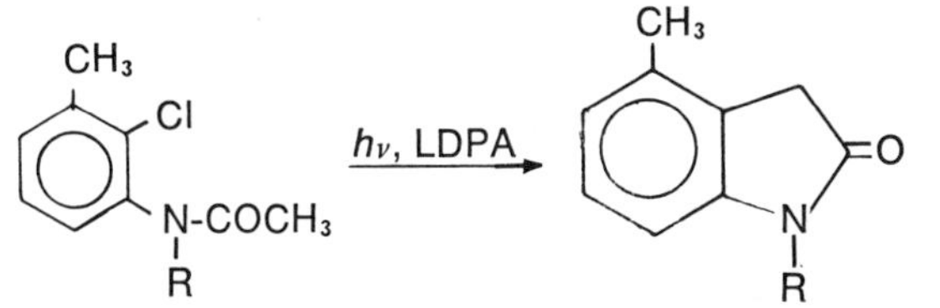

93a, R = CH₃   **94a, R = CH₃ (87% yield)**
  b, R = H   **94b, R = H   (76% yield)**

are unsatisfactory in liquid ammonia. The photostimulated reaction of **88d** with potassium amide afforded 57% yield of oxindole **90d**, along with an 8% yield of aryne-derived *N*-benzyl-3-aminoacetanilide. Attempted cyclizations of other substrates under the same reaction conditions led to much lower yields of the desired oxindoles than obtained with LDPA in tetrahydrofuran–hexane. The photostimulated reaction of *N*-(2-bromobenzyl)-*N*-methylacetamide enolate ion in liquid ammonia also failed to cyclize to give the isoquinoline derivative (*38*). The insolubility of the substrate, together with the formation of reduction and tarry products, do not make this reaction seem promising. However, the reaction has not yet been tried in tetrahydrofuran–hexane where the oxindoles were obtained in very good yields.

Other Stabilized Carbanions.   1-PICOLYL ANION.   Under conditions conducive to the $S_{RN}1$ mechanism, 2- and 4-picolyl anions have been arylated. Chlorobenzene and phenyltrimethylammonium iodide are unreactive toward this nucleophile in the dark, but good yields of the arylated products were obtained in potassium metal or light stimulated reactions in ammonia (Reaction 63) (*39*).

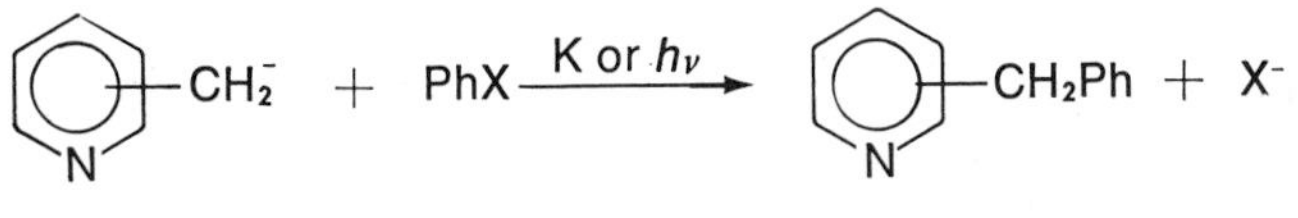

95a, 2-picolyl   **96a, X = Cl**   **97a, 2-benzylpyridine**
  b, 4-picolyl   **b, X = ⁺NMe₃**   **b, 4-benzylpyridine**
                        c, X = Br
                        d, X = I

Substrates **96c** and **96d** react with **95a** in the dark, probably by a benzyne mechanism. The approximate pK_a's of 2- and 4-picoline are about 31 and 29 respectively (*40*), not far from the pK_a of 32.5 for ammonia (*1*). Therefore it is possible that picolyl anions, or the amide ions they are in equilibrium with, could bring about benzyne formation with iodo- or bromobenzene. The yield of substitution product **97a** is somewhat improved by solvated electron or light stimulation, but the mechanism is considered uncertain for **96c** and **96d** (*39*).

On the other hand, 2-bromomesitylene **98**, which is unable to form aryne upon reaction with strong bases, underwent photostimulated reaction with **95a** and **95b** to give excellent yields of the corresponding arylated products **99** and **100** (55% and 89% respectively) (*39*).

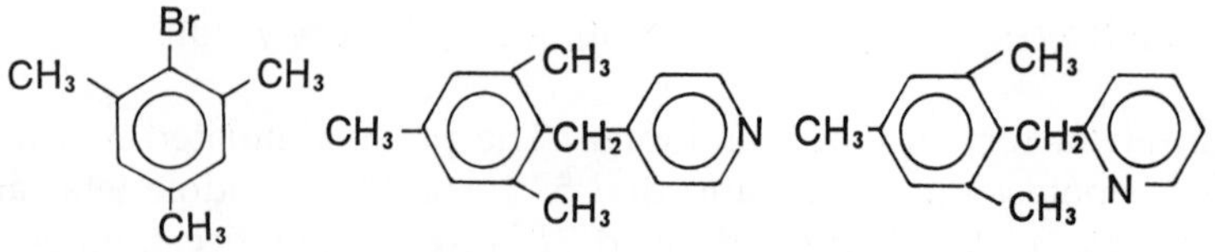

CYANIDE ION.   The cyclic voltammogram of *p*-iodonitrobenzene **101** in dimethyl sulfoxide shows a cathodic peak near −1 V followed by another peak at slightly greater negative potential (cathodic peak near −1.09 V), which corresponds to the reduction of nitrobenzene to its radical anion (*41*).

In the presence of cyanide ion, the cyclic voltammogram of **101** changes, and only the reduction wave attributed to the one-electron reduction of **101** to its anion radical is readily discernible on the first cathodic sweep. After reversal of the potential scan, an anodic wave is observed at −0.72 V. Subsequent cathodic sweeps show an additional reduction process at −0.78 V. These waves are the result of the one-electron reduction of *p*-nitrobenzonitrile following Reactions 64–66 (*41*). Confirmation of *p*-nitrobenzonitrile radical anion (**103**) formation was obtained by cyclic voltammetry of **104** and by electron spin resonance (ESR) spectroscopy.

In addition, 2- and 3-iodonitrobenzenes gave the corresponding substitution products (*41*).

During cyclic voltammetry of 4-bromobenzophenone **105** and 1-bromonaphthalene, substitution of aryl radicals by cyanide ion was observed. However, with the latter compound the wave corresponding to the substitution product was much lower. Electrolysis of **105** in acetonitrile with cyanide ions gave a 95% yield of 4-cyanobenzophenone (**106**) with only 0.25 Faraday/mol of electricity consumed. In contrast, no 1-cyanonaphthalene was isolated from the electrolysis of 1-bromonaphthalene and cyanide ion in dimethyl sulfoxide (Reaction 67) (*42*).

In an attempt to study the cyanide nucleophile in liquid ammonia, it was allowed to react with bromobenzene and 1-chloronaphthalene, and with *p*-iodonitrobenzene in reactions stimulated by light or solvated electrons (*4*). However, no substitution products were found. The elec-

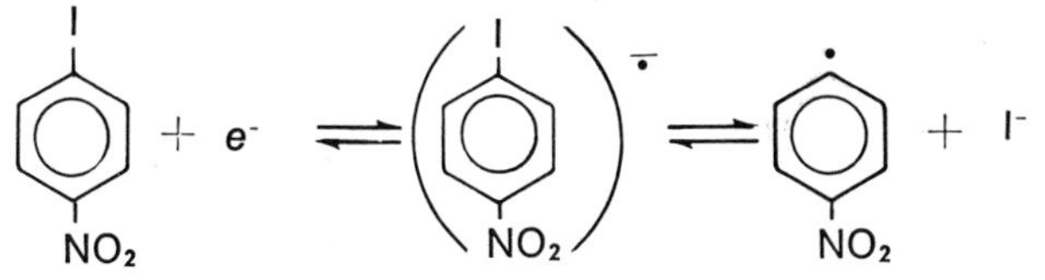

$$ (64) $$

101   102

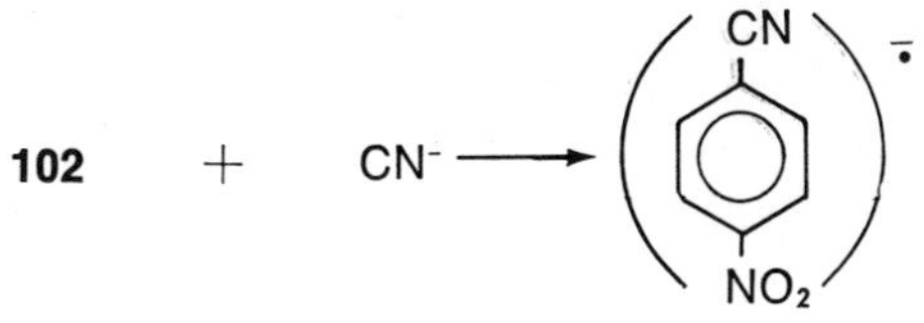

$$ (65) $$

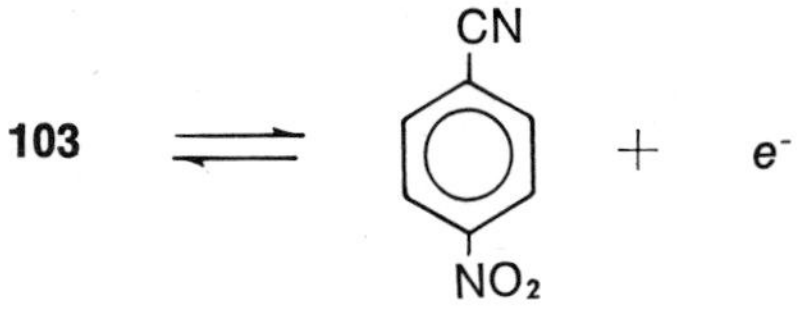

103

$$ (66) $$

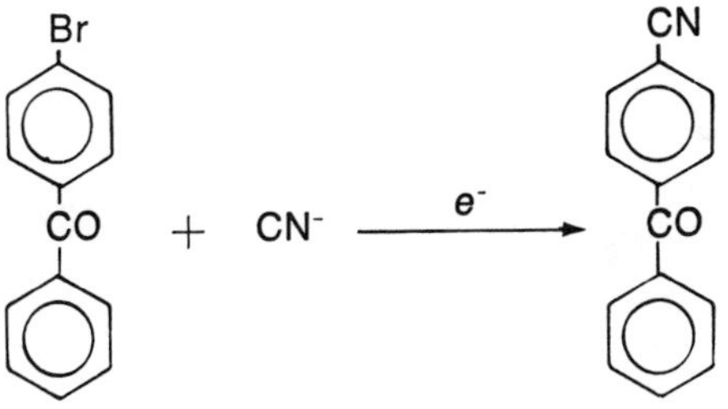

104

$$ (67) $$

105   106

trochemically induced reaction of cyanide ions with 2-chloroquinoline in liquid ammonia was also unsuccessful. The lack of reactivity of cyanide ion in liquid ammonia was attributed to its poor solubility, rather than its low intrinsic reactivity (30).

α-CYANO CARBANIONS.   *Stimulation with Solvated Electrons.* Nitrile α-hydrogens are acidic enough to form carbanions in liquid ammonia. Reaction with monosubstituted benzenes promoted by solvated electrons gives several products (Reaction 68) (43).

$$ PhX + RCHCN^- \xrightarrow[NH_3]{K} PhCHRCN + PhCH_2R + PhH \qquad (68) $$
$$ + Ph_2CHR + (PhCHR)_2 + PhNH_2 $$

The relative amount of products obtained depends on the PhX employed, the temperature, the proportion of reactants, and the way the experiment is conducted.

In reactions with cyanomethyl anion **107** (R = H, Reaction 68), toluene is the major product in yields of 14–50%, benzene in 8–44%, 1,2-diphenylethane **111** in 3–40%, phenylacetonitrile **109** in 2–31%, and diphenylmethane and aniline to minor extents. These results were explained by the reactions of Scheme I (43).

Reactions 69–71 are obvious and need no further explanation. Major differences in the other examples of $S_{RN}1$ reactions are that the radical anion **108** (formed when the phenyl radical couples with **107**, (Reaction 72) undergoes two competitive reactions: electron transfer to PhX, to give phenylacetonitrile **109** (Reaction 73), or C–CN bond fragmentation to give benzyl radical **110** and cyanide ion (Reaction 74).

Benzyl radical **110** can be reduced further with another solvated electron to give benzyl anion, which is protonated by ammonia to give toluene and amide ions (Reaction 75), or **110** can dimerize forming 1,2-diphenylethane **111** (Reaction 76).

$$PhX + e^- \rightarrow (PhX)^- \qquad (69)$$

$$(PhX)^- \rightarrow Ph^\cdot + X^- \qquad (70)$$

$$Ph^\cdot + e^- \xrightarrow{NH_3} Ph^- \rightarrow PhH + NH_2^- \qquad (71)$$

$$Ph^\cdot + {}^-CH_2CN \rightarrow (PhCH_2CN)^- \qquad (72)$$
$$\quad \mathbf{107} \qquad\qquad \mathbf{108}$$

$$\mathbf{108} + PhX \rightarrow (PhX)^- + PhCH_2CN \qquad (73)$$
$$\qquad\qquad\qquad\qquad \mathbf{109}$$

$$\mathbf{108} \rightarrow PhCH_2^\cdot + CN^- \qquad (74)$$
$$\qquad\qquad \mathbf{110}$$

$$\mathbf{110} \xrightarrow{e^-,\ NH_3} PhCH_3 \qquad (75)$$

$$2\ \mathbf{110} \rightarrow PhCH_2CH_2Ph \qquad (76)$$
$$\qquad\qquad\qquad \mathbf{111}$$

Scheme I

Because of its relative acidity, the phenylacetonitrile product **109** will react with the stronger base **107** almost completely (Reaction 77), and the phenylacetonitrile anion **112** produced can react further, by Reactions 78–80, giving the other products observed.

$$109 + 107 \longrightarrow CH_3CN + Ph\overline{C}HCN \tag{77}$$
$$\textbf{112}$$

$$112 + Ph^{\cdot} \longrightarrow (Ph_2CHCN)^{\overline{\cdot}} \tag{78}$$
$$\textbf{113}$$

$$113 \longrightarrow Ph_2CH^{\cdot} + CN^{-} \tag{79}$$
$$\textbf{114}$$

$$114 \xrightarrow{e^{-},\ NH_3} Ph_2CH_2 \tag{80}$$

This is a rather complex system, with several reactive intermediates involved, each of which give rise to different products. However, it has been shown to be an useful method for introducing an alkyl group into a benzene ring, replacing the nucleofugal group.

Examples were reported concerning the introduction of methyl, ethyl, propyl, isopropyl, butyl, isobutyl and benzyl groups (Table VI). Although the yields reported are less than 50%, this method offers several advantages compared with others such as Friedel–Crafts alkylation or benzyne mechanisms. Attempts to find optimum conditions for this system could possibly improve the yields obtained.

The fact that the relative amount of substitution and decyanation product was thought to be governed by the competitive reactions of the radical anion **108** (Reaction 73 vs. 74) stimulated an investigation of this reaction in the presence of good electron acceptors which might increase the electron transfer rate in Reaction 73. However, in the reaction of PhCl or PhBr with acetonitrile anion **107** in the presence of a saturated solution of naphthalene as electron acceptor, the amount of **109** drops to zero. It was suggested that the electron transfer reaction of Reaction 73 occurs through an adduct formed between **108** and an electron acceptor (EA) present in the reaction solution, including the halobenzene substrate (Reaction 81) (44). Formation of this adduct increases the stability of the

$$108 + EA \rightleftharpoons (108-EA)^{\overline{\cdot}} \begin{array}{c} \nearrow (EA)^{\overline{\cdot}} \ + \ \textbf{109} \\[2mm] \searrow EA \ + \ CN^{-} \ + \ \textbf{110} \end{array} \tag{81}$$

radical anion. This gives it time to react with another electron (Reaction 82), thereby forming mainly toluene at the expense of **109**.

$$(108 - EA)^{\overline{\cdot}} \xrightarrow{e^{-}} (108-EA)^{=} \longrightarrow PhCH_2^{-} + CN^{-} + EA \tag{82}$$

Table VI

Reaction of Monosubstituted Benzenes with α-Cyanocarbanions Stimulated by Potassium Metal in Liquid Ammonia

PhX X	⁻CHR–CN R	PhCH$_2$R %	PhH %
F	H	30–49	11–31
Cl	H	14–46	8–30
Br	H	26	43
I	H	14	32
OP(O)(OEt)$_2$	H	43	11
$^+$NMe$_3$	H	26	21
Cl	CH$_3$	34	38
OP(O)(OEt)$_2$	CH$_3$	14	35
$^+$NMe$_3$	CH$_3$	8	36
F	CH$_2$CH$_3$	28	35
Cl	CH$_2$CH$_3$	37	27
I	CH$_2$CH$_3$	6	27
OP(O)(OEt)$_2$	CH$_2$CH$_3$	28	20
Br	(CH$_3$)$_2$[a]	27	31
Cl	n-C$_3$H$_7$	56	38
OP(O)(OEt)$_2$	n-C$_3$H$_7$	38	27
Cl	(CH$_3$)$_2$CH	37	31
OP(O)(OEt)$_2$	(CH$_3$)$_2$CH	31	32
OP(O)(OEt)$_2$	Ph	12	25
$^+$NMe$_3$	Ph	17	43

[a]Isobutyronitrile
Source: Reproduced from Ref. 43, Copyright 1973, American Chemical Society.

The reaction of 3-bromothiophene with **107** at −68°C stimulated with sodium metal in ammonia afforded 3-cyanomethylthiophene **115** (15%), along with 3,3′-(1,2-ethanediyl)bisthiophene **116** (4.5%), and an unspecified amount of starting material, thiophene, and 3-methyl-

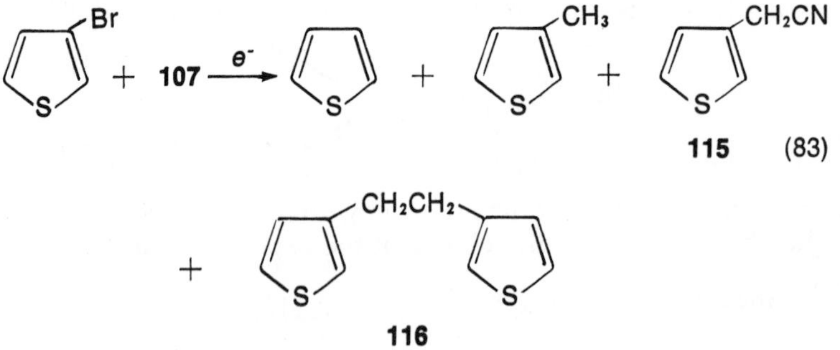

115   (83)

116

thiophene (Reaction 83) (45). These products suggest that the reaction pathway is the same as for halobenzenes.

Photostimulated Reactions.   Bromobenzene and cyanomethyl anions do not react in the dark. However, after 2 h of irradiation there was 38% reaction, and the products obtained were **109, 111** and very low yields of diphenylmethane, toluene and benzene (43).

The low overall reactivity of cyanomethyl anions **107** as compared with acetone enolate anions (which give 100% reaction with bromobenzene in 10 min of irradiation) is because the intermediate radical anion **108** expels cyanide ion forming benzyl radicals (Reaction 74). In this system, the benzyl radical is ineffective at propagating a radical chain, and accumulates until it dimerizes (Reaction 76) or is reduced by accepting an electron (Reaction 75), both of which are termination steps. In fact, **107** seems to be more reactive toward phenyl radicals than ketone enolate ions because, in competitive reactions, **109** or products derived from **109** are predominantly formed (43).

The product pattern observed in photostimulated reactions resembles that from potassium metal stimulated reactions, except that the yields of benzene and toluene are almost zero, as expected according to the genesis of these products proposed in Reactions 71 and 75.

An exhaustive investigation of the reaction of halobenzenes with **107** showed the presence of other products along with those cited above, albeit in very low yields. They include 1,1,2-triphenylethane **117** and 1,1,2,2-tetraphenylethane **118**. These can be accounted for by Reactions 84 and 85 which represent the coupling of diphenylmethyl radicals **114** with benzyl radical **110** to give **117**, and with itself to give **118** (44).

The photostimulated reaction of cyanomethyl anions was studied with the halobenzenes, and also in the presence of several electron acceptors. Overall reactivity was found to be similar to the known order for other $S_{RN}1$ reactions, namely PhI>PhBr>PhCl>PhF, but the product

$$Ph_2CH^{\cdot} + PhCH_2^{-} \rightarrow Ph_2CHCH_2Ph \qquad (84)$$
$$\textbf{114} \qquad \textbf{110} \qquad \textbf{117}$$

$$\textbf{114} + \textbf{114} \rightarrow Ph_2CHCHPh_2 \qquad (85)$$
$$\textbf{118}$$

distribution was almost independent of the halobenzene substrate, or other substances added.

According to the proposed mechanism, the product distribution is determined by the relative rates of Reactions 73 and 74, where the radical anion **108** either transfers its odd electron or decomposes into benzyl radical and cyanide ion. These reactions are of different molecularity, two for the former and one for the latter, and the fact that the product distribution was not affected by changes in the halobenzene substrate, its concentration, or the presence of good electron acceptors (which should increase the rate of Reaction 73 leaving the rate of Reaction 74 unaffected) indicates that some aspects of the mechanism should be changed.

It is clear that the substitution products as well as the decyanation products should come from a common intermediate, and two mechanistic models have been suggested (*44*). One of these has been mentioned in regard to the potassium metal stimulated reaction (Reaction 81). This was based on the knowledge that dianions and electron acceptors form adducts before the actual electron transfer reaction (*46*) and that charge transfer complexes are formed as intermediates in other electron transfer reactions.

The second mechanism is the possibility of electron release from **108** to the solvent ammonia. This mode of formation of **109** has the same concentration dependence as **110**, and should be independent of the substrate and electron acceptors (Reactions 86–87).

In the photostimulated reaction of **107** with 1-chloronaphthalene **119**, the only product obtained was 1-cyanomethylnaphthalene **120**. Products derived from C–CN bond fragmentation were not observed (Reaction 88) (*47*).

$$(PhCH_2CN)^{\overline{\cdot}}\ \textbf{108} \longrightarrow PhCH_2CN\ \textbf{109} + e^- \quad (86)$$

$$(PhCH_2CN)^{\overline{\cdot}}\ \textbf{108} \longrightarrow PhCH_2^{\bullet}\ \textbf{110} + CN^- \quad (87)$$

107

119 120 (88)

The completely different behavior of **119** compared to that of the halobenzenes was suggested to result from the different localizations of the odd electron in the radical anion intermediates.

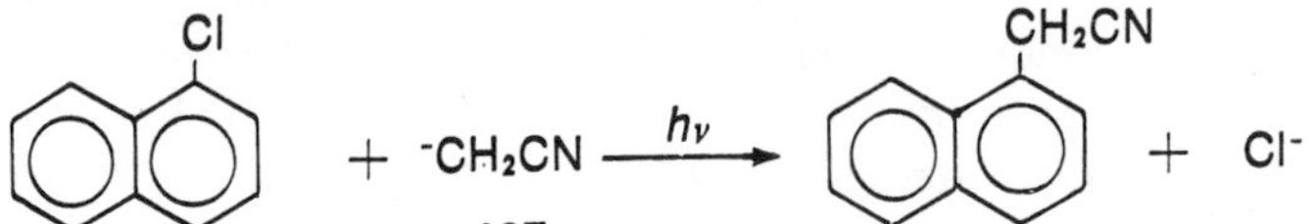

When C–CN bond fragmentation occurs (this is a termination step in the chain propagation cycle of the $S_{RN}1$ mechanism), the overall reactivity of halobenzenes with **107** is sluggish. But in the photostimulated reaction of **119** with **107** (where this termination step does not exist) **119** reacts quantitatively with less than 15 min of irradiation. This is further evidence that when there is an increase in the rate of termination steps in the $S_{RN}1$ mechanism, the overall reactivity is low.

The suggestion that aromatic compounds in which the aromatic moiety has a low energy antibonding molecular orbital (LUMO) will give substitution products without fragmentation, was supported when it was found that haloderivatives of phenanthrene, biphenyl, benzophenone and pyridine gave very good yields of the expected substitution products (Table VII) (47). Moreover, 2-chloropyrazine reacted with phenylacetonitrile anion in liquid ammonia in the dark, giving up to 78% of the substitution product. Inhibition by di-t-butyl nitroxide provided evidence that the reaction was mainly a thermal $S_{RN}1$ substitution (33).

The photostimulated reaction of 3-bromothiophene gave 3-cyanomethylthiophene **115** (37.5%) and 3,3'-(1,2-ethanediyl)bisthiophene **116** (6.5%). This result suggests .decomposition of the radical anion in the sense of Reactions 73 and 74 (45), and is consistent with the fact that the LUMO of the thiophene moiety is higher than that of benzene itself.

The photostimulated reaction of 2-bromothiophene and **107** afforded **115** (35%) and **116** (6%) most likely by way of 3-bromothiophene (Reaction 89) (45). The base catalyzed isomerization of 2-bromothiophene to 3-bromothiophene is a well known fact (48).

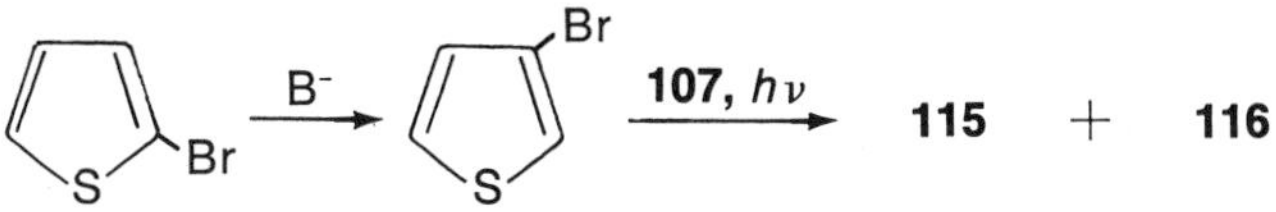

$$\text{(89)} \qquad \underset{S}{\bigcirc}\!\!-Br \xrightarrow{B^-} \underset{S}{\bigcirc}\!\!-Br \xrightarrow{\textbf{107}, h\nu} \textbf{115} \ + \ \textbf{116}$$

Table VII

Photostimulated Arylation of the Cyanomethyl Anion in Ammonia

ArX	Irradiation Time (min)	ArCH$_2$CN (%)	Ref.
Bromobenzene	120[a]	—	43
Bromobenzene	120	8[b]	43
1-Chloronaphthalene	100	89	47
2-Chloronaphthalene	45	93	47
9-Bromophenanthrene	120	70	47
4-Chlorobiphenyl	20	94	47
4-Chlorobenzophenone	60	97	47
2-Chloropyridine	30	56	47
3-Bromothiophene	45	37	45
2-Bromothiophene	60	35[c]	45

[a]Dark reaction.
[b]18% yield of 1,2-diphenylethane.
[c]Product obtained is 3-thienylacetonitrile.

Silicon Nucleophiles

Although the reaction of triorganosilyl anions with haloarenes and haloheteroarenes to give substitution products is a known reaction, mechanistic studies are scarce (Reaction 90) (49).

$$ArX + R_3Si^- \rightarrow ArSiR_3 + X^- \qquad (90)$$

From the work of Shippey and Dervan (50) some features of this reaction are now known. Trimethylsilyl anions, formed in the reaction of hexamethyldisilane and sodium or potassium methoxide in hexamethylphosphoramide (HMPT), reacted with p-halotoluenes (Cl, Br, and I) to give mainly substituted product (63–92%), and toluene as the minor product (4–26%) (Reaction 91) (50). Also, 2-bromopyridine reacted with trimethylsilyl anions to give the substituted product in high yield (80%).

$$X = Cl, Br, I$$

To account for these results, four alternative nucleophilic substitution reactions were considered: 1, an aryne mechanism that was excluded based on regiochemical results; 2, a direct nucleophilic substitution and while not completely discarded, this mechanism was considered unlikely because it is seldom observed with unactivated aryl halides, and it is not possible to explain the appearance of toluene as a product; 3, hydrogen metal exchange to afford an aryl–metal (metal = K, Na, Li) intermediate; and 4, the $S_{RN}1$ reaction.

There was 30% deuterium incorporation when the reaction mixture of iodobenzene and potassium trimethylsilyl was quenched with D_2O, indicating that at least part of the reaction may proceed via the phenylpotassium intermediate. On the other hand, replacement of CH_3O^- by CD_3O^- in the reaction of iodobenzene with hexamethyldisilane affords the reduction product with 64% of deuterium incorporation, consistent with a mechanism involving phenyl radicals. It appears that at least two intermediates are involved in this reaction, and further investigation is required if we are to fully understand the reaction.

Trimethylsilyl anions have been used to reduce a series of cyclic unsaturated ketones in HMPT to their corresponding radical anions, and their ESR spectra were recorded (51). These results suggest that the trimethylsilyl anion is prone to transfer an electron to good electron acceptors, thus giving some support to its involvement in $S_{RN}1$ reactions.

Germanium Nucleophiles

The reaction of haloarenes with triphenylgermyllithium and triethyl-germyllithium in tetrahydrofuran, and in hexamethylphosphoramide has been reported (52). Bromo- and iodotoluenes gave the straight-forward substitution product (Reaction 92) uncontaminated with isomeric (cine) substitution products.

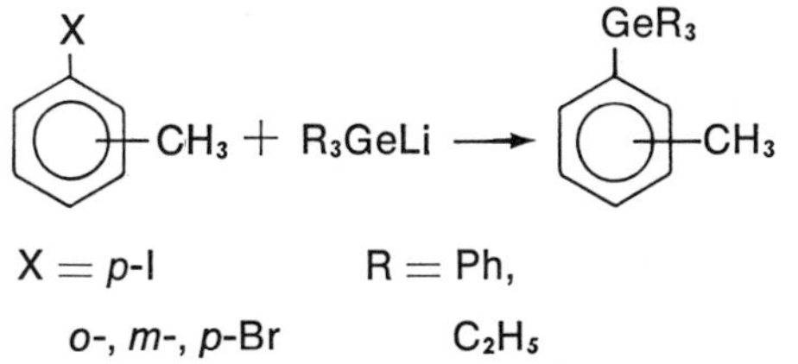

$$X = p\text{-}I \qquad R = Ph,$$
$$o\text{-}, m\text{-}, p\text{-}Br \qquad C_2H_5$$

With p-chloro and p-fluorotoluenes cine substitution of the aryne type was observed, but when the reaction was light stimulated a slight increase of the unrearranged substituted product was found. There is no compelling evidence as to the actual mechanism or mechanisms of these reactions. They may involve metal–halogen exchange, the $S_{RN}1$ mechanism, aryne intermediates, or a mixture of mechanisms that operate simultaneously. It appears that, with bromo or iodo derivatives, metal–halogen exchange and/or $S_{RN}1$ mechanisms are possible. With chloro and fluoro derivatives in tetrahydrofuran, $S_{RN}1$ and aryne mechanisms are consistent with the results (Table VIII).

Tin Nucleophiles

Although the reaction of triorganostannyl ions as nucleophiles with halobenzenes has been known for more than sixty years, systematic studies were not carried out until the last five years. These reactions were carried out in liquid ammonia with bromobenzene (53) and p-dichloro-benzene (54). Other solvents used were ether and tetrahydrofuran (55, 56, 57).

In 1978, Wursthorn, Kuivila, and Smith reported on a detailed study of the reaction of trimethylstannylsodium with halo- and dihaloben-zenes in tetraglyme (TG) solvent (58). The products obtained in this solvent, and with chloro, bromo and iodobenzene, are indicated in Reaction 93 and in Table IX.

$$PhX + Me_3Sn^- \longrightarrow PhH + PhSnMe_3 + Ph_2SnMe_2 + Me_4Sn \qquad (93)$$
$$\mathbf{121} \qquad\qquad \mathbf{122} \qquad \mathbf{123} \qquad \mathbf{124}$$

Table VIII

Reactions of Halotoluenes with Triethylgermyllithium in Tetrahydrofuran

Halotoluenes	Tolyltriethylgermanes[a]		
	Ortho-	Meta-	Para-
p-Iodo[b]			100
o-Bromo	100		
m-Bromo		100	
p-Bromo			100
o-Chloro	82	18	
m-Chloro	2	89	9
p-Chloro		25	75
p-Chloro[c]		15	85
p-Fluoro		33	67

[a]Relative yields.
[b]Solvent HMPT.
[c]Under illumination.
Source: Reproduced, with permission, from Ref. 52. Copyright 1978, *C. R. Acad. Sci. Paris.*

Table IX

Reaction of Halobenzenes with Trimethylstannylsodium in Tetraglyme at 0°C

PhX	PhH	Me$_3$SnPh %	Ph$_2$SnMe$_2$ %	Me$_4$Sn %
PhCl	4.2	64	0.5	19
PhBr	4.0	87	1	4
PhI	7	76	7	10
PhBr[a]	33	60	<1	6

[a]With t-butyl alcohol added.
Source: Reproduced from Ref. 58. Copyright 1978, American Chemical Society.

The various products obtained, along with the observations that t-butyl alcohol inhibited the formation of **123** while decreasing **122** and **124** and increasing substantially the amount of benzene formed, led the authors to suggest a mechanism for Reaction 93 as shown in Scheme II.

The key step in the formation of the products **123** and **124** is the dissociation of **121** to form Me$^-$Na$^+$ and dimethyltin **125** (Reaction 94). Evidence for Reaction 94 came from the observation that a freshly prepared solution of **121** in tetraglyme, containing piperidine as a proton source, evolved about 6% of methane in two days.

Formation of phenylsodium was verified by the increased formation of benzene in the presence of t-butyl alcohol (the reaction conditions otherwise being the same). The amount of benzene formed increased rapidly with increasing amounts of t-butyl alcohol, but approached a plateau when the ratio of alcohol to initial stannylsodium is greater than three. This fact was rationalized by the halogen–metal exchange of Reaction 95, which initially is thought to result in the formation of phenylsodium and bromotrimethylstannane in the solvent cage. A certain fraction of these products react within the cage to give **122**, and the remainder diffuse apart to give phenylsodium and bromotrimethylstannane **126**. These two species may react to form **122**, or phenylsodium can be protonated by the solvent. The added t-butyl alcohol reacts with free phenylsodium, and the saturation in the amount of trapping suggests a cage effect. Experiments in mixtures of tetraglyme and dimethoxyethane, with viscocities from 2.93 cp to 0.6 cp, and a constant amount of t-butyl alcohol, showed that the amount of trapping increases as the viscosity decreases, giving additional support to the cage mechanism (*58*).

In synthetic experiments it was shown that o-dibromobenzene reacted with **121** to give 42% (isolated yield) of o-bis(trimethylstannyl)-benzene (Reaction 100) (*59–60*). o-Bromophenyl anion was evidently an intermediate in this reaction, because it decomposed in part to give benzyne. Benzyne can then be trapped with furan to give the Diels–Alder adduct (Reaction 99).

With the monohalobenzenes p-chloro and p-bromotoluene, benzyne is not an intermediate. Of the products obtained, none were cine-

$$Me_3SN^-Na^+ \rightleftharpoons Me^-Na^+ + (Me_2Sn) \qquad (94)$$
$$\textbf{125}$$

$$\textbf{121} + PhBr \rightleftharpoons (Ph^-Na^+ + Me_3SnBr) \searrow PhSnMe_3 \qquad (95)$$
$$\textbf{126} \qquad \textbf{122}$$

$$\text{diffusion} \downarrow$$

$$PhH \xleftarrow{\text{SH}} Ph^-Na^+ + Me_3SnBr \nearrow$$
$$\textbf{126}$$

$$Me_4Sn$$
$$Me^-Na^+ + \textbf{126} \longrightarrow \textbf{124} \qquad (96)$$

$$Ph^-Na^+ + \textbf{125} \longrightarrow PhMe_2SN^-Na^+ \qquad (97)$$
$$\textbf{127}$$

$$\textbf{127} + PhBr \longrightarrow Ph_2SnMe_3 \qquad (98)$$
$$\textbf{123}$$

Scheme II

(99)

(100)

(101)

substituted (that is, there were no meta substituted products). With p-fluorotoluene however, both para and meta isomers were found, indicating that, with this substrate, at least part of the mechanism occurs by way of a benzyne intermediate. It was suggested that the formation of methylsodium may be responsible for the generation of the aryne. From the experimental evidence, it was concluded that the reaction in tetraglyme occurs by halogen–metal exchange, and not by the $S_{RN}1$ mechanism.

The products obtained in the reaction of tributylstannyllithium 130 in tetrahydrofuran with halobenzenes depends on the nature of the leaving group. With o-, m- and p-bromotoluenes as substrates in the reaction with 130, the straightforward substitution product was obtained, uncontamined with products coming from cine substitution. However, cine-substitution products were found with p-chloro- and p-fluorotoluenes (61, 62).

The reaction of bromomesitylene with 130 was studied under various experimental conditions in order to determine if the reaction with aryl bromides was a halogen–metal exchange reaction or a $S_{RN}1$ reaction. The reaction of bromomesitylene with 130 in tetrahydrofuran and t-BuOD, gave mesitylene and the substitution product. The mesitylene formed had 70% (mole) incorporation of deuterium, which suggests that the reaction occurs mainly through a mesityl anion intermediate. However, in the reaction of bromomesitylene with 130, under the same experimental conditions, in 2,2,5,5-tetradeutero tetrahydrofuran 132, there was 20% incorporation of deuterium in the mesitylene formed. This suggests that, in part, the mesityl radical 131 is an intermediate in this reaction (Reaction 101) (61, 62).

The reaction of p-fluoro- and p-chlorotoluenes with 130 in tetrahydrofuran gave a mixture of m- and p-tolyltributylstannanes, but the relative yield of these products varied as the "age" of the nucleophile varied. The ratio of the products varied also if the reaction was carried out in the dark, under irradiation, or in the presence of added substances.

The effect of the "aging" of the nucleophile was tentatively attributed to the presence of a suspension of unreacted lithium metal in the solution of the nucleophile formed by reaction of lithium metal and hexabutyldistannane. When the reactant is used soon after being prepared in this manner, the unreacted lithium metal catalyzes the reaction by the $S_{RN}1$ mechanism. By aging the solution of the nucleophile, lithium metal is deposited and more cine-substitution, via a benzyne mechanism, is found (Table X). Indeed, if lithium metal is added, more of the straightforward substitution product is found. The catalytic action of the lithium metal is, of course, to initiate the $S_{RN}1$ reaction (Reaction 102).

$$ArX + Li \longrightarrow (ArX)^{\cdot -} \, Li^+ \longrightarrow Ar^{\cdot} + XLi \qquad (102)$$

From these results it was suggested that there is a competition between at least three different mechanisms of reaction. With aryl bromides the halogen–metal exchange mechanism seems to be the principal route to the substitution product. With aryl chlorides and fluorides, there are two competitive mechanisms, radical substitution and aryne mechanisms. Light and lithium metal enhance the radical substitution pathway, and butyllithium enhances the aryne pathway (61, 62).

Similar results were obtained with hexamethylphosphoramide as solvent (63). An improved procedure for formation of the nucleophile free of lithium metal was reported, and the competitive mechanisms with aryl halides were similar to those reported in tetrahydrofuran (63).

Table X

Reaction of p-Halotoluenes with Tributylstannyllithium in THF

$p\text{-}XC_6H_4CH_3$ X	Age of the Nucleophile (h)	Experimental Conditions	Relative Yields	
			m-Isomer	p-Isomer
Cl	24	dark	12	88
Cl	24	light[a]	8	92
Cl	24	UV[b]	8	92
Cl	24	light–perylene	11	89
Cl	96	dark	42	58
Cl	96	light	20	80
Cl	96	light–butyllithium	43	57
F	48	light	50	50
F	48	light–lithium metal	22	78
F	144	dark	50	50
F	144	light	12	88

[a]Ambient light.
[b]High pressure Hg Lamps.
Source: Reproduced with permission from Reference 62. Copyright 1978,
J. Organomet. Chem.

Literature Cited

1. Coulter, L. V.; Sinclair, J. R.; Cole, A. G.; Roper, G. C. *J. Amer. Chem. Soc.* **1959,** *81*, 2986.
2. Rossi, R. A.; Bunnett, J. F. *J. Org. Chem.* **1973,** *38*, 3020.
3. Bunnett, J. F.; Rossi, R. A.; unpublished data.
4. López, A. F. Ph.D. Thesis, Universidad Nacional de Córdoba, 1979.
5. Fox, M. A. unpublished data.

6. Semmelhack, M. F.; Bargar, T. *J. Am. Chem. Soc.* **1980**, *102*, 7765.
7. Rossi, R. A.; Bunnett, J. F. *J. Am. Chem. Soc.* **1972**, *94*, 683.
8. Bard, R. R.; Bunnett, J. F.; Creary, X.; Tremelling, M. J. *J. Am. Chem. Soc.* **1980**, *102*, 2852.
9. Rossi, R. A.; de Rossi, R. H.; Lōpez, A. F. *J. Am. Chem. Soc.* **1976**, *98*, 1252.
10. Rossi, R. A.; de Rossi, R. H.; Lōpez, A. F. *J. Org. Chem.* **1976**, *41*, 3367.
11. Bunnett, J. F.; Gloor, B. F. *Heterocycles* **1976**, *5*, 377.
12. Komin, A. P.; Wolfe, J. F. *J. Org. Chem.* **1977**, *42*, 2481.
13. Oostveen, E. A.; van der Plas, H. C. *Recl. Trav. Chim. Pays-Bas* **1979**, *98*, 441.
14. Rossi, R. A.; Bunnett, J. F. *J. Org. Chem.* **1973**, *38*, 1407.
15. Bunnett, J. F.; Sundberg, J. E. *Chem. Pharm. Bull.* **1975**, *23*, 2620.
16. Bunnett, J. F.; Wamser, C. C. *J. Am. Chem. Soc.* **1967**, *89*, 6712. Bunnett, J. F.; Takayama, H.; *J. Am. Chem. Soc.* **1968**, *90*, 5173.
17. Wolfe, J. F.; Moon, M. P.; Sleevi, M. C.; Bunnett, J. F.; Bard, R. R. *J. Org. Chem.* **1978**, *43*, 1019.
18. Semmelhack, M. F.; Bargar, T. M. *J. Org. Chem.* **1977**, *42*, 1481.
19. Bunnett, J. F.; Sundberg, J. E.; *J. Org. Chem.* **1976**, *41*, 1702.
20. Hay, J. V.; Wolfe, J. F. *J. Am. Chem. Soc.* **1975**, *97*, 3702.
21. Hay, J. V.; Hudlicky, T.; Wolfe, J. F. *J. Am. Chem. Soc.* **1975**, *97*, 374.
22. Wolfe, J. F.; Greene, J. C.; Hudlicky, T. *J. Org. Chem.* **1972**, *37*, 3199.
23. Beugelmans, R.; Roussi, G. *Tetrahedron* **1981**, *37*, 393.
24. House, H. O.; Kramer, V. *J. Org. Chem.* **1963**, *28*, 3362; House, H. O.; Trost, B. *J. Org. Chem.* **1965**, *30*, 1341.
25. Beugelmans, R.; Ginsburg, H. *J. Chem. Soc., Chem. Comm.* **1980**, 508.
26. Brown, C. A. *J. Org. Chem.* **1974**, *39*, 1324.
27. Bunnett, J. F.; Singh, P. *J. Org. Chem.* **1981**, *46*, 5022.
28. Alonso, R. A.; Rossi, R. A. *J. Org. Chem.* **1980**, *45*, 4760.
29. Weinreb, S. M.; Semmelhack, M. F. *Acc. Chem. Res.* **1975**, *8*, 158.
30. Amatore, C.; Chaussard, J.; Pinson, J.; Saveant, J. M.; Thiebault, A. *J. Am. Chem. Soc.* **1979**, *101*, 6012.
31. Scamehorn, R. G.; Bunnett, J. F. *J. Org. Chem.* **1977**, *42*, 1449.
32. Cram, D. J.; Cram, A. C. *J. Org. Chem.* **1971**, *36*, 184.
33. Carver, D. R.; Komin, A. P.; Hubbard, J. S.; Wolfe, J. F. *J. Org. Chem.* **1981**, *46*, 294.
34. Jockman, L. M.; Lange, B. C. *Tetrahedron* **1977**, *33*, 2737 and references cited therein.
35. Beugelmans, R.; Roussi, G. *J. Chem. Soc., Chem. Comm.* **1979**, 950.
36. Rossi, R. A.; Alonso, R. A. *J. Org. Chem.* **1980**, *45*, 1239.
37. Wolfe, J. F.; Sleevi, M. C.; Goehring, R. R. *J. Am. Chem. Soc.* **1980**, *102*, 3646.
38. Alonso, R. A.; Rossi, R. A. unpublished data.
39. Bunnett, J. F.; Gloor, B. F. *J. Org. Chem.* **1974**, *39*, 382.
40. Zoltewicz, J. A.; Helmick, L. S. *J. Org. Chem.* **1973**, *38*, 658.
41. Bartak, D. E.; Danen, W. C.; Hawley, M. D. *J. Org. Chem.* **1970**, *35*, 1206.
42. Pinson, J.; Saveant, J. M. *J. Am. Chem. Soc.* **1978**, *100*, 1506.
43. Bunnett, J. F.; Gloor, B. F. *J. Org. Chem.* **1973**, *38*, 4156.
44. Rossi, R. A.; de Rossi, R. H.; Pierini, A. B. *J. Org. Chem.* **1979**, *44*, 2662.
45. Goldfarb, I. L.; Ikubov, A. P.; Belenki, L. I. *Zhur. Geter. Soedini* **1979**, 1044.
46. Szwarc, M. *Acc. Chem. Res.* **1972**, *5*, 169.
47. Rossi, R. A.; de Rossi, R. H.; Lōpez, A. F. *J. Org. Chem.* **1976**, *41*, 3371.
48. van der Plas, H. C.; de Bie, D. A.; Geurtsen, G.; Reinecke, M. G.; Adickes, H. W. *Recl. Trav. Chim. Pays-Bas* **1974**, *93*, 33.
49. Häbich, D.; Effenberger, F. *Synthesis* **1979**, 841.
50. Shippey, M. A.; Dervan, P. B. *J. Org. Chem.* **1977**, *42*, 2654.
51. Russell, G. A.; Malatesta, V.; Morita, T.; Osuch, C.; Blankespoor, R. L.; Trahanovsky, K. D.; Goettert, E. *J. Am. Chem. Soc.* **1979**, *101*, 2112.
52. Quintard, J. P.; Hauvette–Frey, S.; Pereyre, M.; Couret, C.; Satgé, J. C. R. *Acad. Sci. Paris*, **1978**, *287*, 247.
53. Bullard, R. H.; Robinson, W. B. *J. Am. Chem. Soc.* **1927**, *49*, 1368.
54. Kraus, C. A.; Sessions, W. *J. Am. Chem. Soc.* **1925**, *47*, 2361.

55. Gilman, H.; Rosenberg, S. D. *J. Am. Chem. Soc.* **1952,** *74,* 531.
56. Gillman, H.; Rosenberg, S. D. *J. Org. Chem.* **1953,** *18,* 630.
57. Tamborski, C.; Ford, F. E.; Sokolski, E. J. *J. Org. Chem.* **1963,** *28,* 181.
58. Wursthorn, K. R.; Kuivila, H. G.; Smith, G. F. *J. Am. Chem. Soc.* **1978,** *100,* 2789.
59. Kuivila, H. G.; Wursthorn, K. R. *J. Organomet. Chem.* **1976,** *105,* C 6.
60. Wursthorn, K. R.; Kuivila, H. G. *J. Organomet. Chem.* **1977,** *140,* 29.
61. Quintard, J. P.; Hauvette–Frey, S.; Pereyre, M. *J. Organomet. Chem.* **1976,** *112,* C 11.
62. *Ibid.,* **1978,** *159,* 147.
63. Quintard, J. P.; Hauvette–Frey, S.; Pereyre, M. *Bull. Soc. Chim. Belg.* **1978,** *87,* 505.
64. Bard, R. R.; Bunnett, J. F. *J. Org. Chem.* **1980,** *45,* 1546.

Nucleophiles Derived from the VA Group of Elements

3

Nitrogen Nucleophiles

Amide Ions. Although the aromatic $S_{RN}1$ reaction was discovered during the amination of iodotrimethylbenzenes **1a** and **1b** (*1*), relatively few examples of $S_{RN}1$ aminations of aryl halides have been reported. It is not clear what species initiate the dark reaction between **1a** or **1b** with amide ions, but we know that alkali metals catalyze the reaction, giving only the unrearranged amines. For instance, **1b** gave **2b** in 54% yield when treated with potassium metal and amide ions in liquid ammonia (Reaction 1) (*1*).

The reaction of *o*-haloanisoles with amide ions in ammonia gave *m*-anisidine by an aryne mechanism (Reaction 2), but with potassium metal stimulation only the *o*-anisidine is formed (Reaction 3) (*1*).

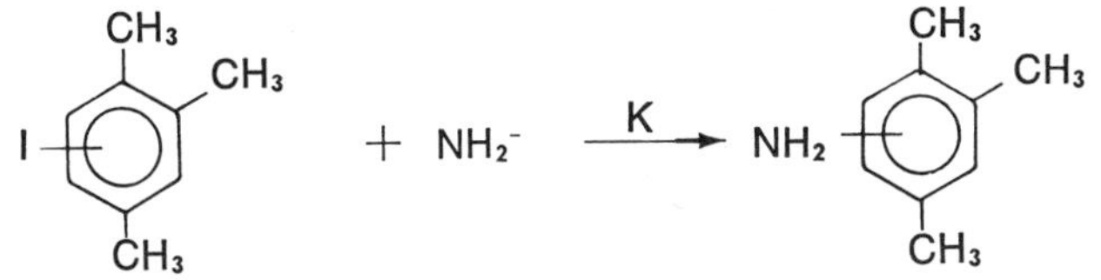

1a: 5-I   2a: 5-NH₂
1b: 6-I   2b: 6-NH₂

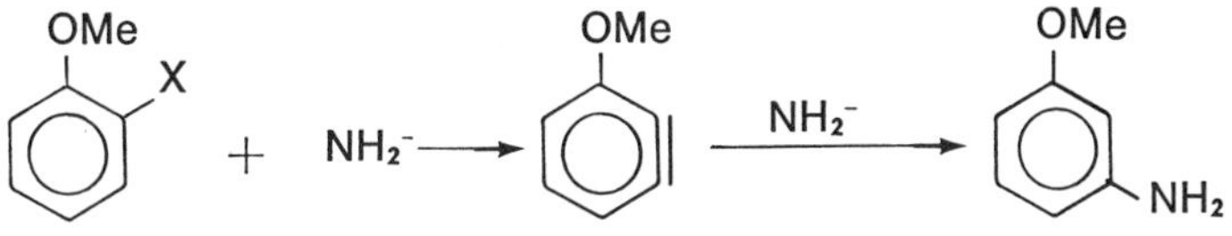

0065-7719/83/0178-0059$06.00/1
© 1983 American Chemical Society

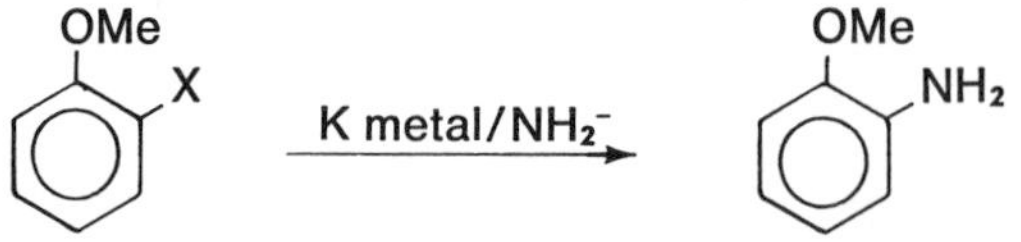

$$(3)$$

2-Iodo-1,3-dimethylbenzene, **3**, is unable to react by the aryne mechanism because it has no hydrogen ortho to iodine. However, it reacts with amide ions and potassium metal to form 2,6-dimethylaniline in 64% yield, together with *m*-xylene (25%) (Reaction 4) (Table I) (*1*).

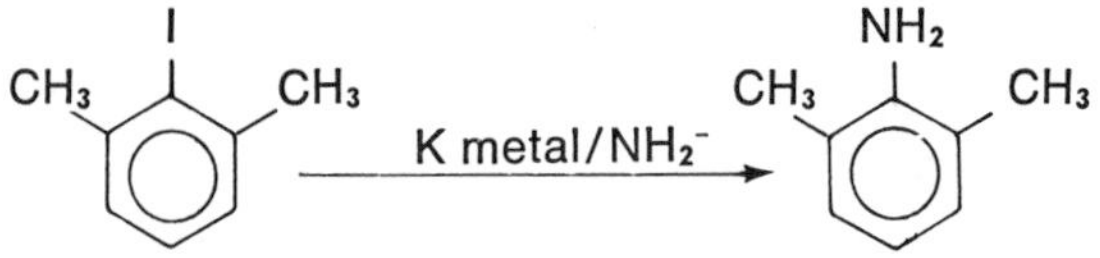

$$(4)$$

A general $S_{RN}1$ amination procedure involves reaction of aryl diethyl phosphates with amide ion and potassium metal in ammonia to form the corresponding anilines in good yields. Since the requisite phosphate esters **4** can be prepared easily from phenols (*2*), the overall sequence provides a convenient method for the conversion of phenols to anilines as shown in Reactions 5–6 (Table I) (*3*).

$$\text{ArOH} + \text{ClP(O)(OR)}_2 \longrightarrow \underset{\textbf{4}}{\text{ArOP(O)(OR)}_2} \qquad (5)$$

$$\textbf{4} \xrightarrow{\text{K metal/NH}_2^-} \text{ArNH}_2 \qquad (6)$$

Table I

Reaction of Aryl Halides with Amide Ion Stimulated by Solvated Electrons

ArX	ArNH$_2$	ArH	Ref.
5-Iodo-1,2,4-trimethylbenzene	50	40	1
6-Iodo-1,2,4-trimethylbenzene	54	30	1
o-Iodoanisole	67	11	1
o-Bromoanisole	64	16	1
2-Iodo-1,3-dimethylbenzene	64	25	1
Diphenyl ether	53	—	1
Phenyl diethyl phosphate	73	—	3
(2,6-Dimethylphenyl) diethyl phosphate	78	—	3
(2-Methoxy-4-methylphenyl) diethyl phosphate	56	—	3

It has been reported that 3-bromothiophene **5** reacts with amide ions to form 3-aminothiophene **6** in 79% yield. However, this reaction does not need potassium metal stimulation to occur (Reaction 7) (4). But, under an atmosphere of air, there is no reaction (only 2% yield of **6** in the same period of time). These results are consistent with the $S_{RN}1$ mechanism. Note that the same reactants are less reactive under illumination than in the dark. It was suggested that the effective initiation process is insensitive to light, but that illumination produces an increase of termination steps rather than initiation steps.

2-Bromothiophene **7** and amide ions under a nitrogen atmosphere furnished 37% of the 3-bromo isomer **5**, which represents bromine migration and 48% yield of **6** (Reaction 8) (Table II).

Potassium anilide **8** in ammonia does not react with iodobenzene, but when potassium metal is present, reaction occurs to form diphenylamine **9** (19%), **10** o-aminobiphenyl and p-aminobiphenyls **11** (11% each) (Reaction 9) (1).

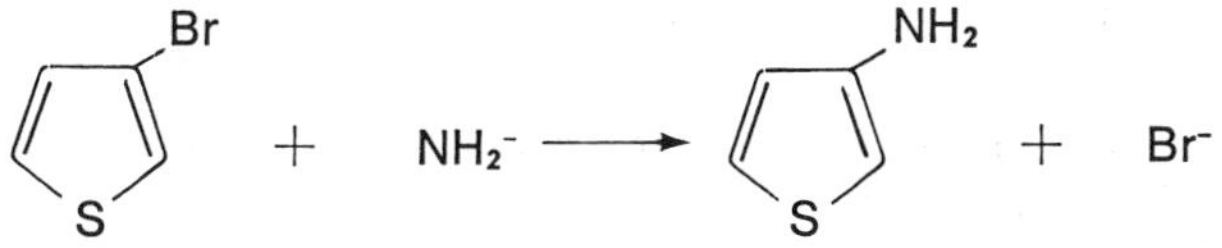

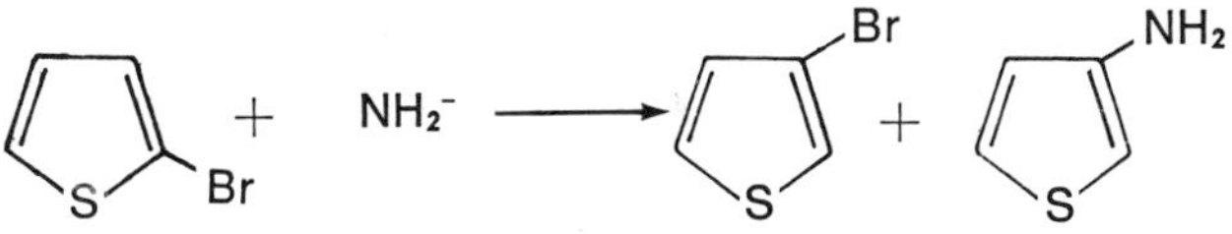

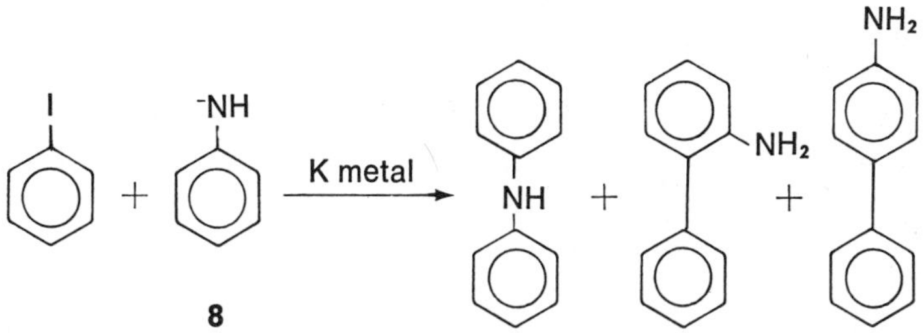

Table II

Reactions of 2- and 3-Bromothiophenes with Potassium Amide in Liquid Ammonia

Substrate	Conditions	3-Amino-thiophene	Thiophene
3-Bromothiophene	dark, nitrogen, 60 min	79	4
3-Bromothiophene	dark, air, 60 min	2	3
3-Bromothiophene	light, nitrogen, 60 min	63	1
2-Bromothiophene	dark, nitrogen, 15 min	48	9
2-Bromothiophene	dark, air, 15 min	31	12

Source: Reproduced, with permission, from Reference 4. Copyright 1976, Sendai Institute of Heterocyclic Chemistry.

Catalysis of this reaction by potassium metal was recognized in the early review of Bunnett (5), although at that time no reference to the possible mechanism was suggested. The knowledge that phenyl radicals are intermediates indicates that the ambient anilide ion nucleophile reacts with the phenyl radical at its ring carbons almost as rapidly as it does at nitrogen. Except for the study of 3-bromothiophene **5** with amide ion under illumination, these reactions have been carried out under potassium metal stimulation, and it is unknown whether they can be light stimulated.

The ring closure reaction of substrate **12** under photostimulation has been attempted, but the desired cyclization product **13** was obtained in only 2% yield. The reduction product **14** was the major product (Reaction 10) (6).

Other Nitrogen Nucleophiles. It is not known whether amines are able to trap aryl radicals because the reactions of these nucleophiles have not been investigated in detail. However, there are indications that certain sterically hindered amines may react by the $S_{RN}1$ mechanism. 2-Methylpiperidine has been reported to react with o-nitroiodobenzene in neat 2-methylpiperidine at 100°C, giving nitrobenzene and N-2-nitrophenyl-2-methylpiperidine (7). The formation of nitrobenzene is decreased in the presence of m-dinitrobenzene, and the aminolysis product is slightly, but measurably, depressed. An indication that this reaction might occur by an electron transfer mechanism (probably the $S_{RN}1$), is that p-iodonitrobenzene is much less reactive than the *ortho*-isomer, in agreement with the 10^5 slower rate of decomposition of the radical anion of the former compound to give aryl radicals (8).

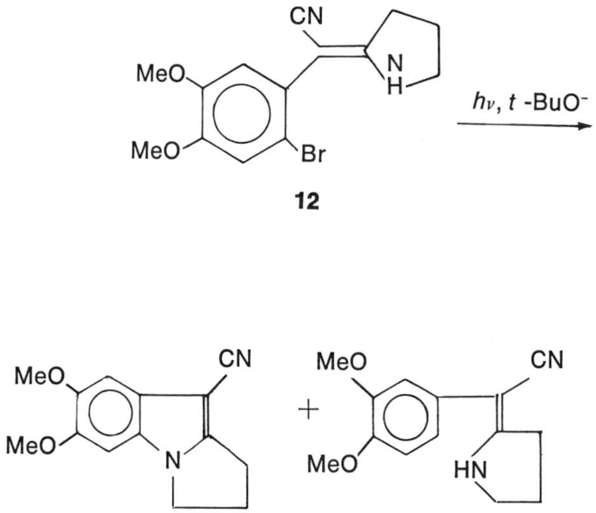

12

$$h\nu, t\text{-BuO}^-$$

13   **14**

(10)

The anion of acetamide **15** formed in liquid ammonia did not react in the photostimulated reaction with 1-chloronaphthalene (*9*); neither did **15** or *N*-methylacetamide ion **16** react when stimulated with electrons from electrodes (*10*).

The pyrrolyl anion **17** was prepared in liquid ammonia and irradiated under the usual conditions of the $S_{RN}1$ mechanism with bromobenzene or 1-chloronaphthalene as substrates. In both cases only the reduced substrate was obtained and substitution products were not detected (*9*).

Chlorobenzene and **17** were allowed to react under potassium metal stimulation to see if the lack of reactivity of **17** was due to the failure of pyrrolyl anion to initiate the reaction. However, no *N*-phenylpyrrole was obtained; only aniline, benzene, and aminobiphenyls were formed, as was reported for the reaction of chlorobenzene with solvated electrons in liquid ammonia (*11*).

The reaction of the azido group with aryl radicals from the decomposition of the diazonium salt **18** has been reported, and the proposed reaction mechanism is shown in Reactions 11–13 (*12*). Although not a chain mechanism, the coupling of the aryl radical with one nitrogen atom in the azido group suggested that azide ions should be able to react similarly with aryl radicals. However, in the photostimulated reaction of sodium azide and bromobenzene there was no reaction at all after 4 hours (*9*).

Nitrite ion belongs to the class of bidentate nucleophiles with a hard

CH_3CONH^- **15**

$CH_3CO\bar{N}CH_3$ **16**

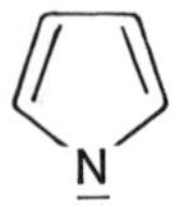

17

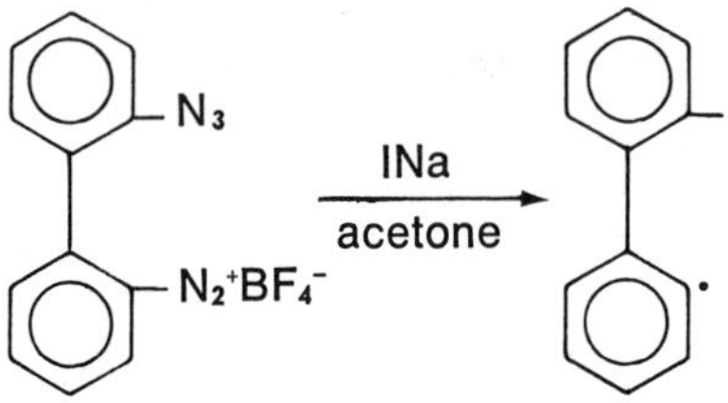

18 **19** (11)

19

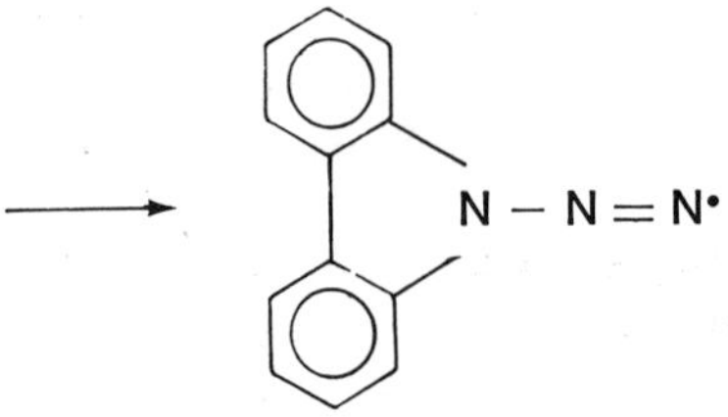

20 (12)

20

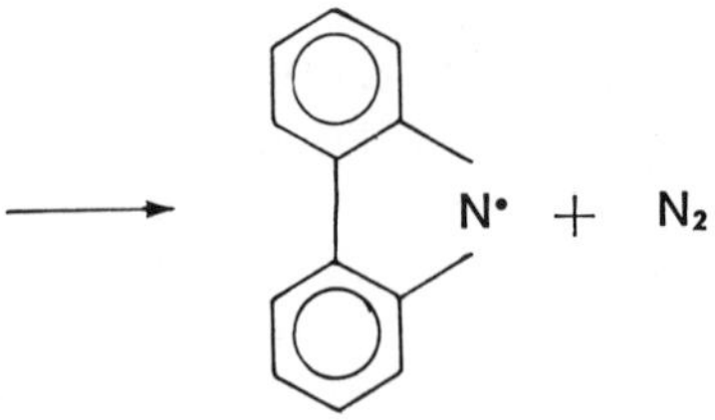

(13)

(oxygen) and soft (nitrogen) part. The formation of nitrobenzene in the
Gatterman process was formulated as in Reactions 14–17 (*13*).

$$ArN_2^+ + Cu^0 \longrightarrow ArN_2^{\cdot} + Cu^+ \tag{14}$$

$$ArN_2^{\cdot} \longrightarrow Ar^{\cdot} + N_2 \tag{15}$$

$$Ar^{\cdot} + NO_2^- \longrightarrow (ArNO_2)^{\overline{\cdot}} \tag{16}$$

$$(ArNO_2)^{\overline{\cdot}} + Cu^+ \longrightarrow ArNO_2 + Cu^0 \tag{17}$$

This mechanism bears some resemblance to the $S_{RN}1$ mechanism, the
difference being that Cu^0 serves as electron donor and Cu^+ as electron
acceptor. However, the fact that the reduction potential of nitrobenzene
is about -1 V (DMF, versus standard calomel electrode) (*14*), and the
reduction potential of benzenediazonium tetrafluoroborate is $+0.3$ V
(sulfolane, versus standard calomel electrode) (*15*) indicates that Reaction 18 is thermodynamically favored, and therefore very fast. Thus, it
seems likely that it competes with Reaction 17 and becomes a chain
propagating step, as in the usual $S_{RN}1$ mechanism. The only function of
the metal then is to promote the reaction (Reaction 14). However, there
is the possibility that this copper catalyzed reaction occurs with the
intermediacy of organocopper compounds.

$$(ArNO_2)^{\overline{\cdot}} + ArN_2^+ \longrightarrow ArNO_2 + ArN_2^{\cdot} \tag{18}$$

The coupling of nitrite ion with aryl radicals by a mechanism very
similar to the $S_{RN}1$ mechanism has also been postulated (*16*).

Phenyl radicals prepared by the thermal decomposition of phenyl-
azotriphenylmethane in DMSO, in the presence of 0.1 M sodium nitrite,
yielded a product which, upon electroreduction, gave a strong ESR spectrum of nitrobenzene radical anions. Gas chromatography showed the
presence of approximately 5% of nitrobenzene and 75% of benzene (*17*).

When sodium dithionite was used to reduce aryldiazonium salts,
aryl radicals such as phenyl, *p*-carboxyphenyl, *m*-carboxyphenyl, *p*-
chlorophenyl, *p*-cyanophenyl, *m*-nitrophenyl, and *p*-nitrophenyl radicals were formed, which could be trapped by nitrite ion to yield the
corresponding nitrobenzene radical anion (Reactions 19 and 20) (*18*).

Coupling of electrochemically generated *p*-nitrophenyl radical **22**
with nitrite ion (Reaction 21) excluded hydrogen atom abstraction as a
reaction pathway, and the radical anion of *p*-dinitrobenzene formed was
detected by cyclic voltammetry in DMSO (Reaction 22) (*17*). Contrary to
these results, in the photostimulated reaction of halobenzenes with nitrite ion in liquid ammonia, there was no reaction by the $S_{RN}1$ mechanism
(*9*).

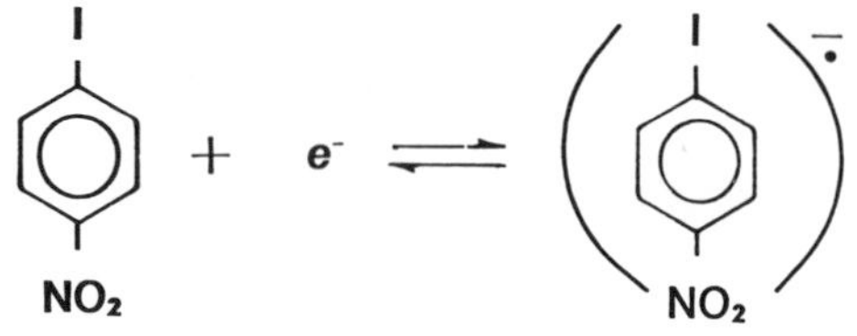

$$\text{(19)}$$

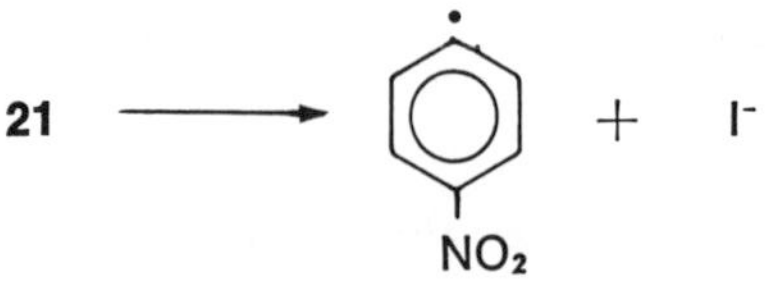

$$\text{(20)}$$

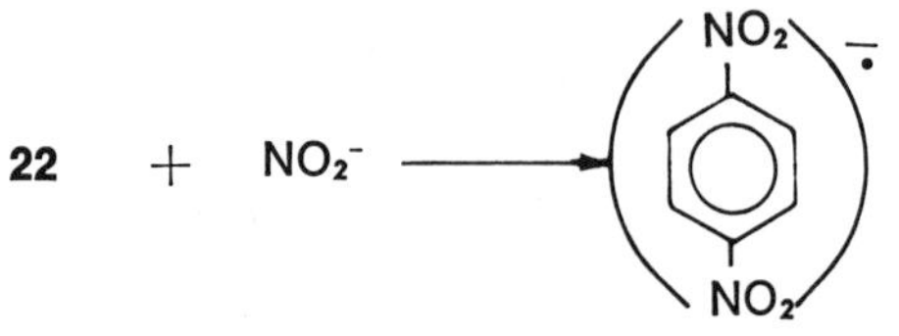

$$\text{(21)}$$

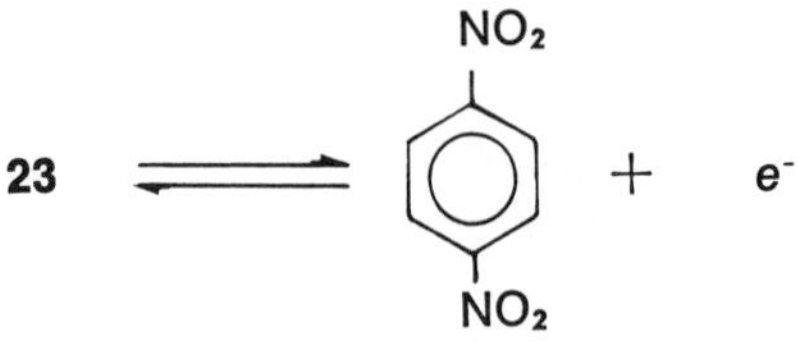

$$\text{(22)}$$

The failure of nitrite ion to act as a nucleophile under these conditions is attributed to the low rate of the propagation steps (such as Reaction 23) rather than to its low intrinsic reactivity.

$$(PhNO_2)^{\overline{\cdot}} + PhX \longrightarrow PhNO_2 + (PhX)^{\overline{\cdot}} \tag{23}$$

The reduction potential of nitrobenzene is about -1 V (-1.06 V in DMF, and -1.1 V in acetonitrile, versus standard calomel electrode) (14), and that of iodobenzene, which has the most positive reduction potential of the four halobenzenes, is -1.6 to -1.8 V in DMSO, versus standard calomel electrode (14), which makes Reaction 23 (X being I) thermodynamically unfavorable by about 0.6 V. Therefore, the reaction is expected to be slow and unable to continue the chain propagation as fast as is necessary for the reaction to proceed.

Phosphorus Nucleophiles

Diphenyl Phosphide Ion. The reaction of lithium diphenylphosphide with p-bromo or p-iodotoluenes in THF was reported by Aguiar and coworkers to give p-tolyldiphenylphosphine, without cine substitution, which rules out an aryne mechanism (Reaction 24). It was suggested that the reaction occurs by a concerted displacement mechanism with a transition state comprised of the aryl halide, lithium diphenylphosphide and a second ion pair (either the nucleophile or lithium chloride) (19).

$$\text{ArI} + \text{LiPPh}_2 \xrightarrow{\text{THF}} \text{ArPPh}_2 + \text{LiI} \tag{24}$$

It has been reported that 8-chloroquinoline reacts with potassium diphenyl phosphide **24** in liquid ammonia in 16 h to give the substitution product, although no mechanistic description was suggested (20). Bunnett et al. reinvestigated Reaction 24 in terms of the $S_{RN}1$ mechanism (21), which can account for the results of Aguiar and coworkers (19). The reaction of m- and p-iodotoluenes with **24** in liquid ammonia or DMSO at 25°C, was found to occur in the dark, and is stimulated by broad spectrum UV light. The product formed is the *ipso* substituted product uncontaminated with cine substituted compounds. This result is strong evidence against an aryne mechanism.

The relative reaction rates of aryl iodides and bromides were determined by direct competition. In ammonia, the iodobenzene:bromobenzene rate ratio is about 400 for the thermal reaction and 200 for the light stimulated reaction, whereas the iodotoluene:bromotoluene rate ratio is roughly 100 for the thermal reaction in ammonia or DMSO (21). Small amounts of electron acceptors, or free radical traps, were found to inhibit the reaction in both solvents.

The reaction of bromotoluene with **24** is slow in DMSO (only 13% of bromide ion is released in 3 h of reaction). The reaction of p-iodotoluene on the other hand is relatively fast (96% of iodide ion released in 2 h).

In a ^{31}P chemically induced dynamic nuclear polarization (CIDNP) (22, 23) study of the reaction of diorganophosphides with alkyl halides, it was found that there is significant radical participation in reactions with alkyl bromides or iodides but not with alkyl chlorides (24). On the other hand, the reaction of potassium dimethylphosphide with bromobenzene occurs without CIDNP, and it was suggested that a metal–halogen exchange may occur (Reactions 25 and 26) (*24*).

$$KPMe_2 + PhBr \rightleftharpoons PhK + BrPMe_2 \xrightarrow{-BrK} PhPMe_2 \qquad (25)$$

$$KPMe_2 + BrPMe_2 \longrightarrow Me_2PPMe_2 + BrK \qquad (26)$$

Although the solvent used (benzene, 1,2-dimethoxyethane) was different from that used by Bunnett et al. (ammonia, DMSO) (*20*), there is a possibility that the reaction occurs by the $S_{RN}1$ mechanism. Because CIDNP detection depends on reactions involving the interaction of two radicals, it is not expected in a chain process unless the chain length is exceedingly short (e.g. 1 to 3) (*25*). Thus, interaction of two radicals in $S_{RN}1$ reactions rarely occurs, and then only as chain-termination steps.

Dialkyl Phosphite Ions. Potassium diethyl phosphite **25** reacts rapidly with aryl iodides in liquid ammonia under irradiation to form diethyl esters of arylphosphonic acid **26** in nearly quantitative yields (Reaction 27) (*26*).

$$\underset{\mathbf{25}}{ArI + (OEt)_2PO^-K^+} \xrightarrow{h\nu} \underset{\mathbf{26}}{ArP(O)(OEt)_2 + I^-} \qquad (27)$$

This reaction has been carried out with iodobenzenes carrying several substituents (*26*), bromobenzene (*27*), and 1-iodonaphthalene (*26*), and can be used as a synthetic route to **26** (*28*). In the dark there is no reaction (Table III).

The four *m*-haloiodobenzenes were studied. *m*-Fluoroiodobenzene and *m*-chloroiodobenzene gave mainly the monosubstitution product, where only iodide is replaced. With *m*-bromoiodobenzene, and *m*-diiodobenzenes and p-diiodobenzenes, the disubstitution product was found (*27*).

These results were explained on the basis of the $S_{RN}1$ mechanism of Scheme I. The *m*-halophenyl radical intermediate **29** formed in Reaction 29 couples with the nucleophile **25** to form the radical anion **30**. Two competitive reactions, electron transfer to the substrate to give the mono-substitution product **31** (Reaction 31), or a C–X bond fragmentation to give a new radical intermediate **32** can then occur. The latter can give ultimately the disubstitution product **34** (Reactions 32–34, Scheme I) (*27*).

Because C–Br and C–I bonds are more readily broken than the C–F or C–Cl bonds, Reaction **32** is favored when X is Br or I, and Reaction **31** predominates when X is F or Cl.

Because the interrupted photostimulated reaction of **35** gave much **34,** much unreacted **35,** but little **36** (Reaction 35), this is further evidence

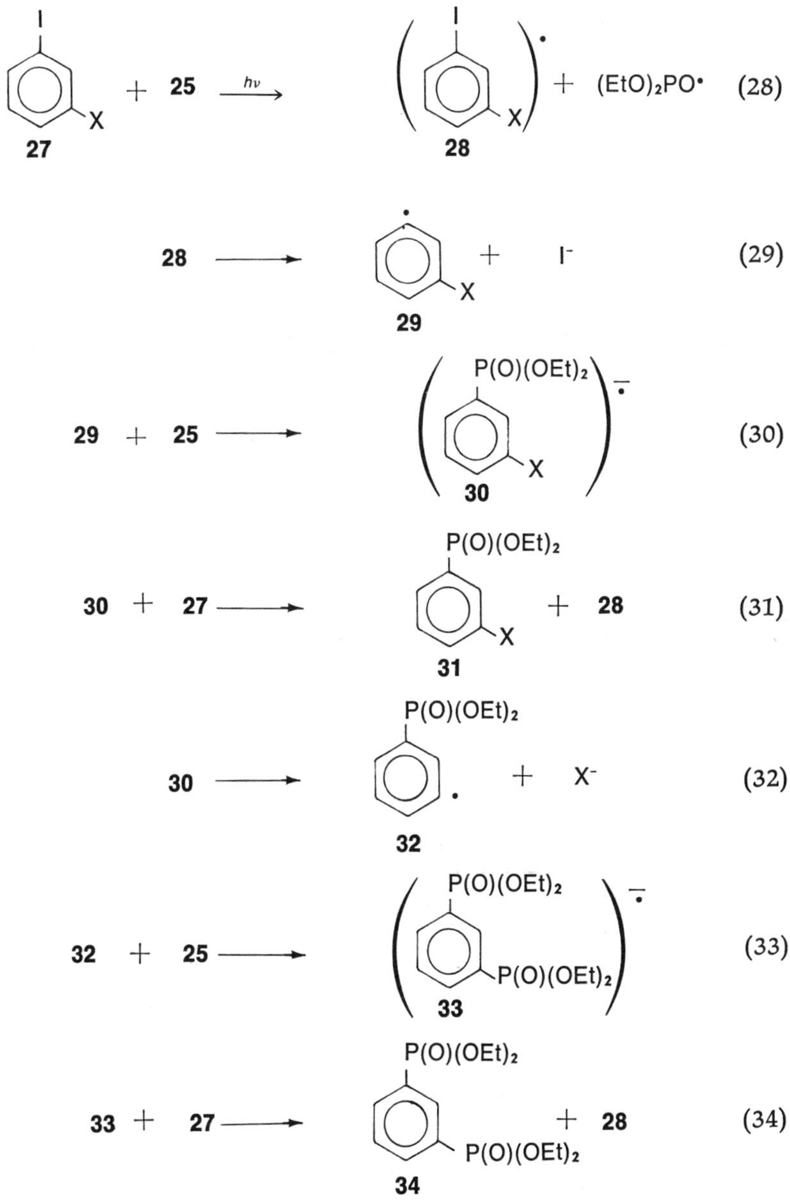

$$27 + 25 \xrightarrow{h\nu} \left(\underset{28}{\bigcirc}_X \right)^{\overset{\bullet}{-}} + (EtO)_2PO^{\bullet} \quad (28)$$

$$28 \longrightarrow \underset{29}{\bigcirc}_X^{\overset{\bullet}{}} + I^- \quad (29)$$

$$29 + 25 \longrightarrow \left(\underset{30}{\overset{P(O)(OEt)_2}{\bigcirc}}_X \right)^{\overset{\bullet}{-}} \quad (30)$$

$$30 + 27 \longrightarrow \underset{31}{\overset{P(O)(OEt)_2}{\bigcirc}}_X + 28 \quad (31)$$

$$30 \longrightarrow \underset{32}{\overset{P(O)(OEt)_2}{\bigcirc}}^{\bullet} + X^- \quad (32)$$

$$32 + 25 \longrightarrow \left(\underset{33}{\overset{P(O)(OEt)_2}{\bigcirc}}_{P(O)(OEt)_2} \right)^{\overset{\bullet}{-}} \quad (33)$$

$$33 + 27 \longrightarrow \underset{34}{\overset{P(O)(OEt)_2}{\bigcirc}}_{P(O)(OEt)_2} + 28 \quad (34)$$

Scheme I

that the sequence of reactions is 28→29→30→32→33→34, rather than through Reaction 35. In addition, 36 is less reactive than 35 with 25 under these conditions, indicating that 36 cannot be an intermediate in the main route from 35 to 34. In 7 min. of irradiation, 35 was obtained in 27.5% yield, 36 in 5.9% yield, and 34 in 62.7% yield (29).

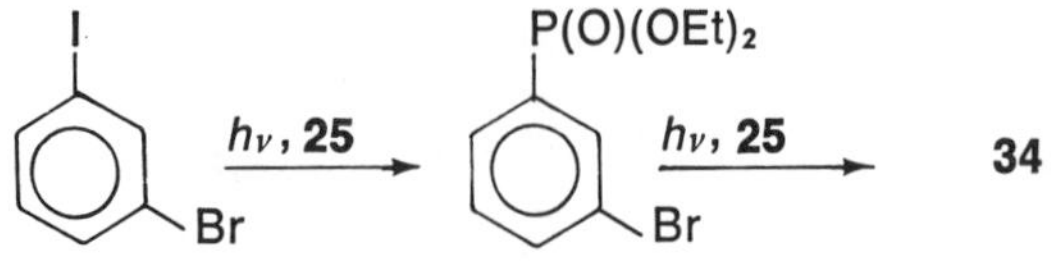

$$(35)$$

Table III

Photostimulated Reactions of Potassium Dialkyl Phosphites with Iodoarenes in Ammonia

ArI	$(RO)_2POK$ R	Irradiation Time (min)	Substitution Product (%)	Ref.
PhI	CH_3	50	93	26
PhI	$n\text{-}C_4H_9$	60	88	26
PhI	C_2H_5	45	96	26
p-Iodotoluene	C_2H_5	75	95	26
p-Iodoanisole	C_2H_5	65	95	26
2-Iodo-m-xylene	C_2H_5	60	87	26
$m\text{-}CF_3C_6H_4I$	C_2H_5	55	95	26
1-Iodonaphthalene	C_2H_5	130	93	26
m-Fluoroiodobenzene	C_2H_5	50	96	26
m-Chloroiodobenzene	C_2H_5	40	89	26
m-Bromoiodobenzene	C_2H_5	60	87[a]	26
m-Diiodobenzene	C_2H_5	90	94[a]	26
p-Bromoiodobenzene	C_2H_5	240	56[a]	27
p-Diiodobenzene	C_2H_5	205	87[a]	27
p-Chloroiodobenzene	C_2H_5	90	59	27
p-Fluoroiodobenzene	C_2H_5	50	96	27

[a]Disubstitution product.

Other Phosphorus Nucleophiles. The potassium salt of n-butyl phenylphosphonite **37** reacted rapidly with iodobenzene in refluxing ammonia during 15 min of irradiation, giving n-butyl diphenylphosphinate in 95% yield (Reaction 36) (30).

$$\text{PhI} + \text{PhP(OBut)O}^- \xrightarrow{h\nu} \text{Ph}_2\text{P(O)(OBut)} + \text{I}^- \qquad (36)$$
$$\phantom{\text{PhI} + }\mathbf{37}\phantom{\text{P(OBut)O}^- \xrightarrow{h\nu} }\mathbf{38}$$

Reaction occurred less rapidly in DMSO solution. The dark reaction in ammonia gave only 15% of iodide elimination after 4 h. Bromobenzene reacts with **37** similarly, but more slowly, furnishing **38** in 72% yield after 75 min of irradiation.

Potassium diphenylphosphinite **39** reacted rapidly under photostimulation to form triphenylphosphine oxide **40** in high yield (Reaction 37). However, the analogous reaction with bromobenzene was undetectable after 80 min of irradiation.

$$\text{PhI} + \text{Ph}_2\text{PO}^- \xrightarrow{h\nu} \text{Ph}_3\text{PO} + \text{I}^- \qquad (37)$$
$$\qquad\quad \mathbf{39} \qquad\qquad \mathbf{40}$$

Potassium O,O-diethyl thiophosphite **41** reacted quickly with iodobenzene under irradiation to form O,O-diethyl phenylthiophosphonate **42** almost quantitatively (Table IV) (Reaction 38). Bromobenzene reacted more slowly.

$$\text{PhI} + (\text{EtO})_2\text{PS}^- \xrightarrow{h\nu} \text{PhP(S)(OEt)}_2 + \text{I}^- \qquad (38)$$
$$\qquad\quad \mathbf{41} \qquad\qquad\quad \mathbf{42}$$

Irradiation of the potassium salt of N,N,N',N'-tetramethylphosphonamide **43** with iodobenzene released 93% of iodide ion after 30 min and gave 64% yield of N,N,N',N'-tetramethylphenylphosphonamide **44** (Reaction 39). Some benzene was also formed, probably by hydrogen atom abstraction from the nucleophile. Bromobenzene underwent a similar, but slower, reaction.

$$\text{PhI} + (\text{Me}_2\text{N})_2\text{PO}^- \xrightarrow{h\nu} \text{PhP(O)(NMe}_2)_2 + \text{I}^- \qquad (39)$$
$$\qquad\quad \mathbf{43} \qquad\qquad\quad \mathbf{44}$$

Table IV

Photostimulated Reaction of Phosphanions with Phenyl Halides in Ammonia

PhX	Nucleophile	Irradiation Time (min)	Substitution Product (%)
PhI	PhP(OBu)OK	15	**38** (95)
PhBr	PhP(OBu)OK	75	**38** (72)
PhI	Ph$_2$POK	20	**40** (95)
PhBr	Ph$_2$POK	80	— (nil)
PhI	(EtO)$_2$PSK	15	**42** (95)
PhI	(Me$_2$N)$_2$POK	30	**44** (64)
PhBr	(Me$_2$N)$_2$POK	90	**44** (65)

Source: Reproduced, from Reference 30, Copyright 1979, American Chemical Society.

The iodobenzene:bromobenzene rate ratio with phosphanion nucleophiles is summarized in Table V.

The relative reactivity of iodobenzene:bromobenzene with phosphanion nucleophiles in separate reactions is very different from that measured when both reactants compete for the same nucleophile. This entrainment effect is very common in reactions by the $S_{RN}1$ mechanism (*see* Chapter 8), although this effect was not observed with **43** as nucleophile (*30*).

Arsenic Nucleophiles

Dialkylarsenide or diarylarsenide ions are known to react readily with halo and dihalobenzenes. For instance, dimethylarsenide ion **45a** (*31*) and methylphenylarsenide ion **45b** (*32*) react with *o*-dichlorobenzene in tetrahydrofuran to give diarsines of type **46** in good yields (Reaction 40). Diphenylarsenide ion **47** reacts with *p*-dibromobenzene to give the *p*-phenylene-bis(diphenylarsine) in 75% yield (*33*).

In the photostimulated reaction of *p*-chloro, *p*-bromo and *p*-iodobenzenes with diphenylarsenide ion **47** in liquid ammonia, the expected *p*-tolyldiphenylarsine **49** was found, and also triphenylarsine **48**, di(*p*-tolyl)phenylarsine **50** and tris(*p*-tolyl)arsine **51** (Reaction 41) (*34*).

Table V

Iodobenzene: Bromobenzene Rate Ratio with Phosphanion Nucleophiles

Nucleophile	PhI/PhBr	Ref.
25	10^3	27
37	10^3	30
43	5×10^2	30
24	4×10^2	21
41	45	30

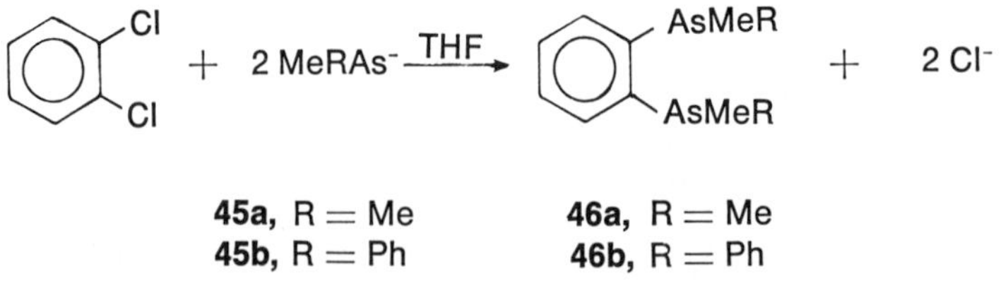

45a, R = Me	**46a**, R = Me
45b, R = Ph	**46b**, R = Ph

$$ToX + Ph_2As^- \xrightarrow{h\nu} Ph_3As + Ph_2ToAs + PhTo_2As + To_3As \qquad (41)$$
$$ \mathbf{47} \phantom{Ph_2As^- \xrightarrow{h\nu} } \mathbf{48} \mathbf{49} \mathbf{50} \mathbf{51}$$

To = p-tolyl
X = Cl, Br, I

In the dark there is no reaction. This reaction was suggested to occur by the photostimulated $S_{RN}1$ mechanism, but additional steps should be added to account for the arsines **48**, **50**, and **51** formed (Scheme II).

$$ToX + \mathbf{47} \xrightarrow{h\nu} (ToX)^{\overline{\cdot}} + Ph_2As^\bullet \qquad (42)$$

$$(ToX)^{\overline{\cdot}} \longrightarrow To^\bullet + X^- \qquad (43)$$
$$\phantom{(ToX)^{\overline{\cdot}} \longrightarrow } \mathbf{52}$$

$$To^\bullet + \mathbf{47} \rightleftharpoons (Ph_2ToAs)^{\overline{\cdot}} \rightleftharpoons PhToAs^- + Ph^\bullet \qquad (44)$$
$$\phantom{To^\bullet + \mathbf{47} \rightleftharpoons (Ph_2ToAs)} -e^- \Big| \mathbf{53} \mathbf{54} \mathbf{55}$$
$$\phantom{To^\bullet + \mathbf{47} \rightleftharpoons } Ph_2ToAs$$
$$\phantom{To^\bullet + \mathbf{47} \rightleftharpoons } \mathbf{49}$$

$$PhToAs^- + To^\bullet \rightleftharpoons (PhTo_2As)^{\overline{\cdot}} \rightleftharpoons To_2As^- + Ph^\bullet \qquad (45)$$
$$\mathbf{54} \mathbf{52} \mathbf{56} \mathbf{57} \mathbf{55}$$
$$ -e^- \Big|$$
$$ PhTo_2As$$
$$ \mathbf{50}$$

$$To_2As^- + To^\bullet \rightleftharpoons (To_3As)^{\overline{\cdot}} \xrightarrow{-e^-} To_3As \qquad (46)$$
$$\mathbf{57} \mathbf{52} \mathbf{58} \phantom{\xrightarrow{-e^-} } \mathbf{51}$$

$$Ph^\bullet + \mathbf{47} \rightleftharpoons (Ph_3As)^{\overline{\cdot}} \xrightarrow{-e^-} Ph_3As \qquad (47)$$
$$\mathbf{55} \phantom{Ph^\bullet + \mathbf{47} \rightleftharpoons (Ph_3As)} \mathbf{59} \phantom{\xrightarrow{-e^-} } \mathbf{48}$$

To = p-tolyl
X = Cl, Br, I

Scheme II

Once the p-halotoluene radical anion is formed (Reaction 42), it decomposes in the usual way to give p-tolyl radical **52** and halide ion (Reaction 43). The p-tolyl radical **52** then couples with **47** to give the intermediate radical anion **53**, which undergoes three competitive reactions: reversion to starting **52** and **47**, decomposition to p-tolylphenylarsenide **54** and phenyl radical **55**, and electron transfer to the substrate to give the product p-tolyldiphenylarsine **49** (Reaction 44).

Reaction 44 forms two new intermediates, **54** and **55**. The arsenide ion **54** competes with **47** as nucleophile for the *p*-tolyl radical **52**. This reaction yields the radical anion **56** (Reaction 45), which can then give the product **50** by electron transfer to the substrate, revert to starting materials **54** and **52**, or decompose to give the intermediate arsenide ion **57** and phenyl radical **55**. The arsenide ion **57** can also react with **52** to give the radical anion intermediate **58** (Reaction 46), which can revert to **57** and **52**, or transfer its extra electron to give triphenylarsine **51**.

Phenyl radicals **55** formed in Reactions 44 and 45 can compete with **52** for the nucleophile **47** to give product **48** ultimately (Reaction 47).

Because **54** and **55** are formed through the same intermediate radical anion **53**, their concentration should be the same. This is roughly the case, with about 60% of **49** and 15–20% of **48** and **50** being formed (Table VI). The ratio of **49** to **48** and **50** indicates that the electron transfer process is about four times faster than bond rupture (Reaction 44). If the same ratio holds for Reaction 45, it is reasonable that the yield of **51** is only in the order of 2–3%.

In the photostimulated reaction of *p*-chloroanisole, *p*-bromoanisole, and *p*-iodoanisole **60** with **47**, the four possible arsines **48**, **61**, **62**, and **63** were found (Reaction 48).

$$\text{AnX} + \mathbf{47} \xrightarrow{h\nu} \mathbf{48} + \underset{\mathbf{61}}{\text{Ph}_2\text{AnAs}} + \underset{\mathbf{62}}{\text{PhAn}_2\text{As}} + \underset{\mathbf{63}}{\text{An}_3\text{As}} \qquad (48)$$

60a, X = Cl
60b, X = Br
60c, X = I

An = *p*-anisyl

With **60b** there was no reaction in the dark, yet reaction was quantitative after 5 min of irradiation. The product distribution was similar for the three *p*-haloanisoles, except that the yield of **61** decreased in the order **60c** > **60b** > **60a**, suggesting that if the trend is real, the rate of electron transfer to the substrate decreases in the same order—consistent with the mechanism proposed in Scheme II. However, in an experiment with tenfold excess of **60b** (to assure a high concentration of electron acceptor throughout the reaction) the yield of **61** did not increase.

In the photostimulated reaction of 4-chlorobenzophenone **64** with **47**, the only arsine product found was the straightforward substitution product **65** (Reaction 49) formed quantitatively in 30 min of irradiation (*34*). During the same period of time, in the dark, there was 63% of reaction, but it was inhibited by *m*-dinitrobenzene and oxygen. The dark reaction was attributed to a thermal electron transfer from **47** to **64**; a similar result was observed between benzophenone and **47** in THF (*35*).

Compared to the scrambling of aryl rings found in the reaction of **47**

Table VI

Photostimulated Reaction of Diphenylarsenide Ion with Haloarenes in Liquid Ammonia

ArX^a	hν (min)	Ph_2AsH	Ph_3As	Ph_2ArAs Yield (%)	$PhAr_2As$	Ar_3As	Ref.
ClTo	60	—	15	61	15	2	34
ClTo	60[b]	75	3	—	—	—	34
BrTo	60	—	21	62	15	1	34
BrTo[c]	60	—	18	60	19	2	34
ITo	60	—	20	50	25	4	34
ClAn	60	—	30	48	19	3	34
BrAn	5	[d]	20	60	22	[d]	34
BrAn	60[b]	83	3	—	—	—	34
BrAn[e]	60	—	25	56	19	[d]	34
IAn	5	—	18	78	5	—	34
4-ClB	30	—	1	100	—	—	34
4-ClB	30[b]	[d]	[d]	63	—	—	34
4-ClB	30[b,f]	[d]	[d]	21	—	—	34
4-ClB	30[b,g]	[d]	[d]	0	—	—	34
1-BrN	10	—	33	40	21	[d]	36
9-BrP	60	—	20	60	15	[d]	36
2-ClQ	5	—	—	60	—	—	36

[a]The abbreviations are as follows: To, p-tolyl; An, p-anisyl; 4-ClB, 4-chlorobenzophenone; 1-BrN, 1-bromonaphthalene; 9-BrP, 9-bromophenanthrene; and 2-ClQ, 2-chloroquinoline.
[b]Dark reaction.
[c]Dissolved in 50 ml DMSO.
[d]Not quantified.
[e]Ten fold excess of the substrate.
[f]20 mole % of m-dinitrobenzene.
[g]Bubbling oxygen through the solution.

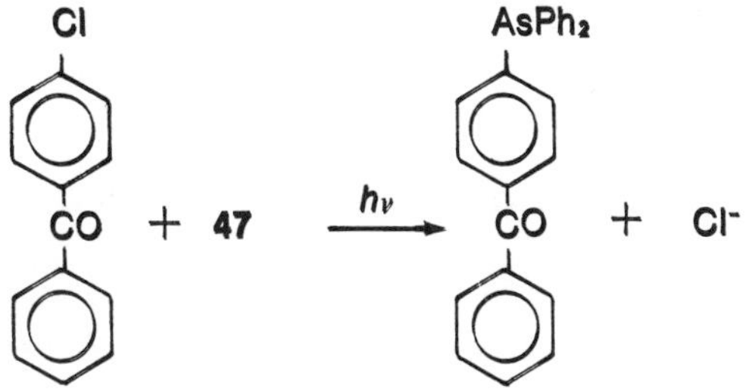

$$ \textbf{64} \qquad\qquad\qquad\qquad \textbf{65} \tag{49}$$

with p-halotoluenes and p-haloanisoles, only the substitution product **65** was found with **64** as substrate.

The reactions of several substrates with **47** were studied in order to determine the factors that produce scrambling of aryl rings (*36*). In the photostimulated reaction of 1-bromonaphthalene and 9-bromophenanthrene with **47** in ammonia scrambling occurred (Table VI), but with 2-chloroquinoline only the straightforward substitution product was formed. It has been reported that **47** reacts with 8-chloroquinoline after 12 h of reaction in liquid ammonia to give (8-quinolyl)diphenylarsine (*20*). This work was published before the photostimulated $S_{RN}1$ reactions were discovered, but the mechanism was thought to be a straightforward nucleophilic substitution. Because 2-chloroquinoline reacts in 5 min under irradiation, the dark reaction of the 8-isomer may also occur by the $S_{RN}1$ mechanism.

These results show contrasting behavior of substrates in photostimulated reactions with **47**. Haloderivatives of toluene, anisole, naphthalene, and phenanthrene, lead to scrambling of aryl rings, but when the substrates have lower reduction potentials, as do derivatives of quinoline and benzophenone, only the straightforward substitution products are formed.

Antimony Nucleophiles

Although dialkyl and diphenylstibides reacted with dichloroalkanes in liquid ammonia to give ditertiary stibines in good yields (*37*) [except for 1,2-dichloroethane, which underwent elimination to give ethylene and tetraphenyldistibine (*38*)], these nucleophiles failed to react with o-dichlorobenzene. Therefore, diphenylstibide ion did not give o-phenylene-bis(diphenylstibine), and dimethylstibide gave only traces of the corresponding distibine (*39*).

The reaction of the dimethylstibide nucleophile with o-bromoiodobenzene gave a 9% yield of the disubstitution product, and even less product with o-dibromobenzene or o-dichlorobenzene (*40*).

The photostimulated reaction of diphenylstibide ion with p-bromoanisole and 4-chlorobenzophenone in liquid ammonia gave four stibines as products of the reaction (Reaction 50) (*36*).

$$ArX + Ph_2Sb^- \longrightarrow Ph_3Sb + Ph_2ArSb + PhAr_2Sb + Ar_3Sb \qquad (50)$$

| ArX = p-Br-anisole | 30 | 45 | 19 | 2 |
| = 4-Cl-benzophenone | 30 | 33 | 15 | — |

Although there is a dark reaction with 4-chlorobenzophenone, the mechanism was ascribed to the $S_{RN}1$ reaction, because the scrambling of aryl rings in the products is hard to explain by known ionic mechanisms of nucleophilic aromatic substitution. However, the $S_{RN}1$ mechanism, which involves the formation of stibine radical anion intermediates that react in part by electron transfer and in part by Ph–Sb bond breaking (the mechanism resembles that of diphenylarsenide ion, *see* Scheme II), gives a good account of the observations.

It is remarkable that 4-chlorobenzophenone with diphenylarsenide ion gave the straightforward substitution product, but with diphenylstibide ion as nucleophile, gave scrambled products.

Literature Cited

1. Kim, J. K.; Bunnett, J. F. *J. Am. Chem. Soc.* **1970**, *92*, 7463–7464.
2. Bliznyuk, N. K.; Kolomiets, A. F.; Kvasha, Z. N.; Levskaya, G. S.; Zhemchuzhin, S. G. *Zh. Obshch. Khim.* **1966**, *36*, 480.
3. Rossi, R. A.; Bunnett, J. F. *J. Org. Chem.* **1972**, *37*, 3570.
4. Bunnett, J. F.; Gloor, B. F. *Heterocycles* **1976**, *5*, 377.
5. Bunnett, J. F.; Zahler, R. E. *Chem. Rev.* **1951**, *49*, 273.
6. Kametani, T.; Takahashi, K.; Ihara, M.; Fukumoto, K. *J. Chem. Soc., Perkin Trans. I* **1976**, 389.
7. Pietra, F.; Bartolozzi, M.; Del Cima, F. *J. Chem. Soc., Chem. Commun.* **1971**, 1232.
8. Danen, W. C.; Kensler, T. T.; Lawless, J. G.; Marcus, M. F.; Hawley, M. D. *J. Phys. Chem.* **1969**, *73*, 4389.
9. López, A. F.; Ph.D. Thesis, Universidad Nacional de Córdoba, 1979.
10. Amatore, C.; Chaussard, J.; Pinson, J.; Saveant, J. M.; Thiebault, A. *J. Am. Chem. Soc.* **1979**, *101*, 6012.
11. Rossi, R. A.; Pierini, A. B.; de Rossi, R. H. *J. Org. Chem.* **1978**, *43*, 1276.
12. Benati, L.; Montevecchi, P. C.; Spagnolo, P. *Tetrahedron Lett.* **1978**, 815.
13. Russell, G. A. *J. Chem. Soc., Esp. Pub. 24*, **1970**, 271.
14. Nelson, R. F.; Carpenter, A. K.; Seo, E. T. *J. Electrochem. Soc.* **1973**, *120*, 206.
15. Elofson, R. M.; Gadallah, F. F. *J. Org. Chem.* **1969**, *34*, 854.
16. Opgenorth, H. J.; Rüchardt, C. *Liebig Ann. Chem.* **1974**, 1333.
17. Bartak, D. E.; Danen, W. C.; Hawley, M. D. *J. Org. Chem.* **1970**, *35*, 1206.
18. Russell, G. A.; Metcalfe, A. R. *J. Am. Chem. Soc.* **1979**, *101*, 2359.
19. Aguiar, A.; Greenberg, H. J.; Rubenstein, K. E. *J. Org. Chem.* **1963**, *28*, 2091.
20. Feltham, R. D.; Metzger, H. G. *J. Organomet. Chem.* **1971**, *33*, 347.
21. Swartz, J. E.; Bunnett, J. F. *J. Org. Chem.* **1979**, *44*, 340.
22. Ward, H. R. *Acc. Chem. Res.* **1972**, *5*, 18.
23. Lawler, R. G. *Acc. Chem. Res.* **1972**, *5*, 25.
24. Bengerter, B. W.; Beatty, R. P.; Kouba, J. K.; Wreford, S. S. *J. Org. Chem.* **1977**, *42*, 3247.
25. Lepley, A. R. "Chemically Induced Magnetic Polarization"; Lepley, A. R.; Closs, G. L.; Eds., Wiley: New York, 1973; p. 327.
26. Bunnett, J. F.; Creary, X. *J. Org. Chem.* **1974**, *39*, 3612.
27. Bunnett, J. F.; Traber, R. P. *J. Org. Chem.* **1978**, *43*, 1867.
28. Bunnett, J. F.; Weiss, R. H. *Org. Synth.* **1978**, *58*, 134.
29. Bunnett, J. F.; Shafer, S. J. *J. Org. Chem.* **1978**, *43*, 1873.
30. Swartz, J. E.; Bunnett, J. F. *J. Org. Chem.* **1979**, *44*, 4673.
31. Felthan, R. D.; Silverthorn, W. *Inorg. Synth.* **1967**, *10*, 159.
32. Henrick, K.; Wild, S. B. *J. Chem. Soc., Dalton Trans.* **1975**, 1506.
33. Zorn, H.; Schindlbauer, H.; Hammer, D. *Monatsh. Chem.* **1967**, *98*, 731.
34. Rossi, R. A.; Alonso, R. A.; Palacios, S. M. *J. Org. Chem.* **1981**, *46*, 2498.

35. Busse, P. J.; Irgolic, K. J.; Dominguez, R. J. G. *J. Organometal. Chem.* **1974,** *81,* 45.
36. Alonso, R. A.; Rossi, R. A. *J. Org. Chem.* **1982,** *47, 77.*
37. Meinema, H. A.; Martens, H. F.; Noltes, J. G. *J. Organometal. Chem.* **1976,** *110,* 183.
38. Meinema, H. A.; Martens, H. F.; Noltes, J. G. *J. Organometal. Chem.* **1973,** *51,* 233.
39. Levason, W.; McAuliffe, C. A.; Murray, S. G. *J. Organometal. Chem.* **1975,** *88,* 171.
40. Shewchuk, E.; Wild, S. B. *J. Organometal. Chem.* **1977,** *128,* 115.

4

Nucleophiles Derived from VIA Group of Elements

Oxygen Nucleophiles

Several ambient nucleophiles such as ketone, N, N-disubstituted amide and ester enolate ions (Reaction 1), where the negative charge is shared by carbon and oxygen, react with aryl radicals only at the carbanionic functionality.

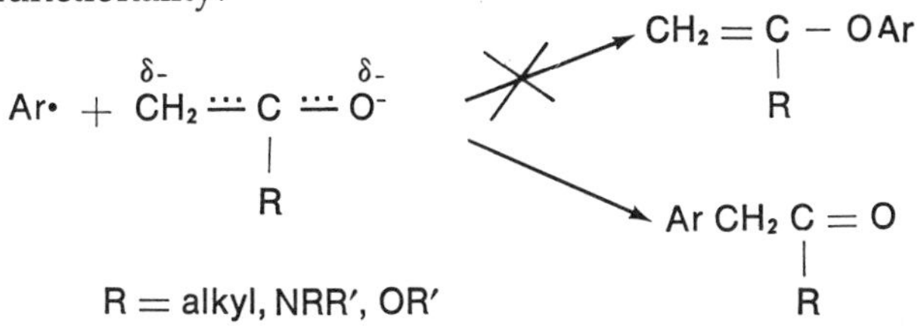

$$Ar\bullet + CH_2 \overset{\delta-}{\cdots} C \overset{\delta-}{\cdots} O^- \underset{R}{\Big|} \quad \xrightarrow{\quad\quad} \begin{cases} CH_2 = \underset{R}{\overset{|}{C}} - OAr \\[2em] Ar\,CH_2\,\underset{R}{\overset{|}{C}} = O \end{cases} \tag{1}$$

$$R = \text{alkyl, NRR', OR'}$$

Alkoxides derived from primary or secondary alcohols do not react as nucleophiles with aryl radicals. Instead it has been shown that methoxide (1) and 2-propoxide (2) ions promote hydrogen atom abstraction (Reactions 2, 3).

$$Ar^{\cdot} + CH_3O^- \longrightarrow ArH + (CH_2O)^{\cdot -} \tag{2}$$

$$Ar^{\cdot} + (CH_3)_2CHO^- \longrightarrow ArH + CH_3 - \overset{\bullet}{\underset{\underset{O\_}{|}}{C}} - CH_3 \tag{3}$$

Although, *tert*-butoxide ion has been used widely as a base in $S_{RN}1$ reactions it has never been found to react with aryl radicals. It is also a poor hydrogen atom donor since it has only hydrogens bonded to primary carbon atoms.

Phenoxide ion behavior is different from the other derivatives of the VIA group of elements, such as benzenethiolate, phenyl selenide, and

0065-7719/83/0178-0079$06.50/1

phenyl telluride ions, because it does not react as a nucleophile in reactions in liquid ammonia stimulated by either solvated electrons (from dissolution of alkali metals (3) or by electrodes (4) or by light (5). The lack of reactivity of p-cresolate ions toward iodobenzene was observed even under unusually intense irradiation in liquid ammonia and, in this case, only the dehalogenation of iodobenzene was observed (6).

Halobenzenes have been reported to react with phenoxide ion in 50% aqueous t-butyl alcohol under stimulation by solvated electrons from dissolution of sodium amalgam, giving good yields of dipheyl ether (7). However, these results could not be reproduced, and the only product formed was benzene (8). Similarly, phenoxide ion did not react with aryl radicals fromed from haloarenes in dimethyl sulfoxide/tert-butoxide ion (9).

In ammonia, alkali phenoxides are slightly soluble, but several derivatives of related functionality, such as 1–3, are soluble. Yet no reaction was observed when they were irradiated in the presence of 1-chloronaphthalene in liquid ammonia (5).

1   **2**   **3**

From these results it was concluded that oxygen nucleophiles do not react with aryl radicals; or, they react slowly in comparison with the rates of other competing reactions.

The lack of reactivity of alkoxide ions toward aryl radicals was also shown in the reaction of triarylsulfonium (10) and triaryliodonium (11) salts with alkoxides. Both reactions produce aryl radicals and products derived from them in high yields, but the small amount of alkyl aryl ethers formed arise from conventional nucleophilic substitution reactions.

Sulfur Nucleophiles

Aryl Sulfide Ions. When solutions of aryl iodides and sodium or potassium benzenethiolate in refluxing ammonia are irradiated under nitrogen, reaction occurs to form aryl phenyl sulfides (Reaction 4). This procedure is attractive for the preparation of diaryl sulfides. The aryl iodide is mixed with benzenethiol in ammonia and, after photo-stimulation with 350-nm light, leads to a substitution product. Aryl bromides also react, but much more slowly than aryl iodides. Several specific reactions are summarized in Table I (*12*).

$$ArI + PhS^- \xrightarrow{h\nu} ArSPh + I^- \qquad (4)$$
$$\mathbf{4}$$

Under the same reaction conditions many dihalobenzenes afford disubstitution products. Typically, little, if any, monosubstitution products are formed. Moreover there is evidence that the monosubstitution product is not an intermediate in the production of the disubstitution product (*13*).

Kinetic analysis of the substitution reaction of *m*-chloroiodobenzene **5** with benzenethiolate **4** shows that the monosubstitution product **9** is not an intermediate in the formation of bis-sulfide **12** (Scheme I). That **5** is more reactive than **9** is demonstrated by the fact that in 105 sec of irradiation, **5** reacted to the extent of 68%, but **9** only to the extent of 15%. Moreover, direct competition between **5** and **9** for **4** showed that **5** was 17 times more reactive than **9**. These reactivity measurements establish that **9** is too unreactive to be an intermediate on the reaction pathway to the disubstitution product. The mechanism of Scheme I gives a good account of these results.

In Scheme I the radical anion **8** has two competing reactions, Reactions 8 and 9. Loss of the odd electron (Reaction 8) by transfer to **5** to give the monosubstitution product **9** is the minor reaction pathway for all the aryl halides except for fluoride (Table I). The major pathway is the loss of halide from **8** (Reaction 9) which leads ultimately to the disubstitution product **12**.

The fact that *m*-fluoroiodobenzene gave only the monosubstitution product indicates that electron transfer from *m*-fluorophenyl phenyl-sulfide radical anion (Reaction 8) is faster than fluoride loss (Reaction 9). This is reasonable considering that C–F bonds are stronger and more difficult to break than C–Cl or C–Br bonds.

The reaction of 4-bromoisoquinoline **13** with **4** in methanol at 147°C to give 4-phenylthioisoquinoline **14** has been reported. But when methoxide ion is present, the rate of reaction is increased and the reduction product, isoquinoline **15,** is formed (Reaction 12, 13) (*14*).

Catalysis by methoxide ions is observed in the time dependence formation of **14** (Figure 1). The formation of **14** in the absence of methoxide ion proceeds by the well known ionic route, involving attack of **4** on **13** to give a σ-complex intermediate that subsequently eliminates bromide to give **14**. But in the presence of methoxide ion, the substitution mechanism follows a $S_{RN}1$ pathway.

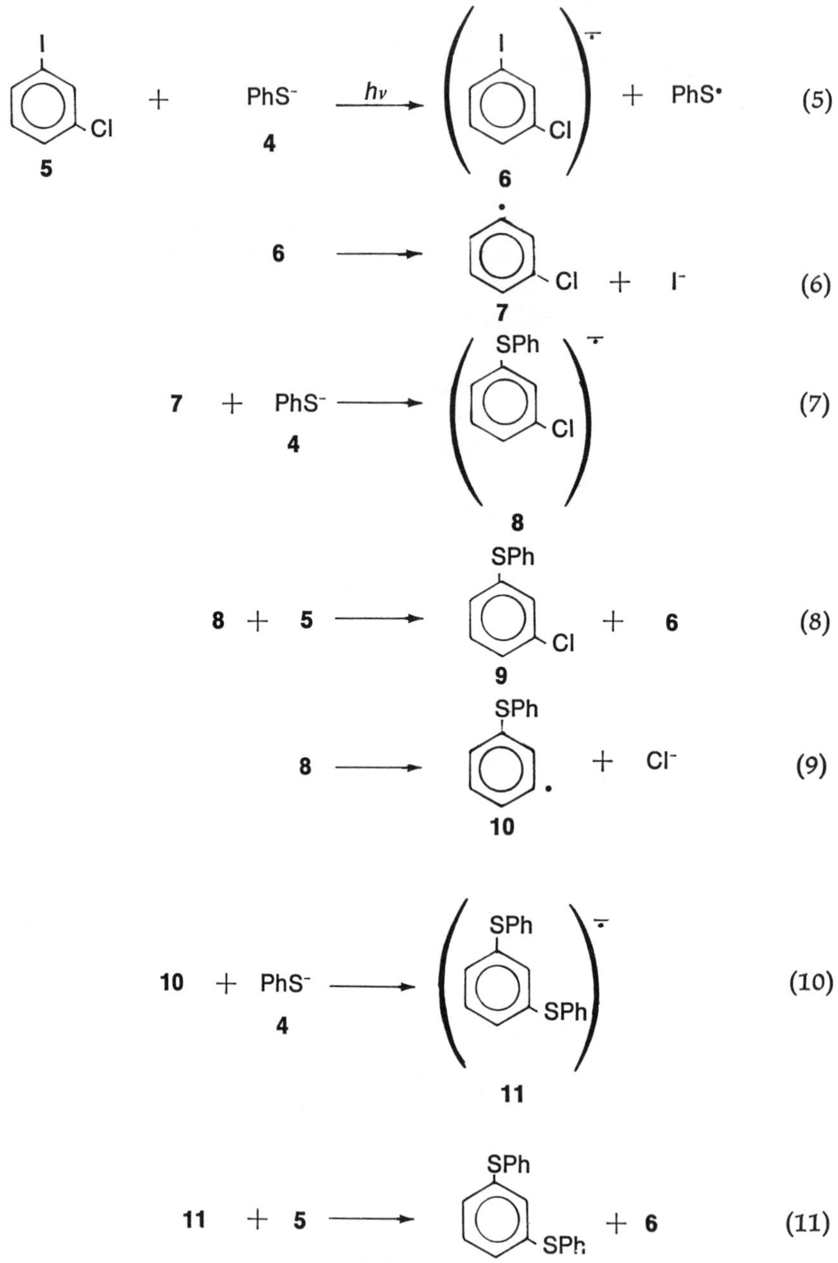

Scheme I

Table I

Photostimulated Reaction of Haloarenes with Benzenethiolate Ion in Liquid Ammonia

ArX	hv (min)	Substitution Product (%)	Ref.
PhI	70	94	12
1-Iodonaphthalene	75	85	12
o-(MeO)C₆H₄I	90	91	12
m-(MeO)C₆H₄I	90	88	12
p-(MeO)C₆H₄I	30	76	12
o-(Me)C₆H₄I	165	68	12
m-(Me)C₆H₄I	135	81	12
p-(Me)C₆H₄I	360	72	12
2-Iodo-1,3-dimethylbenzene	140	19	12
m-Fluoroiodobenzene	100	96	12
p-Iododiphenyl ether	120	92	12
m-(CF₃)C₆H₄I	170	71	12
PhBr	120	23	12
m-Chloroiodobenzene	150	91[a]	13
p-Chloroiodobenzene	120	89[a]	13
m-Dibromobenzene	190	92[a]	13
m-Chlorobromobenzene	180	55[a]	13
p-Dibromobenzene	300	64[a]	13
p-I-C₆H₄NMe₃I	180	95[a]	13
o-Chloroiodobenzene	90	77[a]	13
PhI	120	71[b]	21
p-(MeO)C₆H₄I	100	73[b]	21

[a] Disubstitution product.
[b] The nucleophile was p-methoxybenzenethiolate.

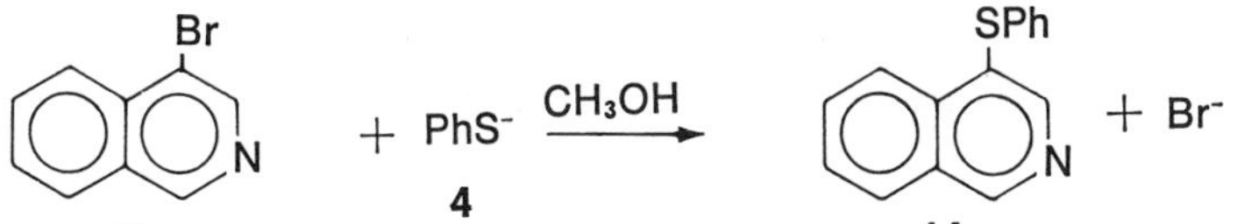

$$13 + PhS^- \xrightarrow{CH_3OH} 14 + Br^- \quad (12)$$

$$13 + PhS^- \xrightarrow{CH_3OH/CH_3O^-} 14 + 15 + Br^- \quad (13)$$

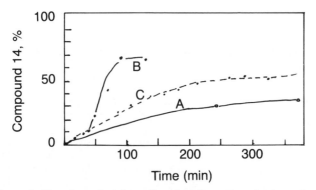

Figure 1. *Time dependent formation of* **14** *from* **13** *and* **4** *in methanol at 147°C.* *(***13***).* = *0.52 M,* *(***4***).* = *0.98 M; A: without additives; B: 0.98 M NaOCH₃ added.* *C* = *0.98 NaOMe and 0.2 M azobenzene (as radical chain inhibitor) added.* *(***14***).*

The 4-bromoisoquinoline radical anion formed, **16,** decomposes to 4-isoquinolyl radical **17** (Reaction 14), which reacts with **4** to give **14** by the $S_{RN}1$ route (Reaction 15), or with methoxide ion to give isoquinoline **15** by hydrogen atom abstraction (Reaction 16).

$$17 + CH_3O^- \longrightarrow 15 + (CH_2O)^{\overline{\cdot}} \qquad (16)$$
$$ 19 \phantom{^{\overline{\cdot}} \qquad}$$

$$19 + 13 \longrightarrow 16 + CH_2O \qquad (17)$$

It has been shown that the electrochemical reduction of 4-bromobenzophenone **20** in acetonitrile and dimethylformamide involves cleavage of the C–Br bond and formation of benzophenone. Decomposition of the initial radical anion **21** gives the aryl radical **22** (Reaction 19) and, in the presence of **4, 22** reacts to give the radical anion of the substitution

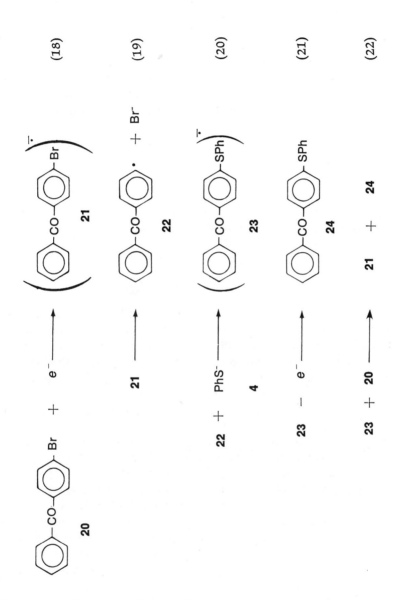

product **23** (Reaction 20). This radical anion **23** can be oxidized by the electrode to the substitution product 4-phenylthiobenzophenone **24** (Reaction 21), or it can transfer its extra electron to **20** in the chain propagation step of the $S_{RN}1$ mechanism (Reaction 22) (*15*).

The catalytic character of electrochemical reduction is clearly shown by the fact that at an electrolysis potential of -1.8 V the yield of **24** is 80% with a consumption of 0.2 F/mol of the substrate **20**.

Electrolysis at -1.9 V yields not only the substitution product **24**, but also the alcohol **25**, and indicates that at this potential the radical anion **23** is further reduced by the electrode to the alcohol **25** (Reaction 23).

$$23 \ + \ e^{-} \longrightarrow$$

(23)

25

Electrochemical reduction offers greater versatility compared to reaction stimulated with solvated electrons from dissolution of alkali metals, because the control of the electrolysis potential provides the equivalent of a continuous series of reducing agents of variable strength (*15*).

In the electrochemical reduction of 4-bromoacetophenone in the presence of **4** as nucleophile, it has been suggested that the principal reaction pathway of the radical anion **23** formed (Reaction 20) is homogeneous electron transfer reaction to the substrate, and not reoxidation by heterogeneous electron transfer to the electrode. Conversion of the substrate increases when the concentration is increased, as required of the bimolecular reaction (Reaction 22) wherein the electrolysis potential is kept constant and 0.2 is passed per mole of substrate (Table II) (*16*).

Several haloaromatic compounds were electrolyzed in the presence of **4**, and the results are shown in Table III. From these results it is concluded that there is a definite relationship between the substituted product yield and the number of F/mol consumed: when the yield is good, consumption of electricity is very low, reflecting the catalytic character of the electrochemical process (*17*).

p-Iodobenzonitrile and 1-bromonaphthalene in acetonitrile gave low yields of substitution products and high consumption of electricity, which is attributed to a competing route for aryl radicals (Reactions 24–25).

$$\text{Ar}^{\cdot} + e^{-} \longrightarrow \text{Ar}^{-} \xrightarrow{\text{SH}} \text{ArH} \qquad (24)$$

$$\text{Ar}^{\cdot} \xrightarrow{\text{SH}} \text{ArH} + \text{S}^{\cdot} \qquad (25)$$

Reaction 24 is more important for p-iodobenzonitrile than for p-bromobenzonitrile because p-iodobenzonitrile radical anion fragments into aryl radicals and the halide ion faster than p-bromobenzonitrile radical anion fragment. Therefore, the aryl radical is formed in a zone close to the electrode, and is reduced before it can diffuse away. Reaction 25 seems to be responsible for the lower yield of substitution product

Table II

Electrolysis of 4-Bromoacetophenone in the Presence of Benzenethiolate Ion at −1.7 V

4-Bromoacetophenone (mmol/L)	Charge passed (F/mol)	Conversion (%)
4.5	0.21	50
9.0	0.20	60
13.5	0.20	90
18.0	0.20	100

Source: Reproduced, with permission, from Reference 16. Copyright 1977, Pergamon Press.

Table III

Electrolysis of Haloarenes in the Presence of Ammonium Benzenethiolate

ArX	Solvent	Potential	F/mol	ArSPh	ArH	Ref.
4-Bromobenzophenone	acetonitrile	−1.8	0.2	95	3	17
p-Bromobenzonitrile	acetonitrile	−2.1	0.2	80	10	17
p-Iodobenzonitrile	acetonitrile	−1.7	1.5	20	80	17
1-Bromonaphthalene	dimethyl sulfoxide	−2.2	0.3	100	0	17
1-Bromonaphthalene	acetonitrile	−2.2	1.2	32	40	17
2-Chloroquinoline[a]	ammonia	−1.43	0.1	96	—	4

[a]Potassium benzenethiolate as nucleophile.

from 1-bromonaphthalene in acetonitrile as compared with the result in dimethyl sulfoxide, which is a poorer hydrogen atom donor (18, 19).

α-Diketones are known to be reduced easily to stable semidiones in basic DMSO (20). Thus, the ESR spectrum obtained from 5-halo-2H,3H-benzo(b) thiophene-2,3-diones **26** with an excess of potassium

t-butoxide in DMSO at room temperature, is attributed to its radical anion **27** (Reaction 26) (*9*).

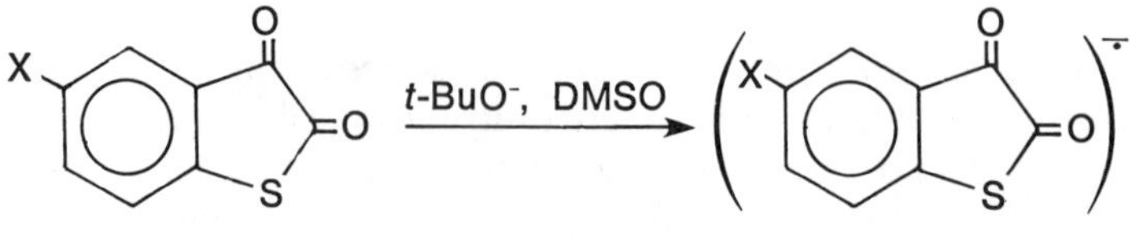

(26)

26   **27**

The spectrum of the radical anion **27** (X = F) remained unchanged after several days, whereas the spectrum of the other three 5-halo derivatives **27** (X = Cl, Br, or I) vanished in a few hours; at the same time another paramagnetic resonance, corresponding to the parent **27** (X = H), became important.

The reduction of **26** to **27** was also achieved with methoxide, hydroxide, phenoxide and methanethiolate ions. However, when **26** (X = Cl, Br) was treated with benzenethiolate ion **4**, the ESR spectra observed had some special features.

Figure 2a shows the ESR first-derivative spectrum of **27** (X = Br) generated with potassium benzenethiolate in DMSO just after mixing. The spectrum gradually changes, and after 30 h the spectrum of Figure 2c is that of the radical anion of the substitution product, which has been unambiguously identified as 5-phenylthio-2*H*,3*H*-benzo[*b*]thiophene-2,3-dione **27** (X = SPh). A possible explanation follows.

When **26** is mixed in DMSO with a base, **27** is formed, which slowly decomposes to a radical **28** and halide ion (X = Cl, Br, I) (Reaction 27). When the base used is *t*-butoxide or hydroxide, the radical **28** does not couple but undergoes hydrogen abstraction to give **29**, that again is reduced to **27** (X = H) (Reaction 28). However, when the base used is benzenethiolate **4**, it reacts with the radical **28** to give the radical anion of the substitution product **27** (X = SPh) (Reaction 29).

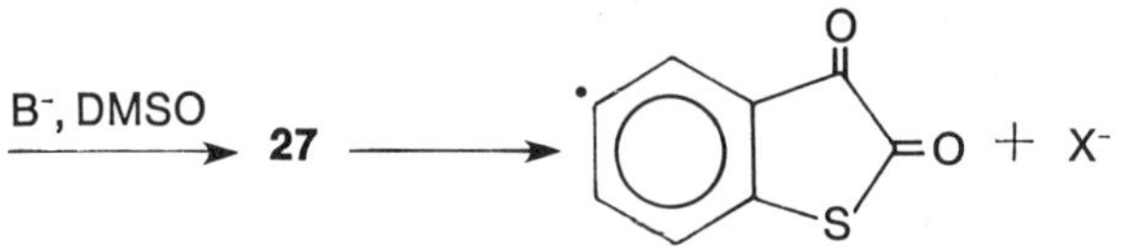

(27)

28

X = Cl, Br, I

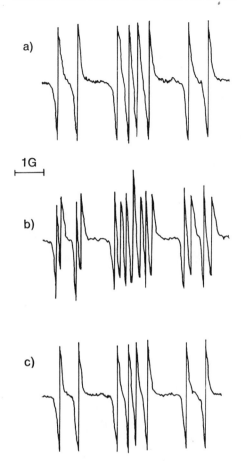

Figure 2. ESR first-derivative spectrum of **27** (X = Br) generated with potassium benzenethiolate in dimethyl sulfoxide just after mixing (a); after 5 h (b), and after 30 h (c). (Reproduced, from Reference 9. Copyright 1978, American Chemical Society.)

28 $\xrightarrow{\text{SH}}$ [structure **29**] $\xrightarrow{\text{B}^-, \text{DMSO}}$ **27** (X = H)   (28)

Alkanethiolate Ions. The photostimulated reaction of iodobenzene with **4** in ammonia is quite fast, and gives a nearly quantitative yield of diphenyl sulfide. However, the photostimulated reaction with ethane-thiolate ion **30** is much slower and, under similar conditions, only 60% of iodide ion is released in 90 min of irradiation. In 200 min of irradiation, and after quenching the reaction with benzyl chloride, there was obtained a 30% yield of phenyl ethyl sulfide **32**, 44% yield of benzyl phenyl sulfide **35**, and 3% yield of diphenyl sulfide **34** (*21*).

By way of explanation, the combination of phenyl radicals with ethane thiolate ions **30** in Scheme II is believed to form the ethyl phenyl

$$28 + PhS^- \longrightarrow 27 \ (X = SPh) \tag{29}$$
$${\bf 4}$$

$$PhI \ + \ EtS^- \xrightarrow{h\nu} (PhI)^{\cdot-} + \ EtS^\bullet \tag{30}$$
$${\bf 30}$$

$$(PhI)^{\cdot-} \longrightarrow Ph^\bullet \ + \ I^- \tag{31}$$

$$Ph^\bullet \ + \ EtS^- \longrightarrow (PhSEt)^{\cdot-} \tag{32}$$
$${\bf 30} {\bf 31}$$

$$PhI \ + \ (PhSEt)^{\cdot-} \longrightarrow (PhI)^{\cdot-} + \ PhSEt \tag{33}$$
$${\bf 31} \phantom{\longrightarrow (PhI)^{\cdot-} + \ } {\bf 32}$$

$$(PhSEt)^{\cdot-} \longrightarrow PhS^- \ + \ Et^\bullet \tag{34}$$
$${\bf 31}$$

$$Ph^\bullet \ + \ PhS^- \longrightarrow (PhSPh)^{\cdot-} \tag{35}$$
$${\bf 33}$$

$$PhI \ + \ {\bf 33} \longrightarrow (PhI)^{\cdot-} + \ PhSPh \tag{36}$$
$$\phantom{PhI \ + \ {\bf 33} \longrightarrow (PhI)^{\cdot-} + \ PhS}{\bf 34}$$

$$PhS^- \ + \ PhCH_2Cl \longrightarrow PhSCH_2Ph \ + \ Cl^- \tag{37}$$
$${\bf 35}$$

Scheme II

sulfide radical anion **31**. These ions react to transfer an electron to iodobenzene (Reaction 33) forming **32** or to expel ethyl radical (Reaction
34), the resulting benzenethiolate ion reacting with iodobenzene by the
$S_{RN}1$ mechanism to form **34**. However, most of the benzenethiolate does
not react, but is later captured by reaction with benzyl chloride to give
35 (Reaction 37) (*21*).

Although the fate of the ethyl radical formed in Reaction 34 is un-
clear, Reaction 34 is a pretermination step and therefore depresses the
overall reactivity compared with the reaction of benzenethiolate ion as
nucleophile.

The photostimulated reaction of the sodium salt of alkanethiolate
ions **36** with 1-chloronaphthalene or 1-bromonaphthalene, gave naphthalene and alkyl-1-naphthyl sulfides **37** in good yields (Table IV) (Reaction 38). Products derived from the decomposition of the radical anion
intermediate, as in Reaction 34, were not observed (*22*).

The different behavior of 1-halonaphthalenes as compared to the
reaction of iodobenzene with alkanethiolate ions stimulated an investi-

Table IV

**Photostimulated Reactions of Haloaromatic Compounds with
Alkanethiolate Ions in Liquid Ammonia**

ArX	R–S⁻ R	hν (min)	ArSR	ArS⁻
PhI	CH_3	180	12	19
PhI[a]	C_2H_5	200	30	44
PhI	n-C_4H_9	180	14	16
PhI	t-C_4H_9	180	36	33
PhI	$PhCH_2$	180	0	4
2-Chloropyridine[b]	n-C_4H_9	180	72–85	14–15
2-Chloropyridine	$PhCH_2$	180	8	31
2-Chloropyridine[c]	$PhCH_2$	180	11	23
1-Chloronaphthalene[d]	n-C_4H_9	170	81	0
1-Iodonaphthalene[d]	CH_2CH_2OH	150	74	0
1-Iodonaphthalene	t-C_4H_9	180	88	0
1-Bromonaphthalene[b]	$PhCH_2$	180	2–4	15–20
9-Bromophenanthrene[b]	$PhCH_2$	180	0–2	5–10
2-Chloroquinoline	$PhCH_2$	90	69	0
2-Chloroquinoline[e]	$PhCH_2$	90	0	0

[a] Ref. 21.
[b] Values of two or more reactions.
[c] Ten fold excess of substrate.
[d] Ref. 22.
[e] Dark reaction.
Source: Reproduced, from Reference 23. Copyright 1981, American Chemical Society.

gation about the behavior of several pairs of substrates and alkane-thiolates ions as nucleophiles. The objective was to determine the factors which affect the delicate balance between electron transfer (Reaction 33) and bond fragmentation (Reaction 34). These results are reported in Table IV and discussed in Chapter 8.

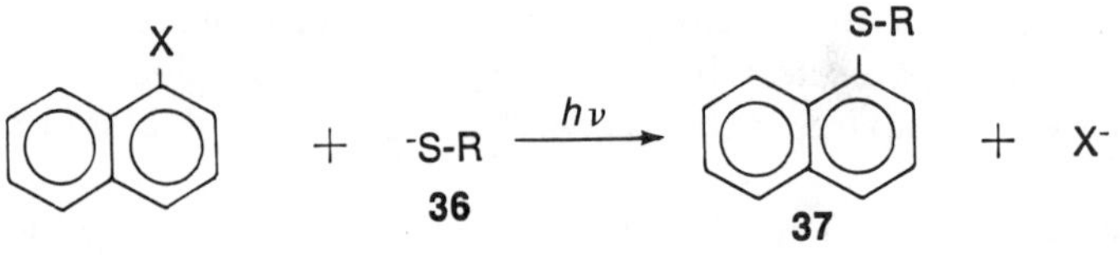

$$X = Cl, Br \qquad R = n\text{-}C_4H_9 \qquad R = n\text{-}C_4H_9 : 81\%$$
$$= CH_2CH_2OH \qquad = CH_2CH_2OH : 74\%$$

Other Sulfur Nucleophiles. Because alkyl and aryl sulfide ions react by the $S_{RN}1$ mechanism, the arylation of disodium or dipotassium salts of sulfide ions (S^{2-}) was attempted. However, these salts were insoluble in liquid ammonia, and no substitution products were found either by photostimulation or stimulation by solvated electrons (24).

Potassium thiocyanate **38**, potassium ethyl xanthate **39**, and sodium dithionite **40** are soluble in liquid ammonia, but they are unreactive with 1-chloronaphthalene by photostimulation (24).

Potassium O,O-diethyl thiophosphite, $(EtO)_2PS^-$, is a bidentate nucleophile, with charges at phosphorus and sulfur, but the only product observed in the photostimulated reaction with iodobenzene is the coupling of phenyl radical at the phosphorus atom (25).

Selenium Nucleophiles

When a solution containing phenyl selenide ion **41** and haloarenes was irradiated in liquid ammonia, the substitution product aryl phenyl selenide **42** was isolated (Reaction 39) (26, 27).

$$ArX + PhSe^- \xrightarrow{h\nu} ArSePh + X^- \qquad (39)$$
$$\qquad\qquad \mathbf{41} \qquad\qquad \mathbf{42}$$

With iodobenzene as the aryl halide, the yield of **42** was good (see also Table V) but bromobenzene gave very small amounts of **42,** and chlorobenzene and p-chlorobenzophenone did not react. It was suggested that the lack of reactivity with p-chlorobenzophenone might be due to a competing addition of the nucleophile to the carbonyl group (Reaction 40) (27).

Dihaloa.enes also react with **41** and give disubstitution products

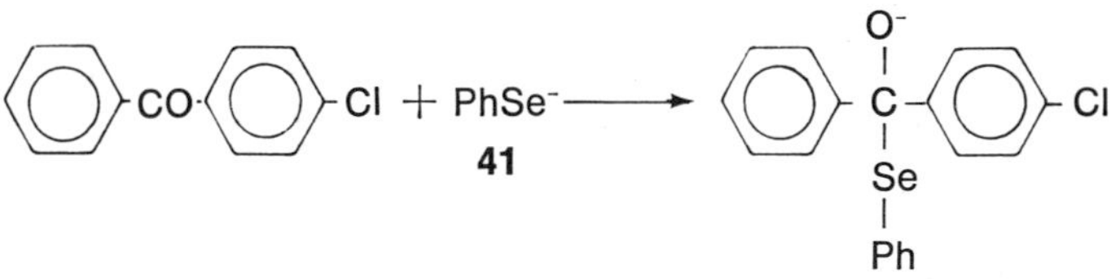

$$\text{Ph}\text{-CO-}\text{C}_6\text{H}_4\text{-Cl} + \text{PhSe}^- \longrightarrow \text{Ph-}\underset{\underset{\text{Ph}}{\overset{|}{\text{Se}}}}{\overset{\overset{\text{O}^-}{|}}{\text{C}}}\text{-}\text{C}_6\text{H}_4\text{-Cl} \qquad (40)$$

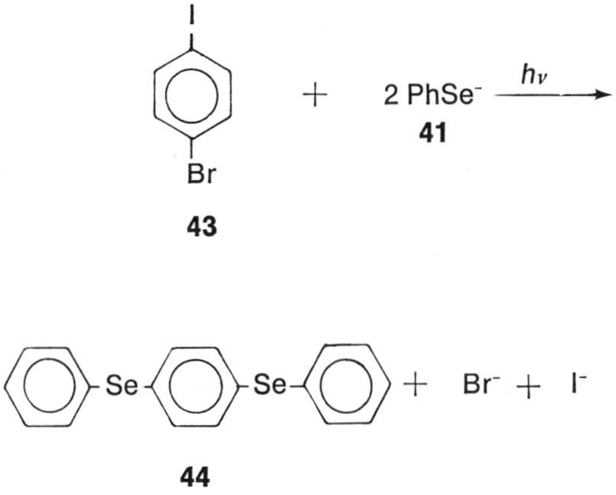

41

43 + 2 PhSe⁻ $\xrightarrow{h\nu}$

41

$$\text{Ph-Se-}\text{C}_6\text{H}_4\text{-Se-Ph} + \text{Br}^- + \text{I}^- \qquad (41)$$

44

when irradiated in liquid ammonia. For instance, p-bromoiodobenzene **43** under photostimulation reacted with **41** to give the disubstitution product **44** in 70% isolated yield (27). The yield of iodide ion (86%) is similar to that of bromide ion (83%) even though iodide is a better leaving group than bromide. Also, about 12% of the substrate **43** was recovered unchanged, and only traces of monosubstitution products were found. These results suggest that the radical anion intermediate **45** formed when p-bromophenyl radical couples with **41,** decomposes (Reaction 42) to p-(phenylselenyl)phenyl radical **46** and bromide ion faster than it reacts by electron transfer to **43,** to give the monosubstituted product **47** (Reaction 44). The radical **46** leads ultimately to **44** (Reaction 45).

These results are comparable with the reaction of benzenethiolate ion with dihaloarenes where there is strong evidence that the disubstitution product is formed directly, without the intervention of the monosubstitution product.

Table V

Photostimulated Reactions of Haloarenes and Dihaloarenes with Phenyl Selenide Ion in Liquid Ammonia

Substrate	hν (min)	Substitution Product (%)
PhCl	180	0
PhBr	100	5
PhI	220	73
1-Chloronaphthalene	220	70
4-Chlorobiphenyl	240	52
9-Bromophenanthrene	220	72
2-Chloroquinoline	220	46
4-Chlorobenzophenone	180	0
4,4'-Dibromobiphenyl	220	23[a]
p-Dibromobenzene	220	13[b]
p-Bromoiodobenzene	220	70[b]

[a] Insoluble in liquid ammonia; the yield reported is of bromide ion.
[b] Disubstitution product.
Source: Reprinted, from Ref. 27. Copyright, 1979, American Chemical Society.

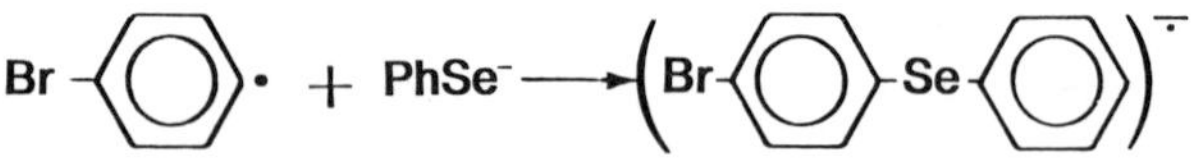

$$Br-\langle\bigcirc\rangle\cdot + PhSe^- \longrightarrow \left(Br-\langle\bigcirc\rangle-Se-\langle\bigcirc\rangle\right)^{\cdot-} \tag{42}$$

41   **45**

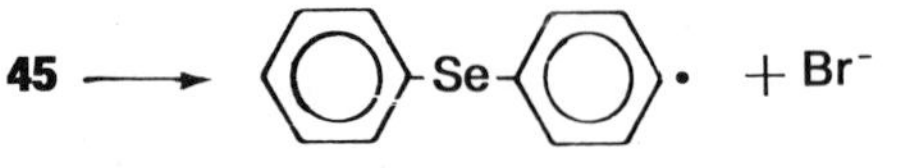

$$\mathbf{45} \longrightarrow \langle\bigcirc\rangle-Se-\langle\bigcirc\rangle\cdot + Br^- \tag{43}$$

46

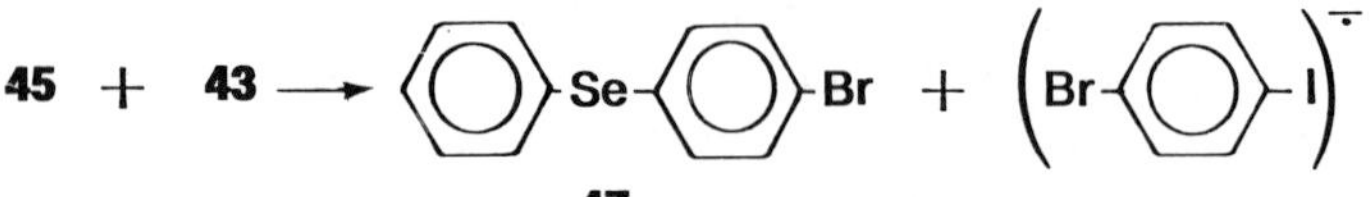

$$\mathbf{45} + \mathbf{43} \longrightarrow \langle\bigcirc\rangle-Se-\langle\bigcirc\rangle-Br + \left(Br-\langle\bigcirc\rangle-I\right)^{\cdot-} \tag{44}$$

47

$$\mathbf{46} + PhSe^- \longrightarrow \longrightarrow \mathbf{44} \tag{45}$$

41

Disodium selenide can be formed in liquid ammonia from the reaction of selenium metal with sodium metal (Reaction 46) (*28*).

$$Se + 2\,Na \xrightarrow{NH_3} Se^{2-} + 2Na^+ \tag{46}$$

When disodium selenide solutions formed in this manner were allowed to react under photostimulation with excess iodobenzene, 12% of diphenyl selenide and 78% of diphenyl diselenide were formed. The fact that there is no dark reaction indicates that the reaction probably occurs by the $S_{RN}1$ mechanism. The formation of diphenyl selenide implies that after phenyl selenide ion is produced (Reaction 48) it competes with Se^{2-} for phenyl radicals. Formation of diphenyl diselenide was shown to arise from oxidation of phenyl selenide ion during work-up. When methyl iodide was added as quencher, the only products formed were phenyl methyl selenide and diphenyl selenide (*29*)

$$Ph\cdot + Se^= \longrightarrow (PhSe^-)^{\overline{\cdot}} \tag{47}$$

$$(PhSe^-)^{\overline{\cdot}} + PhI \longrightarrow (PhI)^{\overline{\cdot}} + PhSe^- \tag{48}$$

$$Ph\cdot + PhSe^- \longrightarrow (PhSePh)^{\overline{\cdot}} \tag{49}$$

$$PhSePh \xrightarrow{e^-,NH_3} PhSe^- + PhH \tag{50}$$

Because diphenyl selenide is transformed easily into phenyl selenide ions by reaction with solvated electrons (Reaction 50), subsequent reaction with aryl or alkylating agents represents a convenient method to form C–Se bonds (Reactions 52 and 54). This method offers the possibility for "one pot synthesis" of several phenylselenium derivatives as indicated in Scheme III (*29*).

By way of example, 1-naphthyl phenyl selenide can be prepared in 98% yield by Reaction 52 (ArX = iodobenzene, Ar'X = 1-iodonaphthalene). Likewise, *o*-methylphenyl methyl selenide is obtained in 87% yield from Reaction 54 (ArX = *o*-iodotoluene, RX = methyl iodide).

Tellurium Nucleophiles

Bromobenzene reacts with phenyl telluride ion **48** under photostimulation in liquid ammonia to give the substitution product diphenyl

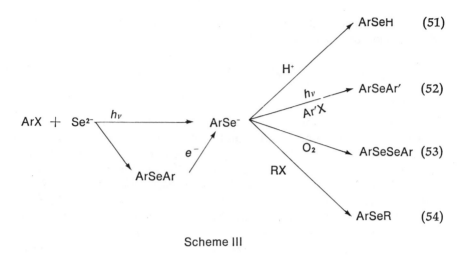

Scheme III

telluride **49** in 20% yield. Iodobenzene gave a 90% yield of **49** under the same reaction conditions (Reaction 55) (Table VI) (27).

$$PhX + PhTe^- \xrightarrow{h\nu} PhTePh + X^-$$
$$X = Br, I$$
$$48 \phantom{+ PhTe^- \xrightarrow{h\nu} Ph}49$$
(55)

Table VI.

Photostimulated Reaction of Haloarenes and Dihaloarenes with Phenyl Telluride Ion in 220 min of Irradiation in Ammonia

Haloarene[a]	Substitution Product (yield, %)
PhBr	PhTePh (20)
PhI	PhTePh (90)
AnI	PhTePh (15)
	AnTePh (73)
	AnTeAn (11)
1-ClNp	PhTePh (9)
	1-NpTePh (41)
	$(1\text{-}Np)_2Te$ (6)
1-BrNp	PhTePh (16)
	1-NpTePh (53)
	$(1\text{-}Np)_2Te$ (10)
$p\text{-}BrC_6H_4I$	PhTePh (20)
	$PhTeC_6H_4TePh$ (40)
	$(p\text{-}BrC_6H_4)TePh$ (7)

[a] An, p-anisyl; Np, naphthyl.
Source: Reprinted, from Ref. 27. Copyright 1979, American Chemical Society.

However, in the photostimulated reaction of p-iodoanisole 50, besides the expected substitution product 51, the symmetrical tellurides 49 and 52 were formed (Reaction 56) (27).

$$\text{AnI} + \text{PhTe}^- \xrightarrow{h\nu} \text{PhTePh} + \text{AnTePh} + \text{AnTeAn} \qquad (56)$$

50   48   49   51   52

$$\text{An} = p\text{-anisyl}$$

This result indicates that the photostimulated reaction between 50 and 48 in ammonia differs from the general scheme for ArS$^-$. Instead, other steps should be added to account for the formation of symmetrical products 49 and 52 of Reaction 56.

It has been suggested that light promotes the reaction in the usual way; that is, electron transfer from 48 to 50. The radical anion 53 formed dissociates to a p-anisyl radical 54 and iodide ion. Then 54 couples with 48 to form p-anisyl phenyl telluride radical anion 55, which may then react by three competitive pathways: regression to starting intermediates 54 and 48, electron transfer to 50 to give the substitution product 51, or by C–Te bond fragmentation to give two new intermediates, p-anisyl telluride ion 56 and phenyl radical 57 (Reaction 59) (Scheme IV) (27).

The new intermediates 56 and 57 formed, can enter the cycle of the $S_{RN}1$ mechanism. The ion 56 competes as a nucleophile with 48 for reaction with 54 (Reaction 60) to give 52 finally, whereas radical 57 can react with 48 to give 49 (Reaction 61).

According to Scheme IV the symmetrical tellurides 49 and 52 come from the decomposition of the radical anion intermediate 55 to form 56 and 57, competing with electron transfer to the substrate. This behavior is not observed with benzenethiolate nucleophile.

Related dissociation of 1-naphthyl phenyl telluride radical anions has been observed. When 1-chloronaphthalene or 1-bromonaphthalene were used as substrates, not only the unsymmetrical 1-naphthyl phenyl telluride 58, but also 49 and bis(1-naphthyl) telluride 59 (Reaction 62) were found (Table VI).

The photostimulated reaction of p-bromoiodobenzene with 48, afforded only 40% yield of p-bis(phenyltelluryl)benzene 60. (Reaction 63).

The nucleophile 48 was prepared by the Te–Te bond fragmentation of diphenyl ditelluride and sodium metal in ammonia. It is interesting to note that when excess alkali metal was used, 48 reacted ultimately to give benzene (30). The behavior of the group VIA anions of structure PhZ$^-$ (Z being O, S, Se, or Te) resembles the behavior of the group VA anions Ph$_2$Y$^-$ (Y being P, As, Sb, or Bi). In both series, the species with the heaviest element (Te and Bi) is cleaved at the metal–carbon bond with alkali metals in ammonia (30, 31).

$$\text{AnI} + \text{PhTe}^- \xrightarrow{h\nu} (\text{AnI})^{\overline{\cdot}} + \text{PhTe}^{\cdot} \qquad (57)$$
$$\quad\;\, \mathbf{50} \qquad \mathbf{48} \qquad\qquad\quad \mathbf{53}$$

$$(\text{AnI})^{\overline{\cdot}} \longrightarrow \text{An}^{\cdot} + \text{I}^- \qquad (58)$$
$$\quad\; \mathbf{53} \qquad\qquad \mathbf{54}$$

$$\text{An}^{\cdot} + \text{PhTe}^- \rightleftharpoons (\text{AnTePh})^{\overline{\cdot}} \rightleftharpoons \text{AnTe}^- + \text{Ph}^{\cdot} \qquad (59)$$
$$\;\, \mathbf{54} \qquad \mathbf{48} \qquad\qquad \mathbf{55} \qquad\qquad \mathbf{56} \qquad \mathbf{57}$$

$$(\text{AnTePh})^{\overline{\cdot}} \xrightarrow{-e^-} \text{AnTePh}$$

$$\text{AnTe}^- + \text{An}^{\cdot} \rightleftharpoons (\text{AnTeAn})^{\overline{\cdot}} \xrightarrow{-e^-} \text{AnTeAn} \qquad (60)$$
$$\quad \mathbf{56} \qquad \mathbf{54} \qquad\qquad\qquad\qquad\qquad\quad \mathbf{52}$$

$$\text{Ph}^{\cdot} + \text{PhTe}^- \rightleftharpoons (\text{PhTePh})^{\overline{\cdot}} \xrightarrow{-e^-} \text{PhTePh} \qquad (61)$$
$$\;\, \mathbf{57} \qquad \mathbf{48} \qquad\qquad\qquad\qquad\qquad\qquad \mathbf{49}$$

An = *p*-anisyl

Scheme IV

$$1\text{-XNp} + \text{PhTe}^- \xrightarrow{hv} \text{PhTePh} + 1\text{-NpTePh} + (1\text{-Np})_2\text{Te} \qquad (62)$$
$$\qquad\quad 48 \qquad\qquad 49 \qquad\quad 58 \qquad\qquad 59$$

Np = naphthyl
X = Cl, Br

$$(63)$$

$$\qquad\quad 48 \qquad\qquad\qquad\qquad\quad 60$$

Disodium telluride can be formed in liquid ammonia by reaction of tellurium metal and sodium metal, and it reacts with iodobenzene in photostimulated reactions as does disodium selenide (29). However, it is not a very promising reaction then from the synthetic point of view because mixtures of products are obtained. Diphenyl telluride and phenyl telluride ion cannot be converted to a pure phenyl telluride ion by reaction of the former with solvated electrons in excess, because the product (phenyl telluride ion) is thereby destroyed (30).

Literature Cited

1. Bunnett, J. F.; Wamser, C. C. *J. Am. Chem. Soc.* **1967**, *89*, 6712. Bunnett, J. F.; Takayama, H. *J. Am. Chem. Soc.* **1968**, *90*, 5173.
2. Rossi, R. A.; Bunnett, J. F. *J. Org. Chem.* **1973**, *38*, 1407.
3. Rossi, R. A.; Bunnett, J. F. *J. Org. Chem.* **1973**, *38*, 3020.
4. Amatore, C.; Chaussard, J.; Pinson, J.; Saveant, J. M.; Thiebault, A. *J. Am. Chem. Soc.* **1979**, *101*, 6012.
5. Rossi, R. A.; de Rossi, R. H. unpublished data.
6. Semmelhack, M. F.; Bargar, T. *J. Am. Chem. Soc.* **1980**, *102*, 7765.
7. Rajan, S.; Sridaran, P. *Tetrahedron Lett.* **1977**, 2177.
8. Rossi, R. A.; Pierini, A. B. *J. Org. Chem.* **1980**, *45*, 2914.
9. Ciminale, F.; Bruno, G.; Testaferri, L.; Tiecco, M. *J. Org. Chem.* **1978**, *43*, 4509.
10. Lai, C. C.; McEven, W. E. *Tetrahedron Lett.* **1971**, 3271.
11. McEven, W. E.; Lubinkowski, J. J.; Knapczyk, J. W. *Tetrahedron Lett.* **1972**, 3301.
12. Bunnett, J. F.; Creary, X. *J. Org. Chem.* **1974**, *39*, 3173.
13. Bunnett, J. F.; Creary, X. *J. Org. Chem.* **1974**, *39*, 3611.
14. Zoltewicz, J. A.; Oestreich, T. *J. Am. Chem. Soc.* **1973**, *95*, 6863.
15. Pinson, J.; Saveant, J. M. *J. Chem. Soc., Chem. Commun.* **1974**, 933.
16. van Tilborg, W. J. M.; Smit, C. J.; Scheele, J. J. *Tetrahedron Lett.* **1977**, 2113.
17. Pinson, J.; Saveant, J. M. *J. Am. Chem. Soc.* **1978**, *100*, 1506.
18. Bartak, D. E.; Houser, K. J.; Rudy, B. C.; Hawley, M. D. *J. Am. Chem. Soc.* **1972**, *94*, 7526.
19. Bartak, D. E.; Danen, W. C.; Hawley, M. D. *J. Org. Chem.* **1970**, *35*, 1206.
20. Russell, G. A. "Radical Ions"; Kaiser, E. T.; Keran, C., Eds.; Interscience, New York, 1968.
21. Bunnett, J. F.; Creary, X. *J. Org. Chem.* **1975**, *40*, 3740.
22. Rossi, R. A.; de Rossi, R. H.; López, A. F. *J. Am. Chem. Soc.* **1976**, *98*, 1252.
23. Rossi, R. A.; Palacios, S. M. *J. Org. Chem.*, **1981**, *46*, 5300.

24. López, A. F., Ph.D. Thesis, Universidad Nacional de Córdoba, **1979.**
25. Swartz, J. E.; Bunnett, J. F. *J. Org. Chem.* **1979,** *44,* 4673.
26. Pierini, A. B.; Rossi, R. A. *J. Organomet. Chem.* **1978,** *144,* C 12.
27. Pierini, A. B.; Rossi, R. A. *J. Org. Chem.* **1979,** *44,* 4667.
28. Bogolyubov, G. M.; Shlyk, Y. N.; Petrov, A. A. *Zh. Obshch. Khim.* **1969,** *39,* 1804.
29. Rossi, R. A.; Peñeñory, A. B. *J. Org. Chem.,* **1981,** *46,* 4580.
30. Pierini, A. B.; Rossi, R. A. *J. Organomet. Chem.* **1979,** *168,* 163.
31. Rossi, R. A.; Bunnett, J. F. *J. Am. Chem. Soc.* **1974,** *96,* 112.

5

Participating Substrates

Two important characteristics of the $S_{RN}1$ mechanism are that an appropriate substituent can be installed in a desired position of the aromatic moiety (the position originally occupied by the leaving group), and that the reaction can take place even when substrates carry electron-donating substituents. Due to the multi-step nature of the mechanism, and because it is a chain reaction, the structure of the substrates will change the overall reactivity by affecting the initiation, propagation, or termination steps. The nature of these steps may vary considerably with the reaction conditions making it difficult to predict a reactivity pattern for all $S_{RN}1$ reactions.

The results to date allow some generalizations to be made, and will be discussed in this chapter. The substrates studied are grouped according to the method used to promote the reaction.

Reaction of Benzene Derivatives in Ammonia

Substrates with One Leaving Group. DARK REACTIONS. Carbanionic nucleophiles are able to transfer an electron to iodobenzene and other substrates with low reduction potentials, such as diphenyliodonium chloride. Thus ketone enolate anion reacts slowly with iodobenzene in liquid ammonia, and 8% yield of phenylacetone is obtained after 180 min of reaction in the dark at −33°C (1). On the other hand diphenyliodonium ion gives 33% of the substitution product after 15 min of reaction in the dark at −78°C (2). Chlorobenzene and bromobenzene do not react at all (1). The higher reactivity of diphenyliodonium ion compared to iodobenzene is consistent with the differences in their reduction potentials (diphenyliodonium is reduced at a potential about 1 V more positive than iodobenzene) (3).

The $S_{RN}1$ mechanism was discovered in the reaction of amide ion with 5- and 6-iodotrimethylbenzenes, where the $S_{RN}1$ mechanism competes with the benzyne mechanism; only the chloro and bromo derivatives reacted by the benzyne mechanism (4).

0065-7719/83/0178-0101$11.50/1

Diphenyl phosphide ion reacts in the dark with p- and m-iodo-toluenes in ammonia to give p-tolyldiphenylphosphine and m-tolyldi-phenylphosphine respectively. These reactions were examined for evidence of cine substitution, but there was no detectable contamination of the *meta* and *para* products by their respective *para* or *meta* isomers. p-Bromotoluene reacted in the dark with diphenyl phosphide ion to give the substituted product, although much more slowly than the iodo derivative (5).

Diphenyl arsenide ion did not react in 60 min in the dark with p-chlorotoluene, but it reacted with 4-chlorobenzophenone giving 63% yield in 30 min. There is ample evidence that this reaction occurs by the $S_{RN}1$ mechanism (6). Diphenyl stibide ion also reacted in the dark with 4-chlorobenzophenone (67% yield, 30 min), giving the stibines expected from the scrambling of the aryl rings (7).

SOLVATED ELECTRON STIMULATED REACTIONS. In principle, any substrate with an appropriate leaving group that forms a radical anion by reaction with solvated electrons (Reaction 1), and decomposes to give a free aryl radical according to Reaction 2, is suitable to participate in the $S_{RN}1$ mechanism.

$$ArX + e^- \longrightarrow (ArX)^{\cdot -} \tag{1}$$

$$(ArX)^{\cdot -} \longrightarrow Ar^{\cdot} + X^- \tag{2}$$

In the reaction stimulated by solvated electrons in liquid ammonia, it has been demonstrated that these substrates are suitable (2): PhX (X being F, Cl, Br, I), Ph_2S, Ph_2I^+, Ph_3S^+, Ph_2O, Ph_2Se, $PhOP(O)(OR)_2$, and $PhNMe_3^+$.

The reactions were carried out in the presence of acetone enolate anion to trap any phenyl radical formed, as in Reactions 1–2. The products investigated were phenylacetone **1** and 1-phenyl-2-propanol **2** (Reaction 3) (Table I).

$$PhX + {}^-CH_2COCH_3 \xrightarrow{e^-} PhCH_2COCH_3 + PhCH_2\overset{\overset{\displaystyle OH}{|}}{C}HCH_3 \tag{3}$$
$$\qquad\qquad\qquad\qquad\qquad\quad \mathbf{1} \qquad\qquad\quad \mathbf{2}$$

Substrates like anisole and tetraphenylsilane were unreactive toward solvated electrons, whereas other substrates formed only benzene (Table I). Phenyltrimethylsilane is reduced in ammonia by solvated electrons to give Product **3** (Reaction 4) (8). *meta*-Substituted fluorobenzene compounds **4** are reduced in the ring without loss of the fluorine atom (Reaction 5), (9) whereas the *para*-substituted derivatives **5** eliminate fluoride ion (Reaction 6) (9).

Table I

The Action of Potassium Metal in Liquid Ammonia on Substituted Benzenes in the Presence of Acetone Enolate Ion at −78°C.

Substrate	PhH	PhCH$_2$COCH$_3$[a]
PhF	[b]	51
PhCl	[b]	68
PhBr	5	91
PhI	[b]	71
Ph$_2$I$^+$	[b]	142
PhOMe	0	0
Ph$_2$O	[b]	5
PhOP(O)(OEt)$_2$	[b]	46
(PhO)$_3$PO	[b]	43
Ph$_2$S	[b]	89
Ph$_3$S$^+$	59	55
Ph$_2$SO	89	0
PhSO$_2$CH$_3$	90	0
Ph$_2$SO$_2$	144	0
PhSO$_3^-$	[b]	0
Ph$_2$Se	60	32
PhNMe$_3^+$	[b]	71
Ph$_3$P	70	0
Ph$_3$As	96	0
Ph$_3$Sb	97	0
Ph$_3$Bi	270	0
PhCN	0	0
Ph$_4$Si	0	0
Ph$_3$B	0	0
PhB(OH)$_2$	0	0

[a]Yield listed is sum of phenylacetone, 1-phenyl-2-propanol and 1,1-diphenylacetone.
[b]Not quantified.
Source: Reproduced, from Ref. 2. Copyright 1974, American Chemical Society.

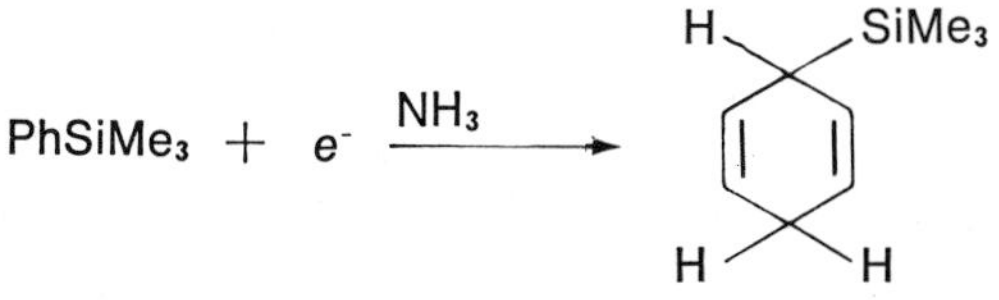

$$\text{PhSiMe}_3 \ + \ e^- \ \xrightarrow{\text{NH}_3} \qquad \qquad \qquad \qquad (4)$$

3

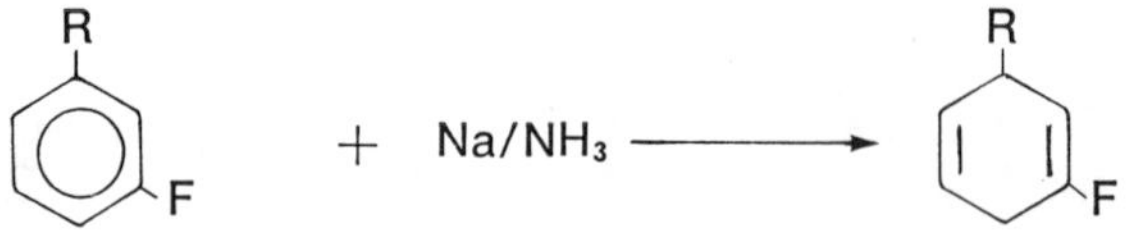

$$\text{4} \quad\quad + \quad Na/NH_3 \quad\longrightarrow\quad \quad\quad\quad (5)$$

$$R = -CO_2^-, Ph, 3\text{-}FC_6H_4, SiMe_3$$

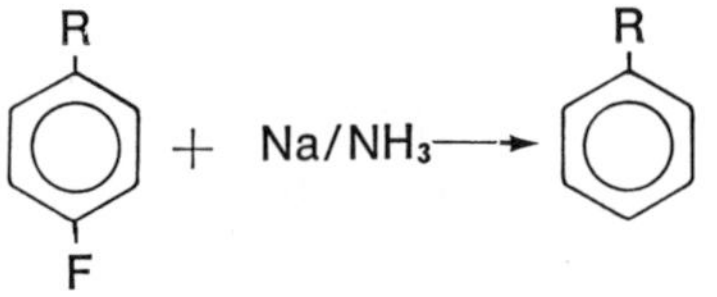

$$\text{5} \quad\quad + \quad Na/NH_3 \longrightarrow \quad\quad\quad\quad (6)$$

$$R = CO_2^-; Ph, 4\text{-}FC_6H_4, OMe, Me$$

That halobenzenes are cleaved by solvated electrons in the sense of Reactions 1 and 2 is widely recognized, and halides are by far the leaving groups most frequently used in $S_{RN}1$ reactions.

The decomposition of diphenyl ether, diphenyl sulfide, and diphenyl selenide radical anions, furnished phenyl radical and the anions of the leaving group (phenoxide, benzenethiolate, and phenyl selenide) (Reaction 7).

$$Ph_2Z + e^- \longrightarrow (Ph_2Z)^{\bar{\cdot}} \longrightarrow Ph^{\cdot} + PhZ^- \quad\quad (7)$$

$$Z = O, S, Se$$

Diaryl sulfides and alkyl aryl sulfides are cleaved cathodically or by alkali metals to form arenethiolate ion and a hydrocarbon (10, 11).

The cleavage of diphenyl telluride with solvated electrons in liquid ammonia was not studied. It is suggested that this substrate decomposes following Reaction 7 because reversible coupling of aryl radical with phenyl telluride ion occurs and leads to the scrambling of aryl rings (12).

Diphenyliodonium reacts with solvated electrons to give iodobenzene and phenyl radical. Iodobenzene also reacts with solvated electrons to give phenyl radical, thus yielding two phenyl radicals per mole of diphenyliodonium ion, as indicated by the amount of 1 and 2 obtained with acetone enolate ion as nucleophile (Reaction 3).

$$Ph_2I^+ + e^- \longrightarrow PhI + Ph^{\cdot} \quad\quad (8)$$

$$PhI + e^- \longrightarrow Ph^{\cdot} + I^- \quad\quad (9)$$

There is electrochemical evidence that the stepwise decomposition of diphenyliodonium occurs as shown by Reactions 8–9 (3).

Triphenylsulfonium ion reacts with solvated electrons to give phenyl radical and diphenyl sulfide. As mentioned before, diphenyl sulfide reacts also with solvated electrons giving phenyl radical and benzenethiolate ion (Reaction 7), so that the overall reaction should produce two moles of phenyl radical per mole of the substrate. However, the actual yield obtained was far from the theoretical yield. A considerable amount of tarry products was obtained, so that this substrate does not seem to be promising for use in $S_{RN}1$ reactions stimulated by solvated electrons.

The reactions of several aryl diethyl phosphate esters with sodium naphthalene and sodium anthracene in tetrahydrofuran and with sodium in liquid ammonia were examined using different combinations of electron donor, concentration, and mode of mixing the reactants (13).

The poorest electron donor, sodium anthracene, gave one cleavage product, the corresponding phenoxide ion. With the other reducing agents the amount of phenoxide ion and arene was highly dependent on the anion radical concentration and mode of mixing. Phenol formation was greater with the poorer reducing agents and at lower concentrations. Interpretation of these results can be set forth as in Reactions 10, 11 (13).

$$ArOP(O)(OEt)_2 \xrightarrow{Ar'H^{-}} ArOP(O)(OEt)_2^{-} \xrightarrow{Ar'H^{-}} ArOP(O)(OEt)^{2-} \quad (10)$$

$$ArO^{-/\cdot} + (EtO)_2PO^{\cdot/-} \qquad Ar^{-/\cdot} + (EtO)_2PO_2^{-/\overline{\cdot}} \quad (11)$$

$$ArOP(O)(OEt)^{2-} \longrightarrow Et^- + ArOPO_2^-$$
$$| \atop OEt$$

In Reaction 10 the radical anion formed initially gave P–O bond fragmentation and only the dianion produced Ar–O bond fragmentation and consequently arene formation.

Previous results for the photostimulated $S_{RN}1$ reaction between phenyl diethyl phosphate ester and acetone enolate ion support this proposed reduction pathway (2). The photostimulated reaction gave only small yields of phenylacetone (13%) and a sizable amount of phenol (71%). The same reaction stimulated by potassium metal gave moderate yields of substitution product. Phenol would be the expected product from this reaction because photochemical electron transfer could easily yield the radical anion, but double photochemical reduction to the dianion would be extremely unlikely (13).

The reaction of triphenylphosphine, arsine, stibine, and bismutine with solvated electrons in the presence of acetone enolate ion, afforded only benzene and diphenyl phosphide, arsenide, and stibide respectively (Reaction 12).

$$Ph_3Y + e^- \longrightarrow PhH + Ph_2Y^- \tag{12}$$
$$Y = P,\ As,\ Sb$$

With triphenylbismutine almost three moles of benzene per mole of substrate were obtained, indicating that all the Ph–Bi bonds were cleaved (2).

The reductive cleavage of aryl alkanesulfonates in the reaction with sodium metal in liquid ammonia has been suggested to give Ph–O bond fragmentation (14), but this substrate has not been tested in S$_{RN}$1 reactions as a source of phenyl or aryl radicals. Other sulfur functionalities fail to give phenyl radicals which can enter in S$_{RN}$1 reactions, such as Ph$_2$SO, PhSO$_2$CH$_3$, Ph$_2$SO$_2$, and PhSO$_3^-$, and all of them are reduced to benzene (Table I) (2).

Benzonitrile did not give 1 or 2 in reactions with solvated electrons in the presence of acetone enolate ion. Other products formed were not identified, but indicated that other reactions did occur (2).

In electrochemical studies of benzonitrile in protic solvents, the nitrile function is reduced, whereas in aprotic solvents, aryl radical and cyanide ions are formed (15). Electrochemical studies of several substrates show that aromatic compounds bearing the nitrile functionality decompose giving cyanide ions (16, 17).

PHOTOSTIMULATED REACTIONS. Halobenzenes, other than iodobenzene, do not react with ketone enolate ions in liquid ammonia in the dark. However, under irradiation with pyrex-filtered light, at 350 nm, they react rapidly to form phenylacetone 1 in high yields (Reaction 13) (1).

$$PhX + {}^-CH_2COCH_3 \xrightarrow{\ h\nu\ } 1 + X^- \tag{13}$$
$$X = F,\ Cl,\ Br,\ I$$

Compounds bearing other leaving groups, such as phenyltrimethyl ammonium ion, triphenylsulfonium, diphenyl selenide, phenyldialkyl phosphate (1), and aryl phenyl sulfides (18) reacted as well as halobenzenes. With aryl phenyl sulfides two modes of fragmentation of the radical anion intermediate 7 are possible (Reactions 14 and 15) (18). Both have been found to occur, but the substituted compounds seem to be favored. The ratio arylacetone 8:phenylacetone 1 varies from 1.2–3.1.

$$(ArSPh)^{\overline{\cdot}} \begin{cases} \nearrow Ar^{\cdot} + PhS^- \xrightarrow{\ ^-CH_2COCH_3\ } ArCH_2COCH_3 \quad (14) \\[4pt] \qquad\qquad\qquad\qquad\qquad\qquad 8 \\[4pt] \searrow ArS^- + Ph^{\cdot} \xrightarrow{\ ^-CH_2COCH_3\ } PhCH_2COCH_3 \quad (15) \end{cases}$$
$$7$$

$$Ar = m\text{-},\ p\text{-tolyl} \qquad\qquad\qquad 1$$
$$ m\text{-},\ p\text{-anisyl}$$

The fact that arylacetone **8** predominates over phenylacetone **1** in all cases indicates that Reaction 14 is favored over Reaction 15 for fragmentation when aryl is m- and p-tolyl, or m- and p-anisyl, but the factors which determine cleavage are not quite clear. If transition state theory applies, the determining factors are not very important because the difference in free energy between the two fragmentation transition states is no greater than 0.5 kcal/mol (18).

From the previous results it is evident that substrates bearing substituents such as methyl and OCH_3 are able to react by the $S_{RN}1$ mechanism. Besides, aryl iodides bearing m-CF_3, m-F, and p-PhO substituents were found to react with benzenethiolate ion giving very good yields of aryl phenyl sulfides (19).

The influence of various substituents on the reaction of ketone enolate ion with several aryl halides was studied (20), and, from the results, several remarkable features of this reaction became evident:

1. The reaction is remarkably insensitive to steric hindrance; thus compounds bearing MeO or Me groups in positions 2 and 6 react very well, and an isopropyl group in the 2 position does not inhibit the reaction. Only when two isopropyl groups, or one t-butyl group are in the o-position, does the yield of arylacetone decrease.

2. The carboxylate group, which has a negative charge, does not inhibit the reaction, but a hydroxy group in the oxyanion form attached to the ring prevents the reaction. In this case only reduction of the aromatic ring is observed. The same observation was found in the photostimulated reaction of 1-bromo-2-naphthoxide with cyanomethyl anion, where 2-naphthol was the only product observed (21). The enolate ion of p-bromoacetophenone is also inadequate as substrate.

3. Contrary to what is found in the traditional S_NAr reaction, strongly electron withdrawing groups, such as nitro and trifluoromethyl groups, are not adequate for this mechanism. With the m-trifluoromethyl substituted benzene ring, a low yield of m-trifluoromethylphenylacetone (34% yield) is formed, but compounds bearing a nitro group are consumed in a dark reaction that forms colored products, but does not lead to the expected substitution products (20).

4. The amino group does not prevent reaction; for instance, 3-bromo-N,N-dimethylaniline and 4-iodo-N,N-dimethylaniline afforded respectively the arylacetones in 82% and 90% yield (22). m-Iodoaniline also reacted (23). These reactions were found to be important for the synthesis of indoles.

The lack of reactivity of compounds bearing O^- as a group may be due to the competition of the nucleophilic reaction with protonation of the radical. In a radiolytic study of the reduction of bromophenols, it was

demonstrated that radicals **9** and **10** are protonated by water and converted into the phenoxyl radical (Reaction 16) (24).

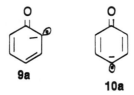

$$\text{(16)}$$

Protonation is of minor importance with the *meta* isomer. The rate
constants for protonation are 1.7×10^5 s^{-1} and 0.5×10^5 s^{-1} for the *para*
and *ortho* isomers respectively. The ease of protonation on the ring was
attributed to the contribution of structures like **9a** and **10a** to the overall
wave function of **9** and **10**.

These radicals were trapped by nitromethane anion at an approxi-
mate rate of $\sim 4 \times 10^7$ $M^{-1}s^{-1}$, indicating that the radical is intrinsically
reactive toward nucleophiles. However, some step of the S$_{RN}$1 mech-
anism is probably slower than with other aryl radicals making other
competing reactions predominant. One step that may be slow is the
electron transfer step, due to electronic repulsion by the negative charge.

RING CLOSURE REACTIONS. An interesting aspect of the S$_{RN}$1 reactions is
that a properly chosen substrate or substrate and nucleophile lead to a
ring closure reaction.

There are some differences however between reactions that occur in
just one step, namely, intramolecular cyclization by the S$_{RN}$1 mechanism,
or reactions which upon nucleophilic substitution by the S$_{RN}$1 mechanism, undergo cyclization in a second step. The former reaction has
already been discussed, the latter will be discussed in the following
section.

Indole Synthesis. The photostimulated reaction of *o*-haloanilines **11** with
ketone or aldehyde enolate ions in liquid ammonia gave, after workup,
indole derivatives **13** in excellent yields in most of the examples studied
(Reaction 17–18) (22, 23).

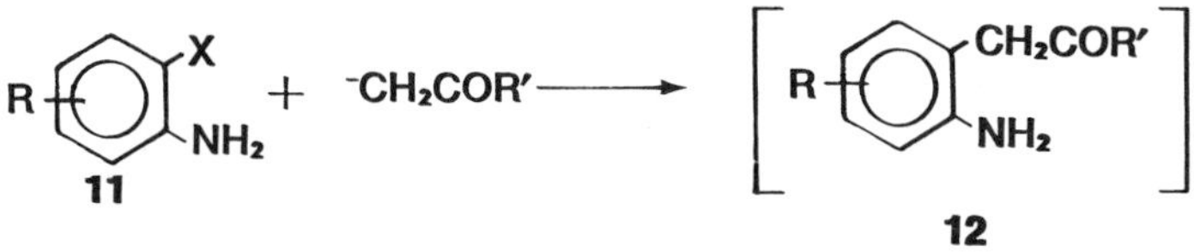

$$R \overbrace{}^{X}_{NH_2} + \ ^-CH_2COR' \longrightarrow \left[R \overbrace{}^{CH_2COR'}_{NH_2} \right] \quad (17)$$

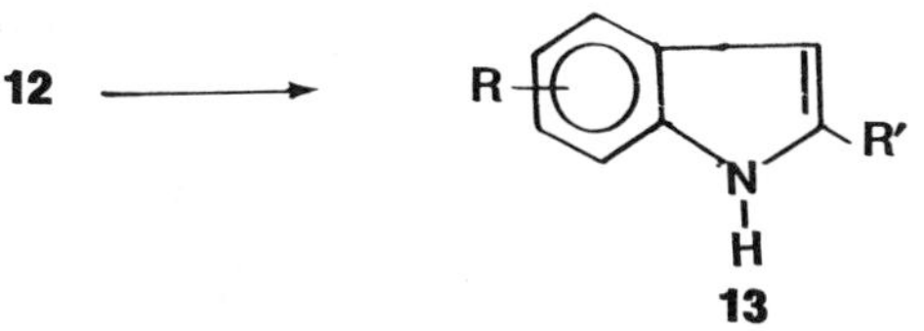

11

12

$$\mathbf{12} \longrightarrow R \overbrace{}^{} \underset{\underset{H}{N}}{\bigcup} R' \quad (18)$$

13

Proposals suggest that this reaction occurs by the $S_{RN}1$ mechanism forming the intermediate **12** which, under strongly basic reaction conditions or after acidification with ammonium salt, undergoes cyclization to indoles **13**. These reactions do not occur in the dark and are totally inhibited by oxygen (23).

Substrates bearing other substituents, besides the *o*-haloamine system necessary for this reaction, such as Me, OMe, Ph, and CO_2^- substituents, produce indoles with the substituent at sites determined by the constitution of the starting *o*-haloaniline. As was pointed out, this method has the advantage of forming only one isomer as opposed to the Fischer indole synthesis and other methods which involve cyclization (Table II) (22).

With cyclohexanone enolate anion, 1,2,3,4-tetrahydrocarbazole **14** was obtained (Reaction 19) (22, 23).

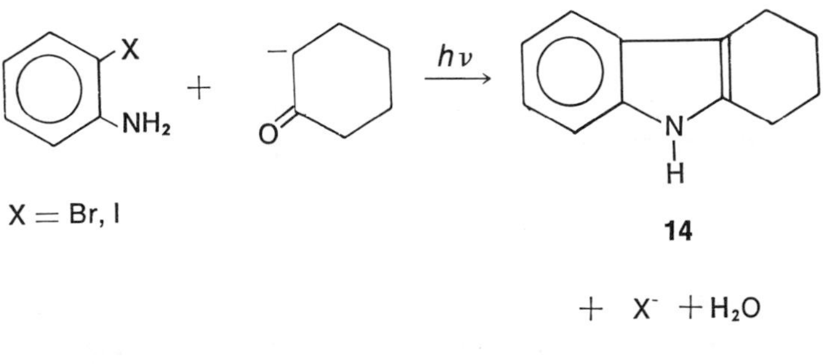

$$\overbrace{}^{X}_{NH_2} + \underset{O}{\bigcirc}^- \xrightarrow{h\nu} \bigcirc\!\!\!\!\bigcirc\!\!\!\!\underset{\underset{H}{N}}{\bigcirc} \quad (19)$$

X = Br, I

14

$$+ \ X^- + H_2O$$

But the yield obtained was modest (Table II) because the ketone enolate ion has β-hydrogen atoms, which allow for another competing reaction (hydrogen atom abstraction). Low yields were observed also with the ketone enolate ion **15**, which gave the derivative **16** in 22% yield (Reaction 20) (25).

Table II

Indole Synthesis by the $S_{RN}1$ Mechanism

Aniline	Enolate from	$h\nu$ (min)	Indole Derivative	Yield (%)	Ref.
2-I	CH_3COH	13	Indole	50	25
2-I	CH_3CH_2COH	20	3-Me	49	25
2-I	$CH_3CH_2CH_2COH$	23	3-Et	33	25
2-I	$(Me)_2CHCH_2COH$	40	$3\text{-}CH_2CH(Me)_2$	26	25
2-Br	CH_3COCH_3	120	2-Me	93	22
2-I	"	16	2-Me	83	25
2-Br-3-Me	"	240	$2,4\text{-}Me_2$	80	22
2-Br-4-Me	"	120	$2,5\text{-}Me_2$	88	22
2-Br-5-Me	"	120	$2,6\text{-}Me_2$	82	22
2-Cl-5-Ph	"	720	2-Me-6-Ph	88	22
2-Cl-5-MeO	"	630	2-Me-6-MeO	42	22
$2\text{-}Cl\text{-}5\text{-}CO_2H$	"	720	$2\text{-}Me\text{-}6\text{-}CO_2H$	89	22
2-Br-N-Me	"	180	$1,2\text{-}Me_2$	79	22
2-I	$(Me)_2CHCOCH_3$	10	$2\text{-}CH(Me)_2$	78	25
2-Br	"	240	$2\text{-}CH(Me)_2$	84	22
2-Br	$(Me)_3CCOCH_3$	180	$2\text{-}C(Me)_3$	94	22
2-I	$CH_3CH_2COCH_3$	25	2-Et	20	25
			$2,3\text{-}Me_2$	40	25
2-I	$CH_3COCH(OMe)_2$	20	2-COH	45	25
2-I	$CH_3COC(OMe)_2CH_3$	25	$2\text{-}COCH_3$	35	25
2-I	cyclohexanone	60	1,2,3,4-tetra-hydrocarbazole	33	25
2-Br	"	240	"	14	22
2-Br-4-Me	3-pentanone	130	$3,5\text{-}Me_2\text{-}2\text{-}Et$	73	22

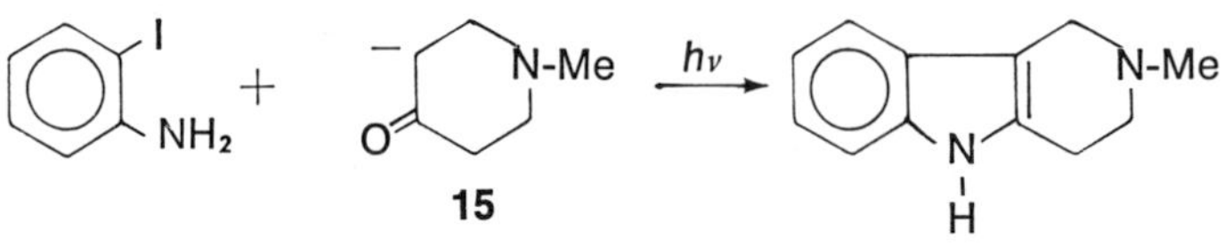

$$(20)$$

2,4-Dihaloanilines showed very low reactivity. 2,4-Dibromoaniline released only 11% of bromide ion during 2 h of irradiation, and 2-iodo-4-chloroaniline released only 9% of iodide ion. The second halogen in the dihaloanilines has been postulated to increase the acidity of the amino group sufficiently so that it exists mainly as an unreactive conjugated base under the usual conditions of the $S_{RN}1$ reactions (potassium t-butoxide used as a base in excess) (22, 26).

With aldehyde enolate ions the yield of indoles is 25–50%, and substantial amounts of aniline are formed (10–40% yield) (25).

The reaction of enolate ions of α-dicarbonyl compounds, when one of the carbonyl groups is protected as the dimethylacetal, give good yields of the 2-substituted indoles (Reaction 21–22) (25).

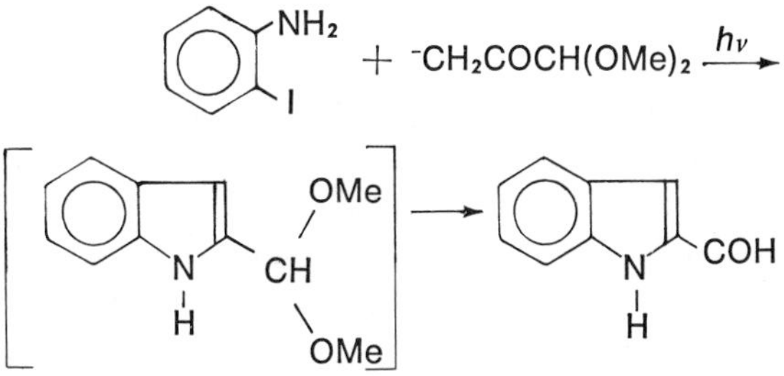

(21)

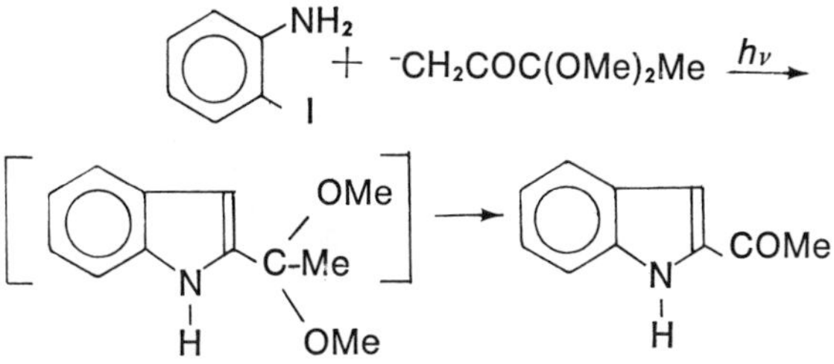

(22)

Pyridine rings with the o-haloamino system also gave these reactions. 2-Chloro-3-aminopyridine **17**, gave the straightforward 4-azaindoles **18** in photostimulated reaction with ketone enolate ions, in good yields (Table III) (Reaction 23) (22, 27). This substrate, like o-haloaniline, gave low yield of the azotetrahydrocarbazole in the photostimulated reaction with cyclohexanone enolate anion, due to β-hydrogen atom abstraction.

Table III

4-Azaindole Synthesis by the $S_{RN}1$ Mechanism from 2-Chloro-3-aminopyridine and Ketone Enolate Ions

Enolate from	hν (min)	4-Azaindole Derivative	Yield (%)	Ref.
Acetone	150	2-Me	23	27
Acetone	630	2-Me	45	22
(CH₃)₂CHCOCH₃	60	2-isopropyl	61	27
(CH₃)CCOCH₃	150	2-t-butyl	100	27
2-Butanone	90	2-ethyl and 2,3-Me₂	45	27
Cyclohexanone	210	1,2,3,4-tetrahydro-5-aza-carbazole	22	27

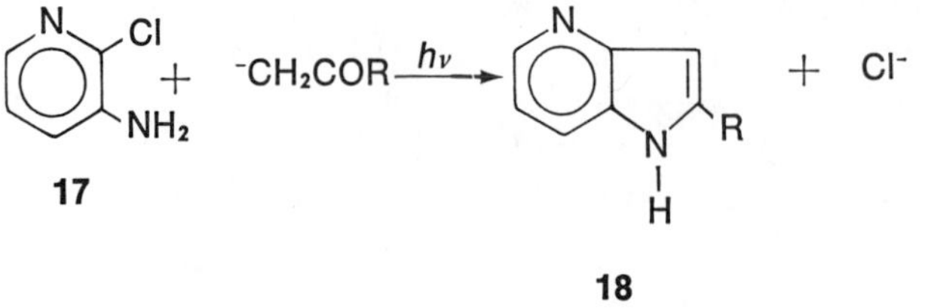

$$(23)$$

Benzo[b]furan Synthesis. In this approach, the overall reaction needs an extra step following the $S_{RN}1$ reaction before the ring closure reaction can occur.

The photostimulated reaction of *o*-iodoanisole with acetaldehyde or ketone enolate ions gave the substitution product **19**, which after deblocking the phenolic group, gave by ring closure the expected benzo[b]furan derivatives **21** (Reaction 24) (*28*). With 2-butanone enolate

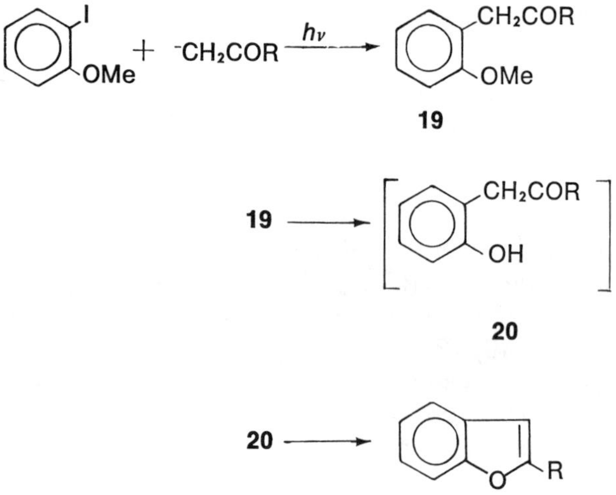

$$(24)$$

21a R = H (40%)
21b R = Me (67%)
21c R = *i*-Pr (66%)
21d R = *t*-Bu (100%)

ion, a 57% yield of a mixture of 2,3-dimethylbenzofuran (60%) and 2-ethylbenzofuran (40%) was obtained (*28*).

3-Acetyl-2-methylindenes. The photostimulated reaction of *o*-dibromoben-zene (**22**) with excess acetone enolate ion afforded 64% yield of a mixture of two acetylmethylindenes, probably **23** and **24**, and 5% yield of 9,10-diacetylanthracene **25** (Reaction 25) (*29*). In the presence of triethyl-amine, only the isomer **23** was formed.

The precursor of **23** and **24** is the straightforward disubstitution product *o*-diacetonylbenzene **26**, that under basic reaction conditions is deprotonated. The resulting enolate ion **27** attacks the carbonyl group of the other side chain, and forms acetylmethylindenes by a normal aldol condensation mechanism (Reaction 26). The by-product **25** is believed to involve the nucleophilic enolate ion **27** in a $S_{RN}1$ reaction with another molecule of **22** (*29*).

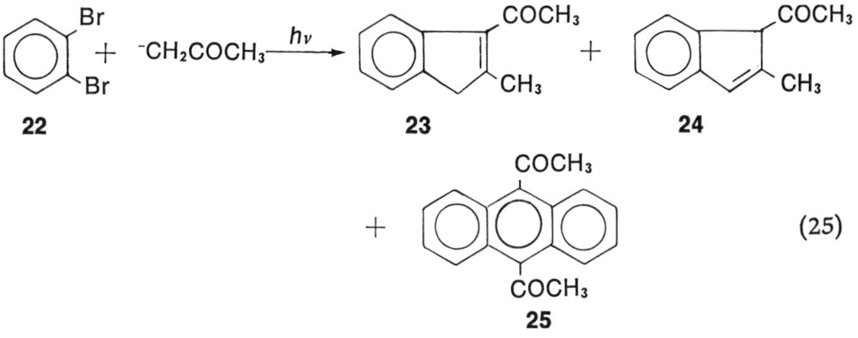

22 + $^-CH_2COCH_3$ $\xrightarrow{h\nu}$ **23** + **24**

22   **23**   **24**

(25)

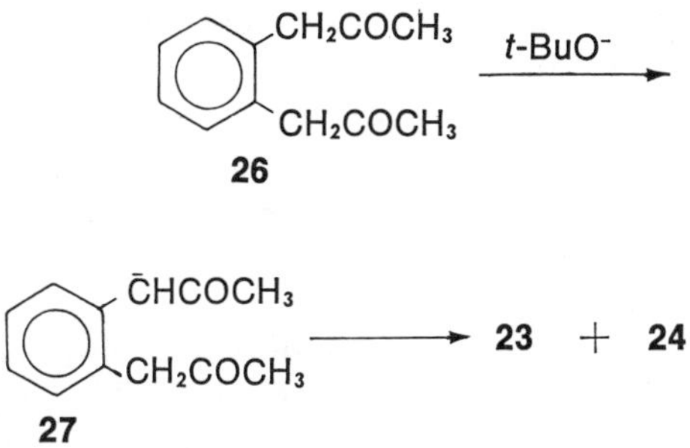

25

26 $\xrightarrow{t\text{-BuO}^-}$

27 $\longrightarrow$ **23** + **24**   (26)

Substrates Bearing Two Leaving Groups. A substrate, such as **28**, with two leaving groups X and Y that reacts by the $S_{RN}1$ mechanism, has several possible reaction pathways not encountered in substrates with only one leaving group.

First, the radical anion formed when the substrate accepts an elec-tron has two possibilities; to form either radical **30a** or **30b** (Reactions 27–28).

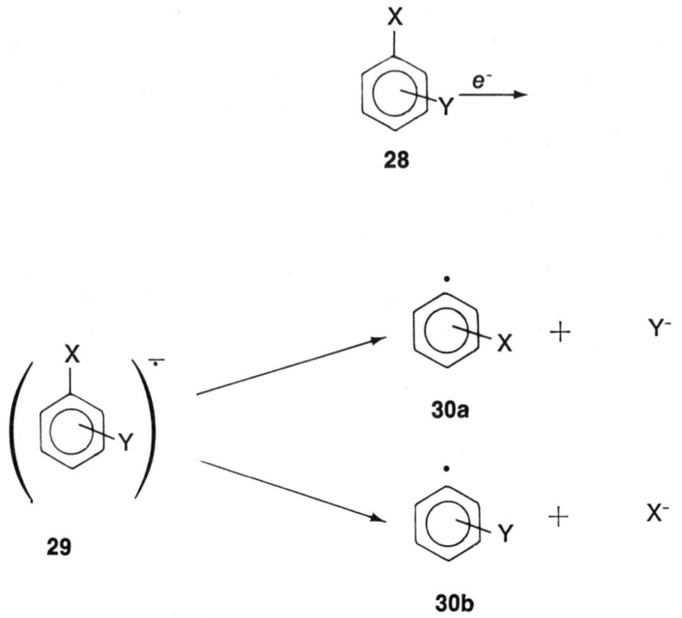

$$\underset{\textbf{28}}{\overset{X}{\bigoplus}Y} \quad \xrightarrow{e^-}$$

$$\left(\underset{\textbf{29}}{\overset{X}{\bigoplus}Y}\right)^{\overline{\cdot}}$$

$$\overset{\cdot}{\underset{\textbf{30a}}{\bigoplus}X} \quad + \quad Y^- \qquad (27)$$

$$\overset{\cdot}{\underset{\textbf{30b}}{\bigoplus}Y} \quad + \quad X^- \qquad (28)$$

Second, the radical anion formed when **30a** or **30b** couples with the nucleophile may eliminate the other leaving group (Reaction 30), or lose the odd electron forming the monosubstituted product **31a** or **31b** (Reactions 29 and 31).

The radical **32** is expected to propagate the chain of the $S_{RN}1$ mechanism as other aryl radicals do, giving ultimately the disubstituted product **33** (Reaction 32).

In principle the disubstitution product can be formed from **31a** or **31b**; however, there is plenty of evidence that the monosubstitution compound is not an intermediate on the main pathway to the disubstitution product.

In studies carried out on substrates with two leaving groups, products from all these processes have been observed. Because Reactions 29–31 represent competing reactions, their relative rates will determine the type of product formed. Halogen mobility is the same as in the monohalobenzenes; $I > Br > Cl > F$.

The photostimulated reaction of *p*-chloroiodobenzene, *p*-bromo-iodobenzene, and *p*-iodophenyltrimethylammonium ion with ben-zenethiolate ion afforded the disubstitution product without the intermediacy of the monosubstituted one, showing that for these reactions $k_f > k_t[\text{ArX}]$. The possibility that the monosubstitution product is formed but consumed very fast was discarded because *m*-chloroiodo-benzene reacts 17 times as fast as the monosubstitution product in

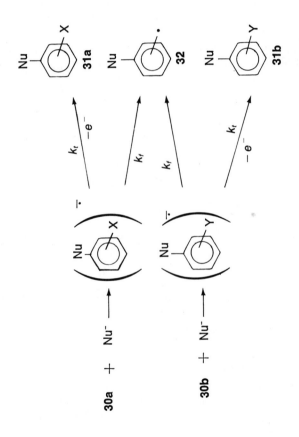

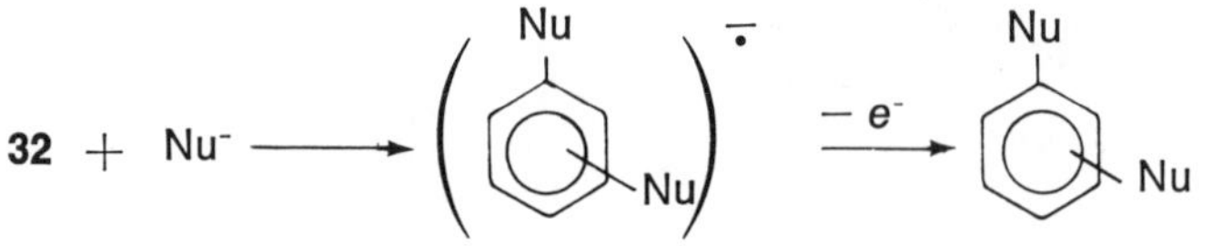

$$\mathbf{32} + Nu^- \longrightarrow \left(\underset{Nu}{\overset{Nu}{\bigcirc}} \right)^{\overline{\cdot}} \xrightarrow{-e^-} \underset{Nu}{\overset{Nu}{\bigcirc}} \quad (32)$$

33

competition experiments (*30*). Besides, in a short period of time, when substantial amounts of substrate are present with the disubstitution product, only mere traces of the monosubstitution product are formed.

The four *p*-haloiodobenzenes react with potassium diethyl phosphite in ammonia. In the case of *p*-bromoiodobenzene and *p*-diiodobenzene, both halogens were replaced to give the disubstituted product. *p*-Chloroiodobenzene gave mainly the disubstituted product accompanied by traces of the monosubstituted one (*31*).

The four *m*-haloiodobenzenes reacted with benzenethiolate ion and diethyl phosphite ion in photostimulated reactions in liquid ammonia (Reactions 33–34).

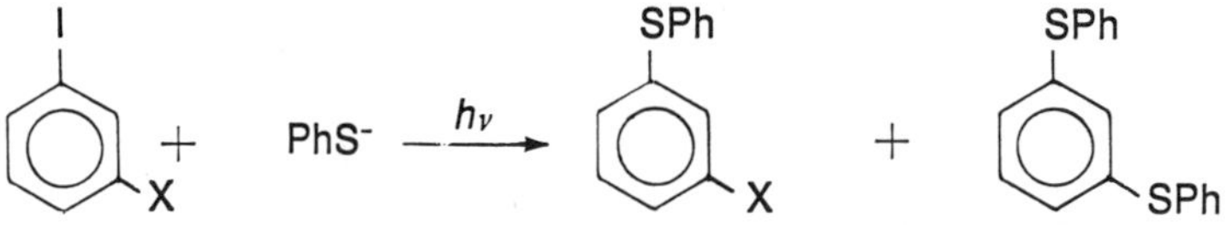

$$\underset{X}{\overset{I}{\bigcirc}} + PhS^- \xrightarrow{h\nu} \underset{X}{\overset{SPh}{\bigcirc}} + \underset{SPh}{\overset{SPh}{\bigcirc}} \quad (33)$$

$$\qquad\qquad\qquad\qquad\qquad \mathbf{34} \qquad\qquad\qquad \mathbf{35}$$

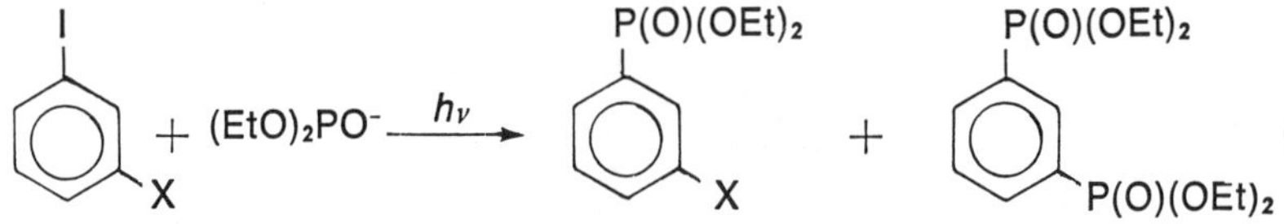

$$\underset{X}{\overset{I}{\bigcirc}} + (EtO)_2PO^- \xrightarrow{h\nu} \underset{X}{\overset{P(O)(OEt)_2}{\bigcirc}} + \underset{P(O)(OEt)_2}{\overset{P(O)(OEt)_2}{\bigcirc}} \quad (34)$$

$$\qquad\qquad\qquad\qquad\qquad\qquad \mathbf{36} \qquad\qquad\qquad \mathbf{37}$$

When X is iodide, both reactions (Reaction 33–34) form the disubstituted products **35** and **37** respectively. When X is bromide, benzenethiolate ion forms **35** exclusively, whereas with diethyl phosphite ion, although **37** is the major product, traces of **36** are found.

Remarkably different behavior is found when X is chloride. In Reaction 33, **35** is formed predominantly, along with traces of **34**, whereas in Reaction 34 the monosubstitution product **36** is formed predominantly, and only 4% of **37** is formed.

p-Fluoro and *m*-fluoro derivatives gave exclusively the monosubstituted product with both nucleophiles.

o-Haloiodobenzenes react with diethyl phosphite ion under photo-stimulation in a manner similar to the *para*-isomer, but there is a dark reaction which takes place by a totally different mechanism. The photo-stimulated reaction is believed to occur by the $S_{RN}1$ mechanism (32).

p-Bromoiodobenzene and p-dibromobenzene were found to react with phenyl selenide ion forming the diselenide in 70% yield. The formation of the diselenide was explained along the same lines (12).

p-Dichlorobenzene reacted with pinacolone enolate ion, affording the corresponding *para*-disubstituted product in excellent yield (33). *m*-Fluoroiodobenzene affords *m*-fluorophenylacetone in a photostimulated reaction with acetone enolate ion (20).

The fact that the reaction of p-dichlorobenzene with pinacolone enolate ion yields the disubstituted product, provides further evidence that the monosubstituted product **39** is not an intermediate in the route to the disubstituted product **42**. In this case if the monosubstituted product **39** were formed, it would be deprotonated to form **41**, which would be unreactive, and only **39** would be observed, as in the case of *m*-fluoroiodobenzene. The latter compound gave *m*-fluorophenylacetone as a consequence of the slow fragmentation of the radical anion intermediate, and the product does not react further, although fluorobenzene reacts under similar reaction conditions (Reaction 35).

Reactions of Polycyclic Hydrocarbons in Ammonia

After the $S_{RN}1$ mechanism was discovered, the possibility of substituting halogens and other leaving groups on other aromatic rings was studied, and several condensed and noncondensed halo-substituted polycyclic aromatic compounds were found to be suitable substrates for this mechanism, giving better yields than the corresponding benzene derivatives in many cases.

1-Chloronaphthalene and 1-iodonaphthalene react with cyano-methyl anion, acetone enolate anion, and butane or 2-hydroxyethane-thiolate ions giving excellent yields of substitution products. The reaction with carbanion nucleophiles was also promoted by potassium metal, but in these cases some reduction of the aromatic ring also occurred (Reaction 36) (34). These reactions call attention to the remarkably different behavior of the naphthalene derivatives and benzene derivatives.

Other polycyclic hydrocarbons were shown to be effective as substrates for photostimulated reactions with several nucleophiles, and these results are summarized in Table IV.

Reactions of Heterocyclic Compounds in Ammonia

Several heterocyclic compounds bearing appropriate leaving groups were found to react by the $S_{RN}1$ mechanism.

3-Bromothiophene reacted with amide ions in the dark under a nitrogen atmosphere to yield 3-aminothiophene (78%) (35). The same

(35)

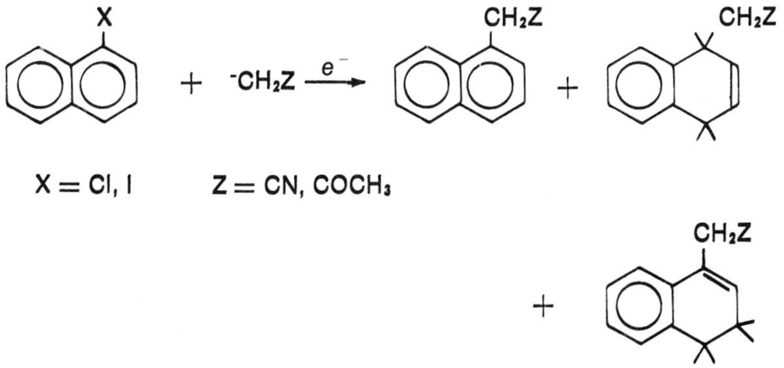

$$X = Cl, I \qquad Z = CN, COCH_3 \tag{36}$$

reaction under air gave different results; 44% of the substrate was recovered, and only 2% of the substitution product and much tar were formed. Under illumination in a nitrogen atmosphere, the reaction behaved as in the dark; but, curiously the reaction was slower.

2-Bromothiophene furnished the 3-amino derivative (48%), together with the isomer 3-bromothiophene (37%) which, represents bromine migration. 3-Bromothiophene was also a good substrate in reactions promoted by alkali metals. On the other hand, 2-bromo and 2-chloro derivatives were less prone to react with ketone enolate ions and mostly gave unreacted material and thiophene under reaction conditions similar to those for the 3-isomer.

In photostimulated reactions with other nucleophiles, such as ammonium benzenethiolate and potassium diethyl phosphite—both very good nucleophiles toward other substrates—2-iodothiophene did not react to give the substitution product. Mostly unreacted starting material was recovered with benzenethiolate ion. With potassium diethyl phosphite there was 91% iodide release, but the only detectable volatile product was thiophene (35).

3-Bromothiophene also reacted with cyanomethyl anion giving products very similar to those derived from halobenzenes in photostimulated and potassium metal stimulated reactions (36).

Six-membered heterocyclic compounds such as pyridine, quinoline, and isoquinoline derivatives appropriately substituted with a nucleofugal group, were also shown to be good substrates for the $S_{RN}1$ mechanism.

In solvated-electron-stimulated reactions six-membered heterocyclic compounds do not seem to behave as well as their homoaromatic partners. The reaction of 2-bromopyridine with acetone enolate ion gave mostly reduction products and unreacted starting material with potassium metal stimulation (Reaction 37). The photostimulated reaction however, gave 2-acetonylpyridine in 95% yield (Reaction 37) (37). The 2-chloro, 2-fluoro, and 3-bromo derivatives also reacted with acetone enolate ion when they were exposed to illumination.

Table IV

Photostimulated Reactions of Polycyclic Hydrocarbons with Nucleophiles by the S$_{RN}$1 Mechanism in Liquid Ammonia

ArX	Nucleophile	hν (min)	Yield (%)	Other Products	Ref.
1-Chloro-naphthalene	CNCH$_2^-$	100	89	4 (ArH)	21
"	n-BuS$^-$	170	81	18 (ArH)	34
"	HOCH$_2$CH$_2$S$^-$	150	74	16 (ArX), 8 (ArH)	34
"	PhSe$^-$	170	73		12
"	PhTe$^-$	220	41	39 (ArX), 9 (Ph$_2$Te), 6 (Ar$_2$Te)	12
"	Ph(Me)NCOCH$_2^-$	120	50		71
1-Bromo-naphthalene	PhTe$^-$	220	53	15 (ArX), 16 (Ph$_2$Te), 10 (Ar$_2$Te)	12
"	Ph$_2$As$^-$	10	40	33 (Ph$_3$As), 21 (PhAr$_2$As)	6
"	PhCH$_2$S$^-$	180	2–4		40
1-Bromo-2-naphthoxide	$^-$CH$_2$CN	60	—	98 (ArH)	21
1-Iodo-naphthalene	CH$_3$COCH$_2^-$	60	76		20
"	t-BuS$^-$	180	88		40
2-Chloro-naphthalene	CNCH$_2^-$	45	98	2 (ArH)	21
2-Iodo-naphthalene	CH$_3$COCH$_2^-$	90	75		20
4-Chloro-biphenyl	CNCH$_2^-$	20	96	2 (ArH)	21
"	PhSe$^-$	180	52		12
9-Bromo-phenanthrene	CNCH$_2^-$	60	60	7 (ArH)	21
"	PhSe$^-$	220	72		12
"	Ph(Me)NCOCH$_2^-$	90	80	5 (ArH)	71
"	CH$_3$COCH$_2^-$	80	62		20
"	Ph$_2$As$^-$	60	60	20 (Ph$_3$As), 15 (PhAr$_2$As)	6
"	PhCH$_2$S$^-$	180	0–2		40
9-Bromo-anthracene	CH$_3$COCH$_2^-$	60	84		20

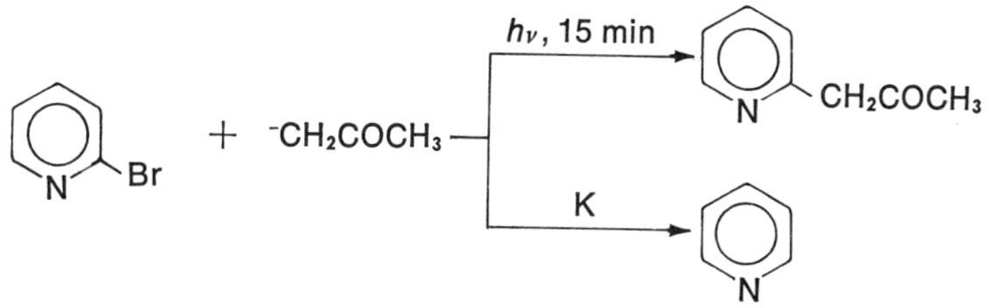

(37)

2,6-Dibromopyridine and 2,6-dichloropyridine under photostimulation reacted with pinacolone enolate ion to give the disubstitution product directly, without the intermediacy of the monosubstitution product (Reaction 38) (37).

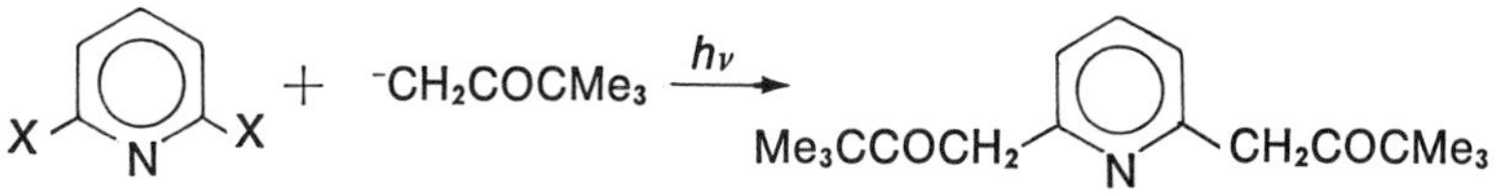

(38)

It seems that the use of an alkali metal catalyst with heteroaromatic compounds results in severe competing reactions. For example, the addition of one equivalent of lithium metal to a reaction mixture of 2-chloroquinoline, three equivalents of acetone, and four equivalents of lithium amide produces a complex mixture of products consisting of 17.9% recovered starting material, 18% quinoline, 15.2% substitution product, 15% 2,3'-biquinoline, and 7.9% 2-aminoquinoline. Under approximately the same reaction conditions, but with stimulation by near ultraviolet light, the substitution product is formed in 80% yield (38).

2-Chloroquinoline has been used as a substrate in other photostimulated reactions, and it gave good yields of substitution products [with phenyl selenide (12), phenyl telluride (39), diphenyl arsenide (6), benzenethiolate ion (40), ketone enolate ions (41), etc.]

Several substituted pyrimidines 43 (a–d) have been shown to react with ketone enolate anions 44 (a–c) (42).

Substrate 43a or 43b reacted with 44a or 44b in the dark, but the reaction is inhibited by di-t-buytl nitroxide. Substrate 43c gives addition to the six position forming 45, besides the substitution product 46 (Reaction 39). Potassium metal stimulation led to the formation of 46 (45–50%) and the reduction product 4-phenylpyrimidine (22%).

There is a sluggish reaction of 43c and 43d in the dark with ketone enolate ions 44, but yields increase to 95% by irradiation with near UV light (42).

2-Chloropyrimidine (47) reacts with acetone and pinacolone enolate ions to give the substitution product 48 and small amounts of the amine

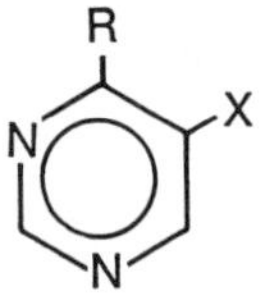

$^-CH_2COR$

	R	X		
43a	Ph	Br	**44a**	R = Me
43b	Ph	Cl	**44b**	R = Ph
43c	*t*-Bu	Br	**44c**	R = *t*-Bu
43d	*t*-Bu	Cl		

43a or **43b** + **44c** ⟶

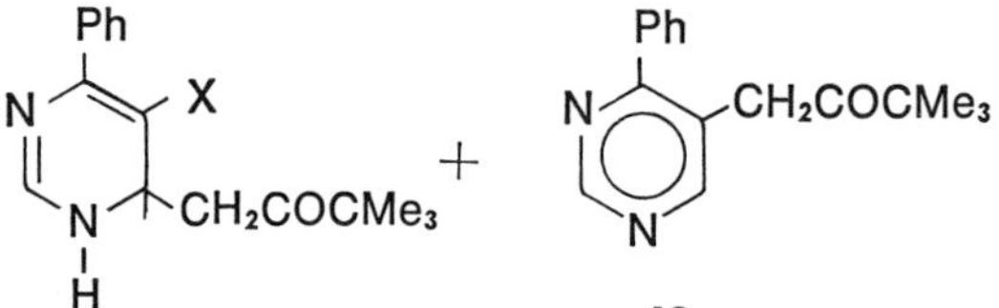

45 **46** (39)

49. In the dark **48** is not formed, and there is an increase in the yield of **49** (Reaction 40) (*43*).

The photostimulated reaction of **47** with diisopropyl ketone enolate ion gave 88% yield of the ketone **48** and 4% yield of **49.** In the dark, or in the photostimulated reaction with di-*t*-butyl nitroxide as inhibitor, only **49** and the addition product **50** (~15%) were obtained.

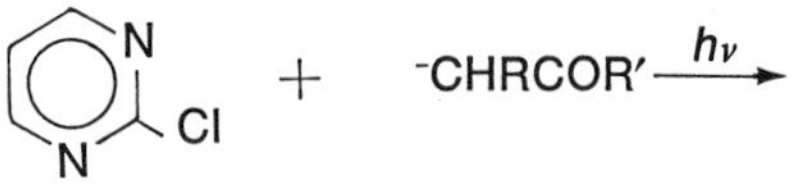

47

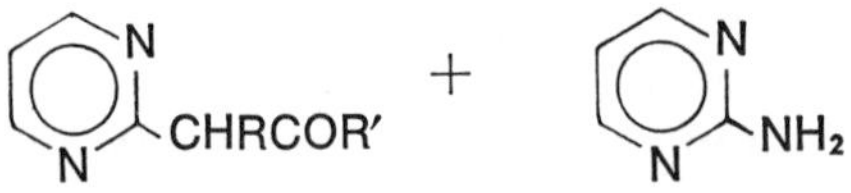

48 **49** (40)

2,6-Dimethoxy-4-chloropyrimidine like the diazine **47,** gave no dark reaction with pinacolone enolate ion, but gave 98% yield of the substitution product under photostimulation (*43*).

3-Chloro-6-methoxypyridazine shows a slow dark reaction with ketone enolate ions, which is catalyzed by light and inhibited with di-*t*-butyl nitroxide. (With diisopropyl ketone enolate ion and in the dark, 8% of the addition product **51** is formed.)

2-Chloropyrazine (**52**) reacts in the dark with acetone, pinacolone, diisopropyl ketone, and acetophenone enolate ions to give 82–98% yield of substitution products **53**. These dark reactions are inhibited by di-*t*-butyl nitroxide (Reaction 41) (*43*).

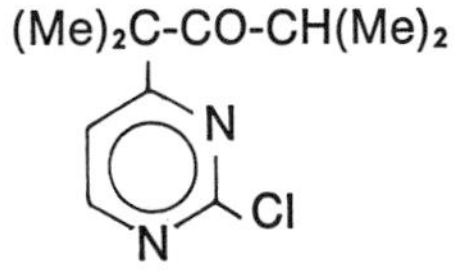

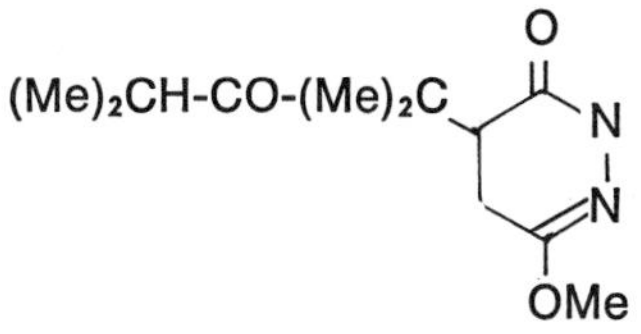

Comparative Reactivity in Ammonia

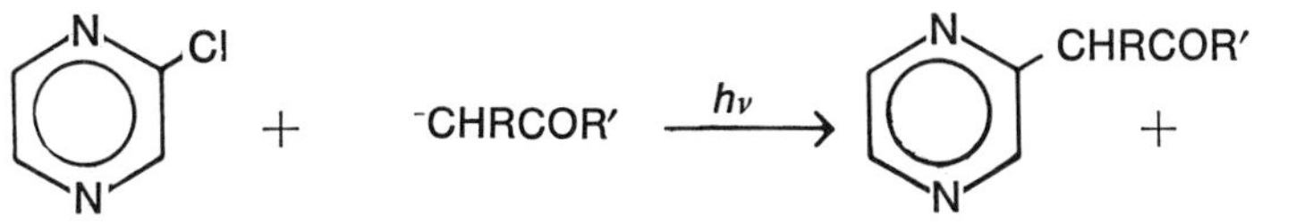

The relative reactivity of substrates depends on the initiator, on the nucleophile, and on the solvent used.

In reactions promoted by solvated electrons it was found that when a pair of halobenzenes (such as iodobenzene and chlorobenzene) were deficient in potassium metal they show equal reactivity (*44*). Similarly, 1-chloronaphthalene was consumed to the same extent as chlorobenzene in reactions carried out with an electron deficiency (*34*). The macroscopic rate of consumption of the substrates seems to be controlled by the rate of mixing (*44*).

In photostimulated reactions, important differences in reactivity are found. In reactions with acetone enolate ion it was demonstrated that the order of reactivity of several monosubstituted benzenes is $I \sim Br > SPh > Cl > F \gg OPh$ (Figure 1) (*1*).

The reactivity of iodo and bromo derivatives initially appeared to be similar because both were consumed totally in the same time. However, further work indicates that iodine is a better leaving group than bromine. For instance, in the reaction of potassium diethyl phosphite with iodobenzene and bromobenzene in ammonia, it was found that iodobenzene

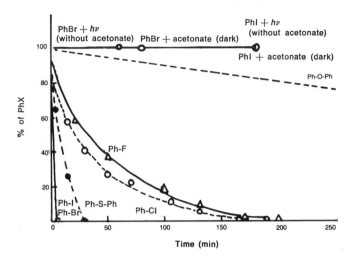

Figure 1. Substrate (%) remaining after various times of exposure to light. Conditions: (Substrate)$_0$, ~0.07 M; ($^-$CH$_2$COCH$_3$)$_0$, ~0.45 M; Key: ◐, bromobenzene and acetone enolate ion in dark; ○, bromobenzene with illumination, but no enolate ion; ⊖, iodobenzene and acetone enolate in the dark; ○, iodobenzene with illumination but no enolate ion. (Reproduced, from Ref. 1. Copyright 1973, American Chemical Society.)

is roughly one thousand times as reactive as bromobenzene in photostimulated reactions (*31*).

The reactivity of substrates with the same leaving group in photostimulated reaction with phenyl selenide ion, indicated that the order of reactivity is 1-chloronaphthalene > 4-chlorobiphenyl > chlorobenzene (*12*). With acetone enolate ion as nucleophile, the order of reactivity found for heterocyclic substrates is 2-haloquinoline > 2-halopyridine > halobenzenes (*37*). With chlorodiazabenzenes as substrates the order of reactivity in reactions with ketone enolate ions is 2-chloro-1,4-diazabenzene > 6-chloro-3-methoxy-1,2-diazabenzene > 2-chloro-1,3-diazabenzene (*43*).

Interesting behavior is found with potassium acetophenone enolate ion and different aromatic substrates. This nucleophile did not react in photostimulated reactions with iodobenzene (*45*) and 2-bromopyridine (*37*) under the usual irradiation conditions (Reaction 42). Substitution product from the reaction of iodobenzene and acetophenone enolate ion was obtained in 67% yield when a Hanovia medium pressure mercury arc positioned in a conventional immersion well was used as the light source (*46*). Homolysis of iodobenzene might occur under these conditions.

With 2-chloroquinoline (*37*) and 1-chloronaphthalene (*34*) a very slow substitution reaction occurred in 60–120 min of irradiation (about 10% yield) (Reaction 43). On the other hand, with 4-phenyl and 4-*t*-butyl-5-halogenopyrimidines, potassium acetophenone enolate ion seems to be an even better nucleophile than other ketone enolate ions (Reaction 44) (*42*), and 2-chloropyrazine reacts even in the dark giving as much as 82% yield of the substitution reaction. This reaction is strongly inhibited by di-*t*-butyl nitroxide (0% yield) (Reaction 45) (*46*).

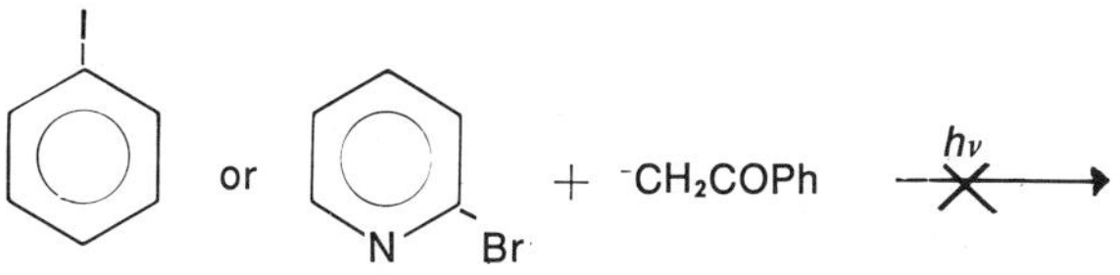

(42)

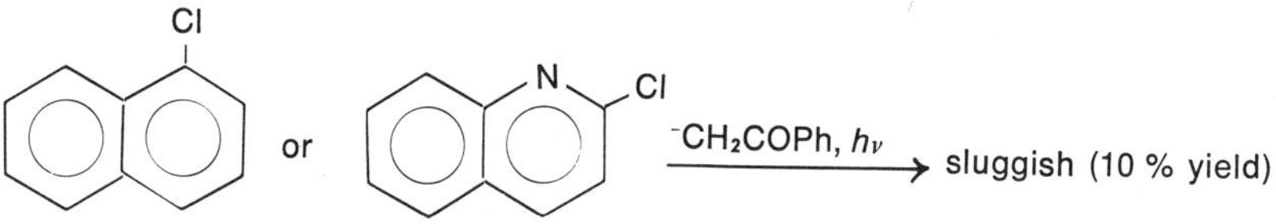

(43)

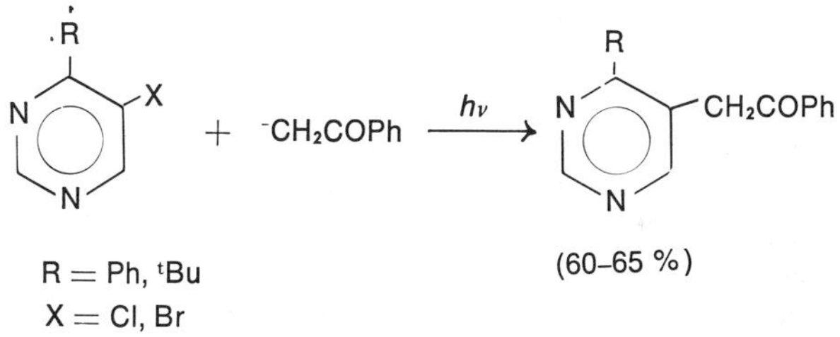

R = Ph, ᵗBu
X = Cl, Br

(44)

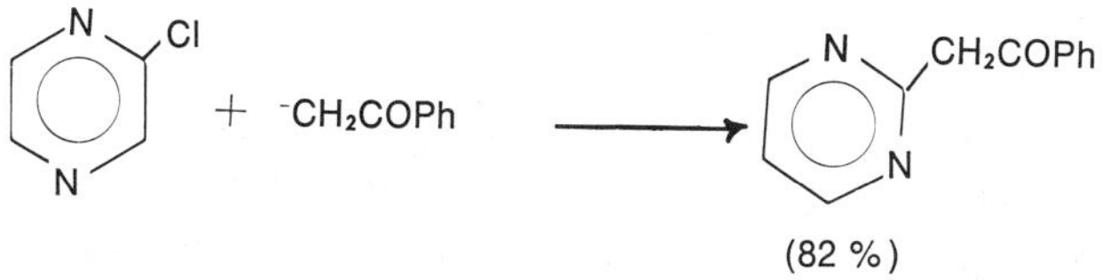

(45)

The position of the leaving group in heterocyclic substrates is important, and different reactivity was found among the isomeric compounds. The reactivity order reported for the conversion of bromopyridines to their respective pyridylacetones after 15 min of irradiation in liquid ammonia is 2-bromopyridine > 3-bromopyridine > 4-bromopyridine.

The halogen mobility found in the pyridine family is similar to that found in the benzene series, as is evident from the data in Table V. With bromothiophenes, it was found that the 3-isomer is much more reactive than the 2-isomer (*35*).

Table V

Halogen Mobility Found in the Pyridine Family

2-Halopyridine	hν (min)	2-Pyridylacetone (%)
Fluoro	120	40
Chloro	60	85
Bromo	15	95

Source: Reproduced, from Ref. 37. Copyright 1977, American Chemical Society.

From these examples it is obvious that the order of reactivity for the same aromatic substrate with different leaving group is $I > Br > Cl > F$ derivatives, and with the same leaving group, the reactivity depends on the reduction potential of the aromatic or heteroaromatic rings. With heterocycles as substrates, reactivity depends on the position of the leaving group.

Reaction in Other Solvents

There are several requirements to be met in order that a solvent for $S_{RN}1$ reactions be adequate. Three obvious requirements are that the solvent dissolve the reactants (usually a nonpolar organic compound and an alkali metal salt of the anionic nucleophile); for photostimulated reactions it should be transparent to the light which provokes the reaction (which is probably in the range of 300–400 nm); and, for solvated electron stimulated reactions, it should not react.

Since most nucleophiles that have been involved successfully in aromatic $S_{RN}1$ reactions are highly basic anions and radical anions, the solvent must be a very weak acid. Because electron transfer steps are involved, solvents which accept electrons readily (nitrobenzene, carbon tetrachloride, etc.) are likely to be inadequate. Solvents which readily react with aryl radicals, especially by hydrogen atom abstraction, present a problem because hydrogen abstraction is a termination step in the chain propagation cycle (Reaction 46). The resulting

$$Ar^{\cdot} + SH \longrightarrow ArH + S \qquad (46)$$

$S^{\cdot}$ radical does not lead back into the propagation sequence and thus inhibits the reaction.

Considering these facts, several solvents were tested using potassium diethyl phosphite, acetone enolate ion, benzenethiolate ion and lithium t-butylamide as nucleophiles. The latter reacts with iodobenzene in t-butylamine more likely by a benzyne mechanism than by the $S_{RN}1$ mechanism (47).

Protic solvents (ammonia, water, methanol, t-butyl alcohol, and t-butylamine) and aprotic solvents (DMSO, HMPT, acetonitrile, and THF) have been used for this reaction.

Among protic solvents, ammonia is the solvent of choice for most of the reactions because it meets most of the requirements already mentioned. It is also easy to purify and to eliminate when the reaction has finished. It has some drawbacks because it has a very low boiling point ($-33°C$), and therefore has to be worked with at very low temperatures. Quantitative studies, such as quantum yield determinations, are difficult to carry out. Another problem is the limited solubility of some substrates and nucleophiles in ammonia. ·

Water was thought to be a good solvent because it offers some advantages, such as it is a poor hydrogen atom donor, and a good solvent for ionic nucleophiles provided they are acidic enough to ionize.

The reactions of water-soluble aromatic substrates bearing polar substituents **54** were studied using benzenethiolate ion, nitromethane anion and, in one case, benzenesulfinate ion, as nucleophiles (47).

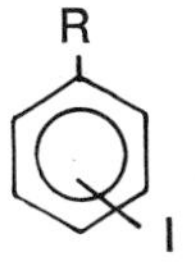

54a R $= p$-$\overset{+}{N}Me_3$

54b R $= p$-CO_2^-

54c R $= p$-$OCH_2CO_2^-$

54d R $= o$-$\overset{+}{N}Me_3$

54e R $= m$-$\overset{+}{N}Me_3$

The reactions of phenyltrimethylammonium iodide and p-iodophenyltrimethylammonium iodide with the anion of nitromethane, photostimulated or using sodium amalgam as a source of solvated electrons, failed to give any substitution products.

The reaction of benzenesulfinate ion with **54a** was also unsuccessful. The only substrate and nucleophile that gave moderate yields of substitution products was **54a** reacting with benzenethiolate ion. The amount of substitution product with all the others did not exceed 6%.

The reaction of **54a** with benzenethiolate ion is interesting for its different behavior compared with its reaction in liquid ammonia (Reaction 47). This difference in behavior is attributed to different relative rates of fragmentation and electron transfer in ammonia and water (47). In ammonia $k_f > k_t[\text{ArX}]$, whereas in water $k_f < k_t[\text{ArX}]$.

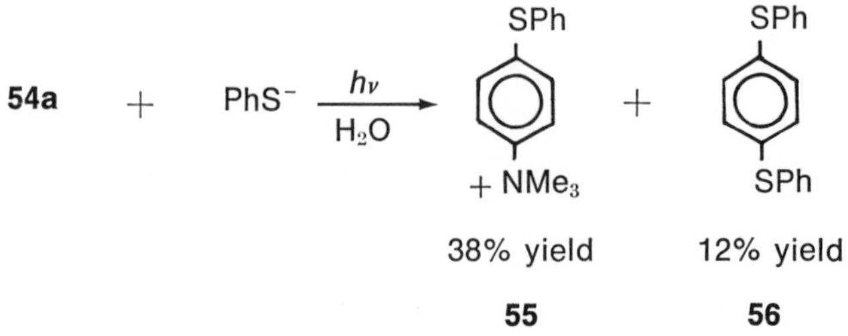

$$\textbf{54a} \quad + \quad \text{PhS}^- \xrightarrow[\text{H}_2\text{O}]{h\nu} \qquad \qquad + \qquad \qquad \qquad (47)$$

55   56

38% yield   12% yield

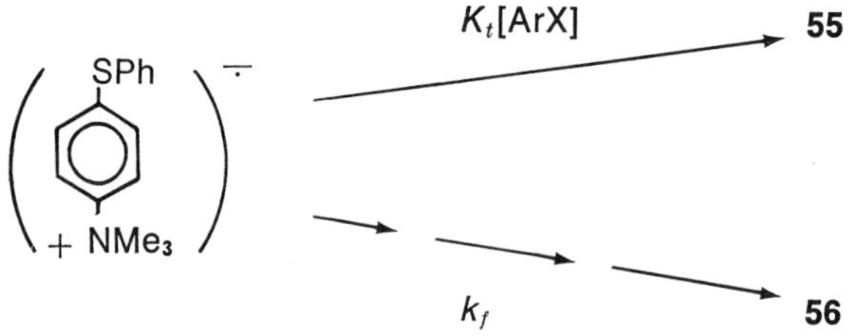

$$\qquad \qquad \qquad (48)$$

Benzenediazonium tetrafluoroborate probably reacts in methanol with potassium iodide by the $S_{RN}1$ mechanism (48, 49), and is also believed to react in water with nitrite ion by this mechanism (50).

The reaction of benzenediazonium ion with sodium dithionite and anions of nitroalkanes (nitromethane, nitroethane, and 1- and 2-nitropropanes) in water, at pH 13–14 gave ESR signals of the aryl substituted nitroalkane radical anions **57** (Reaction 49) from 0.1 s. to several minutes after mixing.

Coupling products were not formed with p-nitro or p-cyanophenyl radicals, and this was attributed to the high electron affinity of these radicals (51), which favors electron transfer (Reaction 50).

p-Nitrophenyl radicals and p-cyanophenyl radicals were trapped by cyanide ion and nitrite ion respectively giving the same ESR signal as the reduced p-cyanonitrobenzene (Reaction 51). These results show that aryl radicals may be trapped by the anion of nitroalkanes, cyanide, and nitrite ions in water (51), although in ammonia these nucleophiles do not react under the usual $S_{RN}1$ conditions.

$$N_2^+ \quad + \quad S_2O_4^= \quad + \quad {}^-CHRNO_2 \longrightarrow \left(\begin{array}{c} R\text{-}CH\text{-}NO_2 \\ \\ S \end{array} \right)^{\overline{\cdot}} \quad (49)$$

57

$$S = H, \ p\text{-}CO_2^-, \ m\text{-}CO_2^- \ p\text{-}Cl, \ p\text{-}MeO$$

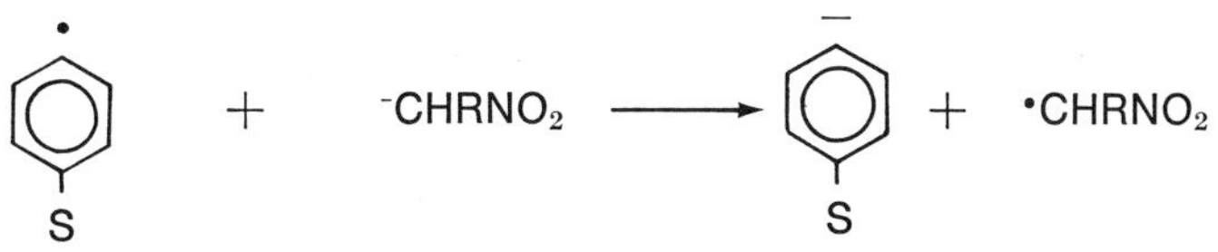

$$\cdot \quad + \quad {}^-CHRNO_2 \longrightarrow \quad {}^- \quad + \quad {}^\cdot CHRNO_2 \quad (50)$$

$$S = NO_2, \ CN$$

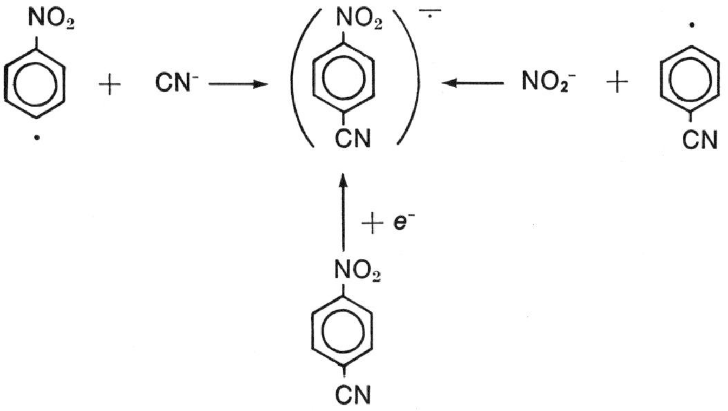

$$NO_2 \quad + \quad CN^- \longrightarrow \left(\begin{array}{c} NO_2 \\ \\ CN \end{array} \right)^{\overline{\cdot}} \longleftarrow NO_2^- \quad + \quad \begin{array}{c} {}^\cdot \\ \\ CN \end{array} \quad (51)$$

$$\Bigg\uparrow + e^-$$

$$\begin{array}{c} NO_2 \\ \\ CN \end{array}$$

3-Bromoisoquinoline reacted with benzenethiolate ion in methanol at 147°C and was strongly accelerated by sodium methoxide which did not react. In the absence of methoxide ion, only 38% of the substitution product was obtained, but with the addition of methoxide ion, the reaction was essentially complete with 65% of the substitution product and 35% of isoquinoline being formed. It was proposed that methoxide ion acted as an electron donor and catalyzed the $S_{RN}1$ mechanism (52).

The reaction of m-dibromobenzene with potassium diethyl phosphite ion as nucleophile in liquid ammonia gave only 32% of the disubstitution product under illumination, and was due to the poor solubility of the substrate in the solvent. The use of 20% of THF or DMSO as cosolvent, reduced the yield to 0–15% with the former, and 15% with the latter (53).

The photostimulated reaction of iodobenzene and potassium di-
ethylphosphite was chosen as a model system to test the behavior of
different solvents in $S_{RN}1$ reactions. The results are shown in Table VI
(47).

The lack of reactivity of a superb nucleophile such as potassium
diethyl phosphite in solvents like N-methyl-2-pyrrolidone, sulfolane,
and HMPT (Table VI), indicates that these solvents are not promising for
$S_{RN}1$ reactions.

Table VI

**Photostimulated Reaction of Iodobenzene with Potassium Diethyl
Phosphite in Diverse Solvents**

| | | | Yield (%) |
Solvent	hν *(h)*	X^-	*Substitution Product*
Ammonia	0.75	99	96
t-BuOH	4.5	81	74
DMSO	4.0	100	68
DMF	4.5	94	63
HMPT	4.5	30	4
Acetonitrile	4.0	98	94
MeCON(Me)$_2$	4.5	72	53
N-Methyl-2-pyrrolidone	4.5	24	10
CH$_3$OCH$_2$CH$_2$OCH$_3$	4.5	73	56
Sulfolane	4.5	28	20
t-BuNH$_2$	0.5	94	76

Source: Reproduced, from Ref. 47. Copyright 1976, American Chemical Society.

The photostimulated reaction of acetone enolate ion with bromobenzene in HMPT failed to give any product, whereas in DMSO, 61%
yield of the substitution product was obtained (47). However, HMPT
was used as solvent in reactions which are believed to occur in part by
the $S_{RN}1$ mechanism (vide infra).

Although the exact mechanism is not yet established, it appears that
part of the reaction of p-halotoluenes with hexamethyldisilane and potassium methoxide (or sodium or lithium methoxide) in anhydrous
HMPT (forming the p-tolyltrimethylsilane in 3h at 25°C) occurs by the
$S_{RN}1$ mechanism. The yield of substitution products obtained was very
good, most substrates giving 95% yield.

2-Bromopyridine and p-bromochlorobenzene also reacted with trimethylsilyl anion under the same reaction conditions. The latter substrate reacted with replacement of the bromine atom only (54–56).

Tetrahydrofuran seems to be a good solvent for photostimulated
reactions with ketone enolate ions, and it was the solvent of choice for

the reaction which leads to oxindoles by the $S_{RN}1$ mechanism (57). 2-Chloro-3-(N-methylacetamido)pyridine also underwent cyclization to afford azaoxindoles (83% yield) (57).

The relative rates of photoinduced consumption of 2-chloroquinoline with acetone enolate ion at ambient temperature, increase with respect to the solvent in the order benzene ≈ diethyl ether ≪ dimethoxyethane < THF < DMF < DMSO. (58).

The photostimulated reaction of bromobenzene with pinacolone enolate ion in THF did not give substitution product (1.3 h irradiation, 35°C) (46), but 2-bromopyridine reacted with acetone enolate ion under photo-stimulation giving 64% yield of the substitution product in 3 h of reaction in THF. Moreover 2-chloroquinoline reacted in THF to give the same amount of substitution product as in liquid ammonia. This latter reaction also proceeded in the dark, and in 1 h yielded 28% of the substitution product (in liquid ammonia only 1% yield was obtained in the dark) (Reaction 52) (58).

$$\text{(quinoline-Cl)} + {}^-CH_2COCH_3 \xrightarrow{\text{THF}} \text{(naphthalene)} CH_2COCH_3 \qquad (52)$$

Because the dark reaction is probably a thermally induced $S_{RN}1$ reaction, the difference in reactivity can probably be attributed to the 50–60°C temperature difference at which the reactions are carried out. The difference in temperature may affect not only the initiation step, but also the propagation steps.

In THF the reactions of halobenzenes (fluoro, chloro, and bromo) with tributylstannyllithium proceeds in part by the $S_{RN}1$ mechanism, although, depending on the reaction conditions and the nature of the halogen, other mechanisms (benzyne or metal–halogen exchange) may predominate (59).

Although $S_{RN}1$ reactions were carried out in several solvents, DMSO was the solvent of choice for quantitative studies. Iodobenzene reacted with potassium diethyl phosphite in DMSO to give the substitution product with a raw quantum yield at 313 nm of 20–50. This was the first quantitative study of the effect of light in the $S_{RN}1$ mechanism (60). It is very important to carefully purify this solvent, because it may contain unknown impurities which inhibit the reaction or give erratic results.

The reaction of m- or p-iodotoluenes or bromotoluenes with potassium diphenylphosphide in DMSO at 25°C gave the corresponding tolyldiphenylphosphines in good yields (5). This reaction occurred in the dark, but was stimulated with irradiation with broad spectrum UV-light. There is no evidence for an aryne mechanism, and azobenzene and m-dinitrobenzene inhibit the reaction, a characteristic feature of the $S_{RN}1$ mechanism. The thermal reaction of potassium diphenylphosphide with

p-bromotoluene in DMSO is quite slow. The reaction with p-iodotoluene, on the other hand, is fast (5).

The photostimulated reaction of ketone enolate ions with bromobenzene is not only affected by the solvent, but also by the counter ion. For instance the potassium salt of pinacolone enolate in liquid ammonia reacted in 1.3 h at $-33°C$ to give 96% of the monophenylated product and 4% of the diphenylated product. The reaction in DMSO, at 35°C gave 84% and 17% of the same products respectively. The lithium salt, under otherwise identical conditions gave only 25% and 3% of the same products (46).

Bromobenzene was unreactive toward acetone enolate ion in the dark in liquid ammonia, but reacted sluggishly in DMSO with potassium pinacolone enolate at about 23°C (61). Iodobenzene was very reactive under these reaction conditions with acetone enolate ion. For example, in a solution of iodobenzene (0.1 M) and acetone enolate ion (0.4 M) at about 23°C, 31% of iodide ion was released in 12 min. (61). The greater reactivity of iodobenzene as compared with bromobenzene in this solvent, resembles their behavior in liquid ammonia.

$$PhX + {}^-CH_2COCR_3 \xrightarrow{DMSO} PhCH_2COCR_3 \qquad (53)$$

$$X = Br, I \qquad R = H, Me$$

Evidence indicates that these reactions occur by the $S_{RN}1$ mechanism, and other possible mechanisms, such as benzyne, can be ruled out. p-Iodotoluene and m-iodotoluene gave the straightforward product without contamination of the other isomer. Furthermore, 2-iodomesitylene, a substrate incapable of being converted to a benzyne derivative, reacts with acetone enolate ion to furnish the arylated ketone.

However, a side reaction by the aryne mechanism was observed in the case of m-iodoanisole. The side reaction forms phenolic products in about 13% yield. Reactions 54–55 depict the mechanism postulated.

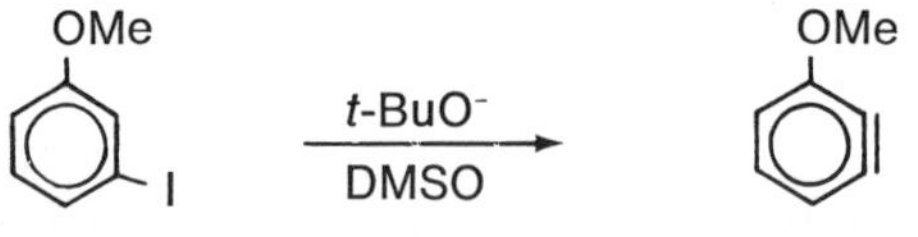

(54)

58

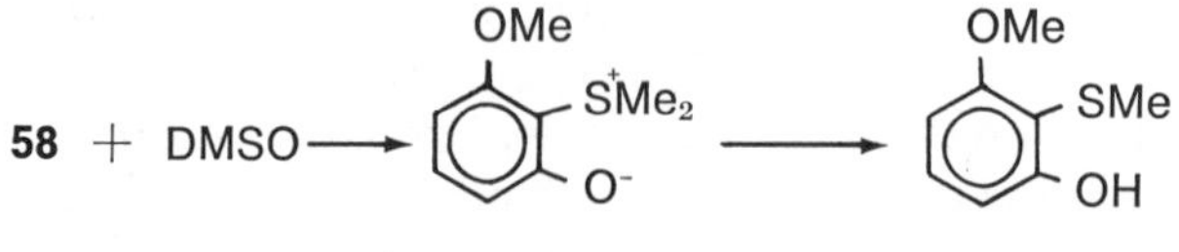

(55)

Similar products were found in the reaction of o-bromoanisole and dimsyl sodium in DMSO (62), and in the reaction of aryl halides with potassium t-butoxide in DMSO (63).

The reactivity of iodobenzene and bromobenzene with pinacolone enolate ion was studied in DMSO. The average iodobenzene : bromobenzene reactivity ratio under dark conditions was 6.7, similar to the value obtained under illumination (5.8), and strongly supports the same mechanism for both reaction systems (64).

The substituent effect in Reaction 56 was studied by two procedures, direct measurement of the reaction rate of the various aryl iodides with pinacolone enolate ion (Table VII), and by determination of the relative reactivity of aryl iodides versus bromobenzene in competition experiments (Table VIII).

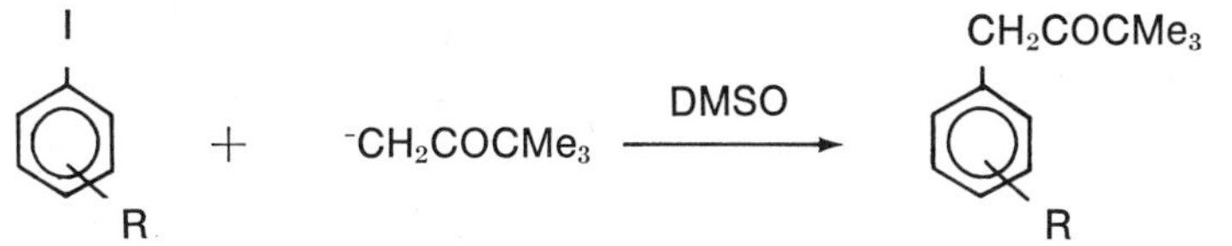

$$\text{(56)}$$

Table VII

Rate Constants for Reactions of Aryl Iodides with Potassium Pinacolone Enolate in DMSO at 25°C in the Dark

Aryl Iodide	$10^7 k,^a (M/s)$	$k_{rel.}$
Iodobenzene	7.0	(1.00)
p-Iodotoluene	0.65	0.093
m-Iodotoluene	0.67	0.096
p-Iodoanisole	1.05	0.15
m-Iodoanisole	260	37
p-Fluoroiodobenzene	23.6	3.4

$^a k$ was obtained from the integrated rate law $dx/dt = k$ (ArI)/X, where X is an unidentified product whose concentration was taken to be equal to iodide ion.
Source: Reproduced, from Ref. 64. Copyright 1979, American Chemical Society.

Tables VII and VIII show that there are enormous differences in the relative reactivity measured by the two methods. This huge difference (from four hundred fold to less than two fold) was interpreted in terms of the $S_{RN}1$ mechanism, which, being a radical chain mechanism, comprises initiation, propagation, and termination steps.

When two substrates react separately, the overall reactivity is determined not only by the propagation cycle, but also by the relative rates of initiation and termination; when they react together, as in the com-

Table VIII

Competition Experiments of Aryl Iodides and Bromobenzene in Reaction with Potassium Pinacolone Enolate in DMSO at 25°C in the Dark

Aryl Iodide	k_{ArI}/k_{PhBr}	k_{ArI} (rel)
Iodobenzene	7.3	(1.00)
p-Iodotoluene	6.9	0.95
m-Iodotoluene	7.4	1.01
p-Iodoanisole	6.4	0.88
m-Iodoanisole	11.8	1.62
p-Fluoroiodobenzene	9.4	1.29

Source: Reproduced, from Ref. 64. Copyright 1979, American Chemical Society.

petition experiments, they suffer the same initiation and termination steps, and the relative rates of electron transfer from the radical anion intermediate **59** of the propagation cycle to the aryl iodide or bromobenzene are measured (Reactions 58–59), provided that both aryl halides react only by fragmentation.

$$Ar^{\cdot} + {}^{-}CH_2COCMe_3 \longrightarrow (ArCH_2COCMe_3)^{-} \qquad (57)$$
$$\textbf{59}$$

$$\textbf{59} \left[\begin{array}{l} \xrightarrow{ArI} (ArI)^{-} + ArCH_2COCMe_3 \qquad (58) \\[2ex] \xrightarrow{PhBr} (PhBr)^{-} + ArCH_2COCMe_3 \qquad (59) \end{array} \right.$$

Correlation of the data from Table VII and Table VIII with substituent parameters σ (Hammett or Wepster) and half wave reduction potential was sought. The rates measured independently seem to be about ten times more sensitive to substituent changes. Although the correlations are not very good in either case, it is remarkable that in both cases (log k and log k_{ArI} (rel)) the correlations are better with the $E_{1/2}$ values than with σ values.

The usefulness of DMSO as solvent for $S_{RN}1$ reactions was ascribed to its low reactivity with aryl radicals. Using the reaction of 1-naphthyl radical with carbon tetrabromide as standard, the absolute rate constants for the reaction of 1-naphthyl radical with DMSO, DMF, acetonitrile, diphenyl disulfide and benzenethiolate ion were obtained (65). The source of 1-naphthyl radicals was 1-naphthalenediazonium tetrafluoroborate which was reduced either by iodide ions or electrochemically. The calculated rate constants are summarized in Table IX.

Table IX

Absolute Rate Constants for Reactions of 1-Naphthyl Radical at 20°C

Reactant/Solvent	$(M^{-1}s^{-1})$
DMSO/DMSO	3 $\times 10^5$
DMF/DMF	8 $\times 10^6$
CH$_3$CN/CH$_3$CN	2.5 $\times 10^5$
PhSSPh/DMSO	1.4 $\times 10^8$
PhSSPh/CH$_3$CN	2 $\times 10^8$
PhS$^-$/DMSO	1.7 $\times 10^8$

Source: Reproduced, with permission, from Ref. 65. Copyright 1980, *Acta Chem. Scand.*

It is seen in Table IX that the rate of hydrogen atom abstraction of 1-naphthyl radical from DMSO is quite high. With this rate, lower limits can be calculated for the rate of reactions where the reduction product is not formed at all in DMSO.

Based on a detection limit of 0.1% for the reduction product derived from hydrogen atom abstraction, it was concluded that the following rate constants are necessary for the reaction of 1-naphthyl radical with the benzenethiolate nucleophile for the $S_{RN}1$ mechanism to appear to be the exclusive pathway: 4.5×10^{10} (DMSO); 4.8×10^{10} (CH$_3$CN), and 1×10^{12} (DMF). The latter rate constant exceeds the diffusion rate, and this may be the reason why DMF is not as good as the other solvents for $S_{RN}1$ reactions.

The hydrogen atom abstraction rate constant reported in Table IX for 1-naphthyl radical in DMSO, $k = 3 \times 10^5$ $M^{-1}s^{-1}$ differs by about three orders of magnitude from the value estimated by Saveant of about 1×10^8 $M^{-1}s^{-1}$ (66).

If we accept that free aryl radicals are involved in $S_{RN}1$ reactions in DMSO, the value given in Table IX is more likely to be correct, otherwise the coupling of the aryl radical with the nucleophile would have to be faster than diffusion to represent the major reaction pathway.

Reaction Stimulated by Electrodes

Evidence that a nucleophile can react with aryl radicals generated by electrons from electrodes was obtained during the study of the cyclic voltammetric behavior of *p*-iodonitrobenzene **60** in DMF (67).

When tetraethylammonium perchlorate was used as supporting electrolyte, two well-defined reduction waves were seen near −1.0 and −1.1 V during the first cathodic sweep, and only a single, small oxidation wave corresponding to reoxidation of the second reduction product appears in the anodic sweep. Each wave corresponds to a one electron reduction process. The mechanism proposed was that p-iodonitrobenzene is reduced to its radical anion **61** in the first cathodic wave (Reaction 60), and is followed by a slow chemical reaction that involves the decomposition of the radical anion **61** forming p-nitrophenyl radical and iodide ion (Reaction 61). Then abstraction of hydrogen atom from the solvent (Reaction 62) yields nitrobenzene. The second wave pertains to the reduction of nitrobenzene to its radical anion **64** (Reaction 63).

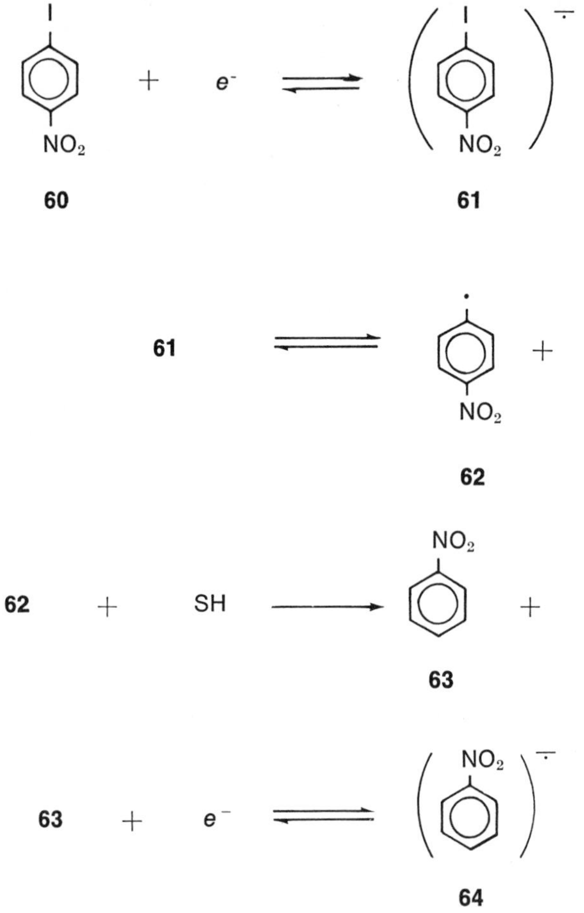

$$\textbf{60} \qquad\qquad\qquad\qquad \textbf{61} \tag{60}$$

$$\textbf{61} \qquad\qquad\qquad\qquad \textbf{62} \tag{61}$$

$$\textbf{62} \quad + \quad \text{SH} \qquad\longrightarrow\qquad \textbf{63} \quad + \quad \text{S} \cdot \tag{62}$$

$$\textbf{63} \quad + \quad e^- \qquad\qquad \textbf{64} \tag{63}$$

When the supporting electrolyte was changed to tetraethylammonium iodide, the rate of formation of nitrobenzene was clearly slowed down, indicating that iodide ion can react with p-nitrophenyl radical, shifting the equilibrium of Reaction 61 to the left.

The pseudo-first-order rate constant for the formation of nitro-benzene decreases from 2.3 s^{-1} (0.02 F of tetraethylammonium iodide plus 0.08 F tetraethylammonium perchlorate) to 0.9 s^{-1} (0.1 F of tet-raethylammonium iodide) (67).

The cyclic voltammogram of p-bromonitrobenzene indicates that there is only one redox process, corresponding to the formation of p-bromonitrobenzene radical anion, and its reoxidation. This result indicates that the radical anion is quite stable. Thus generation of this species by reducing p-iodonitrobenzene electrochemically in the presence of tetraethylammonium bromide was attempted. However, the cyclic voltammetry investigation of this system showed a behavior similar to that obtained in the absence of the bromide ion, and indicates that this ion does not react with p-nitrophenyl radical (67).

The cyclic voltammogram of **60** in DMSO (Figure 2A) changes when sodium cyanide (0.1 M) is present (Figure 2B), and only the reduction wave from the one electron reduction of **60** to its anion radical **61** is readily discernible on the first cathodic sweep. After reversal of the potential scan at −1.2 V, an anodic wave not seen in Figure 2A is observed at −0.72 V. Subsequent cathodic sweeps show an additional reduction process at −0.78 V, that correspond to the couple p-nitro-benzonitrile/p-nitrobenzonitrile radical anion (Figure 2C). Confirmation of p-nitrobenzonitrile radical anion formation was obtained by ESR spectroscopy (68).

These results show that the radical **62** formed in Reactions 60–61 can react with cyanide ions to form p-nitrobenzonitrile radical anion (68).

$$\textbf{62} \quad + \quad CN^- \quad \longrightarrow \quad \underset{NO_2}{\overset{CN}{\bigcirc}} \qquad (64)$$

65

The yield of **65** is greater than 90%, and similar results were obtained with o- and m-iodonitrobenzenes. Nitrite ion also reacts with p-nitro-phenyl radical **62** to form p-dinitrobenzene radical anion (68).

The electrochemical reduction of 4-bromobenzophenone **66** in DMF or acetonitrile in the presence of benzenethiolate ion has been reported to give a high yield of 4-phenylthiobenzophenone **70** (Reactions 65–69) (69).

When the applied electrolysis potential is −1.8 V, **70** is the product formed, but when the electrolysis is carried out at −1.9 V, not only the

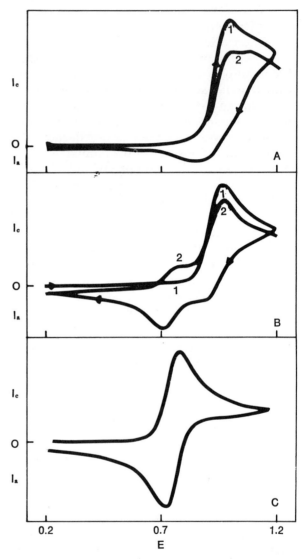

Figure 2. Cyclic voltammograms in DMSO at a scan rate of 80.6 mV/sec. Numbers 1 and 2 represent cycles 1 and 2 respectively. Key: A, 7.32×10^{-3} M p-iodonitrobenzene; B, 7.32×10^{-3} M p-iodonitrobenzene and 0.1 M NaCN; and C, 5.82×10^{-3} M p-nitrobenzonitrile. Reproduced from Ref. 68. Copyright 1970, American Chemical Society.

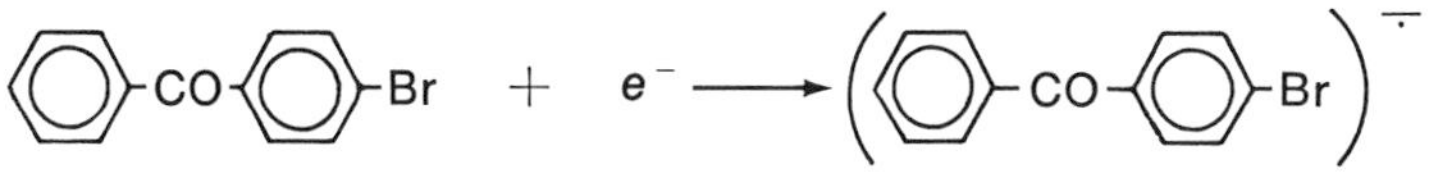

$$\text{[66] } C_6H_5\text{-CO-}C_6H_4\text{-Br} \quad + \quad e^- \longrightarrow \left(C_6H_5\text{-CO-}C_6H_4\text{-Br} \right)^{\overline{\cdot}} \quad \text{(65)}$$

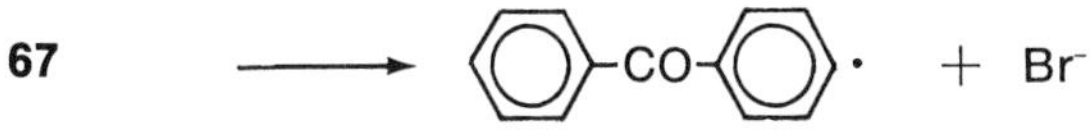

66   67

$$\text{67} \longrightarrow C_6H_5\text{-CO-}C_6H_4\cdot \;\; + \;\; Br^- \quad \text{(66)}$$

68

$$\text{68} \quad + \quad PhS^- \longrightarrow C_6H_5\text{-}\overset{\cdot}{\underset{\overset{\|}{O_-}}{C}}\text{-}C_6H_4\text{-SPh} \quad \text{(67)}$$

69

$$\text{69} \quad + \quad \text{66} \longrightarrow C_6H_5\text{-CO-}C_6H_4\text{-SPh} \;\; + \;\; \text{67} \quad \text{(68)}$$

70

$$\text{69} \xrightarrow{\;-e^-\;} \text{70} \quad \text{(69)}$$

ketone **70** is obtained, but also the alcohol **71** derived from reduction of the radical anion **69** (Reaction 70).

The catalytic nature of the process was evident because only 0.2 F/mol were needed to form 80% of **70** (*69*). Several haloaromatic compounds were shown to be substituted under electrochemical stimulation, and these results are summarized in Table X.

Aromatic radical nucleophilic substitution catalyzed by electrodes can be carried out at a potential more positive than the reduction potential of the aromatic halide, as was observed in the reaction of p-bromobenzonitrile ($E° = -1.66$ V versus Ag/0.01 M Ag$^+$ electrode) with potassium diethyl phosphite (*70*). In a preparative scale experiment, the potential was set at the level of the ArNu/(ArNu)$^{\overline{\cdot}}$ wave which results in an increase of current corresponding to an induction period, followed by a decrease of current featuring the reduction of the substitution product ArNu. The circuit was opened at the time of maximum current and the electricity consumed was extremely low (0.01 electron/molecule) while the yield of the substitution product was virtually 100% (*70*).

69   + e^- $\xrightarrow{\text{SH}}$ ⬡—CH—⬡—SPh   (70)
                                                     |
                                                     OH

71

Table X

$S_{RN}1$ Reactions Stimulated by Electrodes

ArX	Nucleophile	Solvent	Potential	F/mol	ArNu	ArH	Ref.
p-Bromobenzo-phenone	PhS$^-$	CH$_3$CN	−1.8	0.2	95	3	69
"	PhS$^-$	CH$_3$CN	−1.9	2.0	30^a		69
"	MeS$^-$	CH$_3$CN	−1.9	1.7	5	75	72
"	t-BuS$^-$	DMSO	−1.8	0.3	60		72
"	CN$^-$	CH$_3$CN	−1.7	0.25	95	3	72
p-Bromo-benzonitrile	PhS$^-$	CH$_3$CN	−2.1	0.2	80	10	72
p-Iodo-benzonitrile	PhS$^-$	CH$_3$CN	−1.7	1.5	20	80	72
p-Chloro-benzonitrile	(EtO)$_2$PO$^-$	NH$_3^b$	−1.35	0.01	100		70
1-Bromo-naphthalene	PhS$^-$	DMSO	−2.2	0.3	100		72
"	CN$^-$	DMSO	−2.2	0.5	—	55^c	72
p-Bromo-acetophenone	PhS$^-$	CH$_3$CN	−1.7	0.2	95		73
2-Chloro-quinoline	PhS$^-$	NH$_3^b$	−1.43	0.1	96		74
"	p-ClC$_6$H$_4$S$^-$	NH$_3^b$			80		74

^{a}65% of p-PhSC$_4$H$_4$CHOHPh.
^{b}−40°C.
^{c}45% of ArX.

Literature Cited

1. Rossi, R. A.; Bunnett, J. F. *J. Org. Chem* **1973,** *38,* 1407.
2. Rossi, R. A.; Bunnett, J. F. *J. Am. Chem. Soc.* **1974,** *96,* 112.
3. Beringer, F. M.; Messing, S. *J. Org. Chem.* **1972,** *37,* 2484.
4. Kim, J. K.; Bunnett, J. F. *J. Am. Chem. Soc.* **1970,** *92,* 7463, 7464.
5. Swartz, J. E.; Bunnett, J. F. *J. Org. Chem.* **1979,** *44,* 340.
6. Rossi, R. A.; Alonso, R. A.; Palacios, S. M. *J. Org. Chem.* **1981,** *46,* 2498.
7. Alonso, R. A.; Rossi, R. A. *J. Org. Chem.* **1982,** *47,* 77.
8. Benkaser, R. A.; Tincher, C. A. *J. Organomet. Chem.* **1968,** *13,* 139.
9. Jessup, D. W.; Paschal, J. W.; Rabideau, P. W. *J. Org. Chem.* **1977,** *42,* 2620.
10. Truce, W. E.; Tate, D. P.; Burge, D. N. *J. Am. Chem. Soc.* **1960,** *82,* 2872.

11. Gerdil, R. *J. Chem. Soc., B,* **1966,** 1071.
12. Pierini, A. B.; Rossi, R. A. *J. Org. Chem.* **1979,** *44,* 4667.
13. Shafer, S. J.; Closson, W. D.; van Dijk, J. M. F.; Piepers, O.; Buck, H. M. *J. Am. Chem. Soc.* **1977,** *99,* 5118.
14. Carnahan, Jr., J. C.; Closson, W. D.; Ganson, J. R.; Juckett, D.A.; Quaal, K. S. *J. Am. Chem. Soc.* **1976,** *98,* 2526.
15. Romanin, A. M.; Gennaro, A.; Vianello, E. *J. Electroanal. Chem.* **1978,** *88,* 175.
16. Volke, J.; Manousek, O.; Troyepolskaya, T. V. *J. Electroanal. Chem.* **1977,** *85,* 163.
17. Houser, K. J.; Bartak, D. E.; Hawley, M. D. *J. Am. Chem. Soc.* **1973,** *95,* 6033.
18. Bunnett, J. F.; Creary, X. *J. Org. Chem.* **1975,** *40,* 3740.
19. Bunnett, J. F.; Creary, X. *J. Org. Chem.* **1974,** *39,* 3173.
20. Bunnett, J. F.; Sundberg, J. E. *Chem. Pharm. Bull.* **1975,** *23,* 2620.
21. Rossi, R. A.; de Rossi, R. H.; Lopez, A. F. *J. Org. Chem.* **1976,** *41,* 3371.
22. Bard, R. R.; Bunnett, J. F. *J. Org. Chem.* **1980,** *45,* 1546.
23. Beugelmans, R.; Roussi, G. *J. Chem. Soc., Chem. Comm.* **1979,** 950.
24. Schuler, R. H.; Neta, P.; Zemel, H.; Fessenden, R. W. *J. Am. Chem. Soc.* **1976,** *98,* 3825.
25. Beugelmans, R.; Roussi, G.; *Tetrahedron* **1981,** *37,* 393.
26. Bordwell, F. G.; Algrim, D.; Vanier, N. R. *J. Org. Chem.,* **1977,** *42,* 1817.
27. Beugelmans, R.; Boudet, B.; Quintero, L. *Tetrahedron Lett.* **1980,** *21,* 1943.
28. Beugelmans, R.; Ginsburg, H. *J. Chem. Soc., Chem. Commun.* **1980,** 508.
29. Bunnett, J. F.; Singh, P.; *J. Org. Chem.* in press.
30. Bunnett, J. F.; Creary, X. *J. Org. Chem.* **1974,** *39,* 3611.
31. Bunnett, J. F.; Traber, R. P. *J. Org. Chem.* **1978,** *43,* 1867.
32. Bard, R. R.; Bunnett, J. F.; Traber, R. P. *J. Org. Chem.* **1979,** *44,* 4918.
33. Alonso, R. A.; Rossi, R. A. *J. Org. Chem.* **1980,** *45,* 4760.
34. Rossi, R. A.; de Rossi, R. H. Lopez, A. F.; *J. Am. Chem. Soc.* **1976,** *98,* 1252.
35. Bunnett, J. F.; Gloor, B. F. *Heterocycles* **1976,** *5,* 377.
36. Goldfarb, I. L.; Ikubov, A. P.; Belenki, L. I. *Zhur. Geter. Soedini* **1979,** 1044.
37. Komin, A. P.; Wolfe, J. F. *J. Org. Chem.* **1977,** *42,* 2481.
38. Hay, J. V.; Hudlicky, T.; Wolfe, J. F. *J. Am. Chem. Soc.* **1975,** *97,* 374.
39. Rossi, R. A.; Peñeñory, A. B., unpublished results.
40. Rossi, R. A.; Palacios, S. M. *J. Org. Chem.,* **1981,** *46,* 5300.
41. Hay, J. V.; Wolfe, J. F. *J. Am. Chem. Soc.* **1975,** *97,* 3702.
42. Oostveen, E. A.; van der Plas, H. C. *Recl. Trav. Chim. Pays-Bas* **1979,** *98,* 441.
43. Carver, D. R.; Komin, A. P.; Hubbard, J. S.; Wolfe, J. F. *J. Org. Chem.* **1981,** *46,* 294.
44. Bard, R. R.; Bunnett, J. F.; Creary, X.; Tremelling, M. J. *J. Am. Chem. Soc.* **1980,** *102,* 2852.
45. Bunnett, J. F.; Sundberg, J. E. *J. Org. Chem.* **1976,** *41,* 1702.
46. Semmelhack, M. F.; Bargar, T. *J. Am. Chem. Soc.* **1980,** *102,* 7765.
47. Bunnett, J. F.; Scamehorn, R. G.; Traber, R. P. *J. Org. Chem.* **1976,** *41,* 3677.
48. Kumar, R.; Singh, P. R. *Tetrahedron Lett.* **1972,** 613.
49. Singh, P. R.; Kumar, R. *Aust. J. Chem.* **1972,** *25,* 2133.
50. Opgenorth, H. R.; Rüchardt, C. *Justus Liebigs Ann. Chem.* **1974,** 1333.
51. Russell, G. A.; Metcalfe, A. R. *J. Am. Chem. Soc.* **1979,** *101,* 2359.
52. Zoltewics, J. A.; Oestreich, T. *J. Am. Chem. Soc.* **1973,** *95,* 6863.
53. Bunnett, J. F.; Shafer, S. J. *J. Org. Chem.* **1978,** *43,* 1873.
54. Shippey, M. A.; Dervan, P. B. *J. Org. Chem.* **1977,** *42,* 2654.
55. Quintard, J. P.; Hauvette-Frey, S.; Pereyre, M.; Couret, C.; Satgé, J. C. R. *Acad. Sci., Ser C,* **1978,** *287,* 247.
56. Quintard, J. P.; Hauvette-Frey, S.; Pereyre, M. *Bull. Soc. Chim., Belg.* **1978,** *87,* 505.
57. Wolfe, J. F.; Sleevi, M. C.; Goehring, R. R. *J. Am. Chem. Soc.* **1980,** *102,* 3646.
58. Moon, M. P.; Wolfe, J. F. *J. Org. Chem.* **1979,** *44,* 4081.
59. Quintard, J. P.; Hauvette-Frey, S.; Pereyre, M. *J. Organomet. Chem.* **1978,** *159,* 147.

60. Hoz, S.; Bunnett, J. F. *J. Am. Chem. Soc.* **1977,** *99,* 4690.
61. Scamehorn, R. G.; Bunnett, J. F. *J. Org. Chem.* **1977,** *42,* 1449.
62. Birch, A. J.; Chamberlain, K. B.; Oloyede, S. S. *Aust. J. Chem.* **1971,** *24,* 2179.
63. Cram, D. J.; Day, A. C. *J. Org. Chem.* **1966,** *31,* 1227.
64. Scamehorn, R. G.; Bunnett, J. F. *J. Org. Chem.* **1979,** *44,* 2604.
65. Helgeé, B.; Parker, V. D. *Acta Chem. Scand. B,* **1980,** *34,* 129.
66. M'Halla, F.; Pinson, J.; Saveant, J. M. *J. Am. Chem. Soc.* **1980,** *102,* 4120.
67. Lawless, J. G.; Hawley, M. D. *J. Electroanal. Chem.* **1969,** *21,* 365.
68. Bartak, D. E.; Danen, W. C.; Hawley, M. D. *J. Org. Chem.* **1970,** *35,* 1206.
69. Pinson, J.; Saveant, J. M. *J. Chem. Soc., Chem. Comm.* **1974,** 933.
70. Amatore, C.; Pinson, J.; Saveant, J. M.; Thiebault, A. *J. Electroanal. Chem.* **1980,** *107,* 59.
71. Rossi, R. A.; Alonso, R. A. *J. Org. Chem.* **1980,** *45,* 1239.
72. Pinson, J.; Saveant, J. M. *J. Am. Chem. Soc.* **1978,** *100,* 1506.
73. van Tilborg, W. J. M.; Smit, C. J.; Scheele, J. J. *Tetrahedron Lett.* **1977,** 2113.
74. Amatore, C.; Chaussard, J.; Pinson, J.; Saveant, J. M.; Thiebault, A. *J. Am. Chem. Soc.* **1979,** *101,* 6012.

6

Molecular Orbital Considerations

Chemical reactivity is usually discussed in terms of transition state theory. However, the principles of orbital symmetry (1), the perturbational molecular orbital method (PMO) (2, 3), and the frontier orbital approach (4) have been particularly successful in delineating those reactions which can occur and in predicting which reaction path is more favorable.

As two molecules collide, three major forces operate: repulsion from the interaction of filled orbitals of one molecule with the filled orbitals of the other, Coulombic repulsion or attraction due to the polarity or charges of species reacting together, and attractive forces from the interaction of the occupied orbitals of one reactant interacting with the unoccupied orbital of the other (5).

In an expression that takes account of these interactions, the first term is always repulsive, but is of similar value for any approximation of the two reactants. The second term is important in reactions between ions or polar molecules, and the third term is the gain in energy that results from the interaction of occupied–unoccupied orbitals. The third term is important in reactions of neutral molecules (6, 7).

PMO Applied to $S_{RN}1$ Reactions

Aryl radicals are uncharged soft electrophiles and therefore the term that considers electrostatic forces is not important. The relevant term is the interaction of the molecular orbitals (MO's).

The frontier MO approach demonstrates that the strongest interaction between two reacting centers occurs through the frontier orbitals of similar energy (2, 3) so that the singly occupied molecular orbital (SOMO) of the aryl radical will interact with the highest occupied molecular orbital (HOMO) of the nucleophile. This three-electron interaction will generate one two-electron bonding orbital (a σ bond) and an antibonding orbital, which does not necessarily bear the extra electron. The

0065-7719-83/0178-0143$06.00/1

strongest interaction occurs when the two species approach each other along the axes of the new σ bond (Reaction 1).

$$\text{Ar} \uparrow \quad \ldots\ldots\ldots \quad \uparrow\downarrow \text{Nu} \longrightarrow (\text{Ar} \uparrow\downarrow \text{Nu})^{\bar{\cdot}} \qquad (1)$$
$$1\,e^- \qquad\qquad\qquad 2\,e^- \qquad\qquad 3\,e^-$$

Assuming that the energies of the SOMO of the aryl radical and the HOMO of the nucleophile are equal, the change in the π energy, as calculated by the first-order perturbation, is given by Equation 2 where

$$\Delta E_\pi = c_{Ar}^{SOMO}\, c_{r,Nu}^{HOMO}\, \beta \qquad (2)$$

the coefficient of the σ radical, c_{Ar}^{SOMO} is unity; $c_{r,Nu}^{HOMO}$ is the coefficient of the HOMO of the atom r of the nucleophile that will form the bond with the aryl radical; and β is the resonance integral between the two atoms bonded. Because c_{Ar}^{SOMO} is unity, Equation 2 simplifies to Equation 3.

$$\Delta E_\pi = c_{r,Nu}^{HOMO}\, \beta \qquad (3)$$

In the interaction of an aryl radical with a nonconjugated nucleophile with the same energy (SOMO = HOMO), with $c_{r,Nu}^{HOMO}$ unity, Equation 3 simplifies to Equation 4.

$$\Delta E_\pi = \beta \qquad (4)$$

If the SOMO of the aryl radical and the HOMO of the nucleophile are not degenerate, the first-order change in ΔE_π is zero. In this case the change in ΔE_π is given by the second-order perturbation for the interaction of atom r of the nucleophile with the atom s of the electrophile and can be calculated using Equation 5.

$$\Delta E_\pi = 2 \sum_{j}^{occ} \sum_{k}^{unocc} \frac{c_{rj}^2 c_{sk}^2 \beta_{rs}^2}{E_j - E_k} \qquad (5)$$

In our particular case where the electrophile is an aryl radical, we have only one orbital with energy E_k close to α and $c_s=1$. Equation 5 then simplifies to Equation 6, with E_j expressed in β units.

$$\Delta E_\pi = \sum_{j}^{all} \frac{c_{rj}^2 \beta_{rs}^2}{E_j} \qquad (6)$$

Provided that the predominant term of Equation 6 is the one involving the HOMO of the nucleophile, Equation 6 then simplifies to Equation 7.

$$\Delta E_\pi = \frac{(c_r^{HOMO})^2 \beta_{rs}^2}{E_{HOMO}} \tag{7}$$

Consider in detail the interaction of an aryl radical with a non-conjugated nucleophile with the same energy. As we have seen before, the change in energy is given by Equation 4, but this Equation applies when both reacting species are close enough to give the total perturbation. At any point between starting material and products, the changes in energy will be a fraction of β; thus we can rewrite Equation 4 to be Equation 8 (*8*).

$$\Delta E_\pi = \gamma \beta \tag{8}$$

Since γ depends on the distance between the reacting species, it will change between 0 at infinite distance, when no perturbation occurs (Figure 1A), to $\gamma = 1$ for the maximum perturbation (Figure 1C). As γ increases, the gap between σ and σ^* molecular orbitals also increases.

In the coupling of an aryl radical and a nucleophile, not only the σ and σ^* MOs of the new bond are involved, but also the σ, σ^*, π, π^* MOs of the aromatic system, and the nucleophile as well. There are different situations depending on the structure of the aryl radical and the nucleophile, and consequently different reaction pathways are found.

Consider the hypothetical reaction of a phenyl radical with methyl anion (Reaction 9).

$$Ph\cdot + {}^-CH_3 \longrightarrow (PhCH_3)^{\cdot-} \tag{9}$$

In this particular case, we have that the energy of the SOMO of the σ phenyl radical (Figure 2A) is almost equal to the energy of the doubly occupied p orbital of the methyl anion (Figure 2B). The ordering of the π MOs in this Figure is at the Hückel level, while the positioning of the σ MOs is just schematic. The σ MOs of the phenyl radical are not shown in Figure 2A to avoid unnecessary complications.

The product obtained in Reaction 9 is a toluene radical anion, in which the odd electron occupies a π^* MO, because the energy of this MO is lower than the energy of a C–C σ^* MO (*9, 10*).

The MOs involved during the coupling process of a phenyl radical with methyl anion are the σ and σ^* MOs of the bond being formed, which, as the reactants approach each other, will become more and more separated because the energy of the σ MO decreases as the energy of the σ^* MO increases.

As was previously mentioned, the product is a toluene radical anion in which the low lying antibonding MO is the π^* (*9, 10*); thus there must be a point on the reaction coordinate where the energy of the incipient

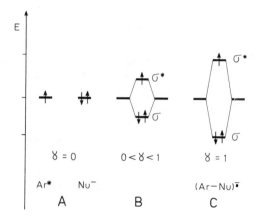

Figure 1. Changes in energy during the coupling of an aryl radical and a nucleophile. Key: A, at infinite distance ($\gamma = 0$); B, at intermediate distance ($0 < \gamma < 1$); and C, in the aryl–nucleophile radical anion product ($\gamma = 1$).

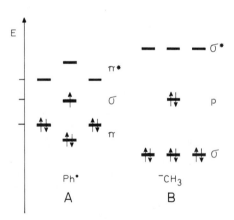

Figure 2: Energy levels of the molecular orbitals. Key: A, phenyl radical; and B, methyl anion.

σ^* equals the energy of the π^* MO, and, thereafter, the odd electron stays in the π^* MO. Figure 3 shows only the MOs involved in this transition.

It can easily be seen that the net gain in energy is higher for this reaction pathway than for a pathway leading to a product with the odd electron in a σ^* MO.

A more quantitative description of this reaction pathway was performed using the INDO method (11), and it was found that at a distance of about 1.9 Å between the phenyl radical and the methyl anion the values of the σ^* MO and π^* MO are equal (degenerate). At shorter distances than the equilibrium distance the π^* MO decreases in energy, whereas the σ^* MO increases. At distances longer than the infinite (about 8 Å) the σ^* MO is lower in energy than the π^* MO (Figure 4).

In the coupling of an aryl radical with a carbanionic nucleophile of $^-CH_2Z$ type, Z being an unsaturated moiety, aryl and Z will not be conjugated in the product because the two moieties are separated by an sp^3 carbon atom. Therefore the odd electron must be located in the lowest unoccupied MO (LUMO) of either the aryl group or the Z group (Reaction 10) (12).

$$Ar^{\cdot} + {}^-CH_2\text{-}Z \underset{\underset{\textbf{2}}{Ar\text{-}CH_2\text{-}(Z)^{\dot{-}}}}{\overset{\overset{(Ar)^{\dot{-}}\text{-}CH_2\text{-}Z}{\textbf{1}}}{\diagup\!\!\!\!\diagdown}} \qquad (10)$$

The radical anion formed predominantly will be the one where the electron is located in the LUMO of lowest energy, which will also be the same as the one formed when the neutral product takes an electron (Reaction 11).

$$Ar\text{-}CH_2\text{-}Z + e^- \underset{\textbf{2}}{\overset{\textbf{1}}{\diagup\!\!\!\!\diagdown}} \qquad (11)$$

In the process of coupling of an aryl radical, which has π and π^* MOs, with a $^-CH_2$-Z nucleophile, which is a conjugated system that also has π and π^* MO's, a new C–C σ bond and a radical anion intermediate are formed. Two possible situations develop, depending on which moiety has the lowest LUMO, the aryl or Z (Figure 5).

As the σ MO decreases in energy, the σ^* MO increases, and there are two conceivable orbital crossings in the reaction coordinate. The odd electron will be located in the first orbital equal to the energy of the σ^* MO and will stay there if the difference in energy between this MO and the other antibonding MO's, σ^* or π^*, is large.

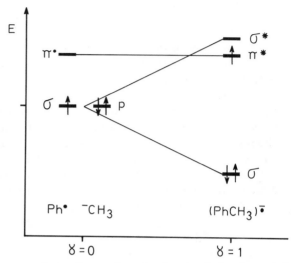

Figure 3. Changes in energy during the coupling of a phenyl radical and a methyl anion at infinite distance ($\gamma = 0$) and in the product toluene radical anion ($\gamma = 1$).

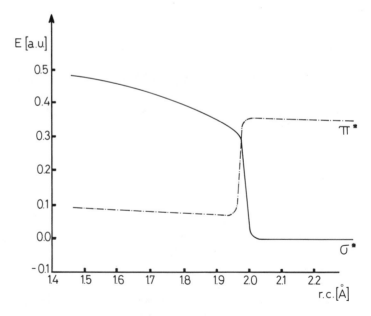

Figure 4. Changes in energy of σ^* and π^* MOs for the reaction of a phenyl radical with a methyl anion when the species approach each other (INDO method).

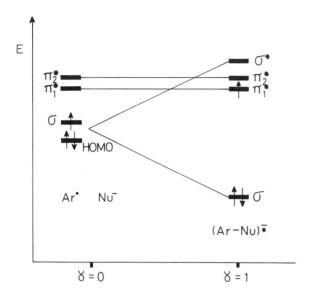

Figure 5. MO energy changes during the coupling of an aryl radical with a nucleophile bearing a π^* *MO.*

If π^*_1 belongs to the aryl moiety and is lower in energy (Figure 5) than π^*_2, belonging to the Z moiety, the structure of the radical anion will be **1**, but if their energies are in the reverse order, the structure will be **2**.

As will be discussed later, the π^* MOs are not always the lowest in energy. In some systems σ^* MOs may be of lower or comparable energy.

The experimental evidence that supports these ideas will be discussed.

Coupling of Phenyl Radicals with Hydrocarbon-Derived Carbanions

It follows from Equations 3 and 7 that the coefficients $c_{r,\ Nu}^{HOMO}$ will determine the position of the coupling. In the HOMO of indene anion, carbon 1 has a higher coefficient (c_1^H) than carbon two (c_2^H), and according to Equation 3, the preferred position for coupling will be position one, in good agreement with the experimental results (Reaction 12).

On the other hand, the coefficients of the HOMO of the pentadienide anion are the same for carbons, 1, 3, and 5 ($c^H = -c_3^H = c_5^H$). Considering only the coefficients c^H, two products should have been formed, with reaction at carbon 1 and carbon 3 (Reactions 13, 14).

$$\text{(12)}$$

$$Ph^{\cdot} + (CH_2 \dot{=} CH \dot{=} CH \dot{=} CH \dot{=} CH_2)^- \rightarrow PhCH_2(CH{=}CH{-}CH{=}CH_2)^{\bar{\cdot}} \quad \text{(13)}$$

$$\mathbf{3}$$

$$Ph{-}CH{\Big\langle}\begin{array}{l}(CH{=}CH_2)^{\bar{\cdot}}\\ CH{=}CH_2\end{array} \quad + (Ph)^{\bar{\cdot}}{-}CH{\Big\langle}\begin{array}{l}CH{=}CH_2\\ CH{=}CH_2\end{array} \quad \text{(14)}$$

$$\mathbf{4} \qquad\qquad\qquad\qquad \mathbf{5}$$

However, the radical anions have different energy in their LUMOs, **3** having lower energy than **4** or **5**.

As the phenyl radical and pentadienide ion approach each other, the π^* energy of the phenyl moiety is not expected to change much because it is not perturbed by the σ bond being formed. However, the π^* MO belonging to the nucleophile changes from the value that it has in the pentadienyl anion to a lower value in the ethylene system, or to a yet lower value in the butadiene system. The fact that only **3** is obtained may indicate that the π^* energy of the latter reaches the value of the incipient σ^* MO before the former. Figure 6 shows a schematic representation of the changes of the MO's energies along the reaction coordinate.

The same arguments explain why the coupling of p-anisylpropenide ion in the 3 position is three times faster than the coupling at the 1 position, although they have almost the same coefficient in the HOMO $(c_1^H = -c_3^H)$. The radical anion formed in the coupling at the 3 position (Reaction 15) has a lower LUMO energy than the radical anion formed in the coupling at the 1 position (Reaction 16), in which any of the three groups bonded to the sp^3 carbon have the same energy in their LUMOs.

From these results it is evident that a major factor in the competition between two different positions of similar HOMO coefficients in the nucleophile is the stability of the radical anion intermediate being formed. The stability of the radical anion seems to be less important in intermolecular competition. The reaction of a methyl radical with a carbanion nucleophile has been shown to be determined totally by the pK_a of the parent hydrocarbon acid (13). The MO following approach has been applied to the interaction of a radical with an anion using the methyl radical and the allyl anion as model reaction (14).

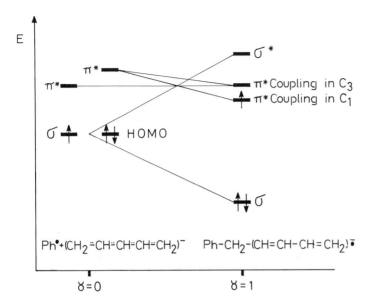

Figure 6. MO energy changes during the coupling of a phenyl radical with a pentadienide ion.

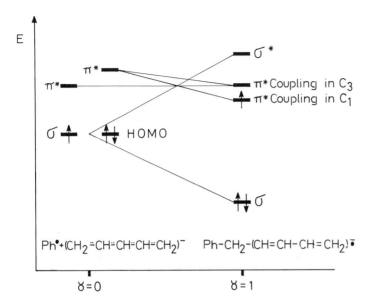

(15)

(16)

Coupling of Aryl Radicals with Ketone Enolate Anions

Allyl anion has the same coefficient on carbon 1 as on carbon 3 in the HOMO, but when carbon 3 is replaced by oxygen there is a modification of the electronic distribution. In Figure 7 the size of the circles can be taken as representing roughly the size of the coefficients. The total

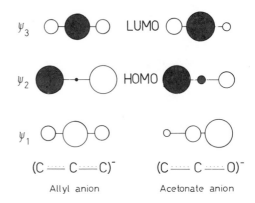

Figure 7. HOMOs of the allyl and acetone enolate anions. Radius and shading of the circles are indicative of the size and sign of the coefficients.

charge in the acetone enolate anion is higher on oxygen than on carbon, but in the HOMO the coefficient is higher on carbon than on oxygen.

In the coupling of a phenyl radical with the acetone enolate anion, which is a nucleophile with a conjugated system that has π and π^* MOs, a new C–C σ bond and a radical anion intermediate are formed (Reaction 17).

$$Ph^\cdot + {}^-CH_2COCH_3 \longrightarrow (PhCH_2COCH_3)^{\overline{\cdot}} \tag{17}$$

Phenylacetone has two π^* MOs separated by an sp^3 carbon atom, one belonging to the phenyl ring and another to the carbonyl moiety, and the latter is lower in energy than the former (Figure 8). Again the LUMO of the carbonyl moiety is lower than the LUMO of the acetone enolate anion because of the reduction of the number of electrons. In Figure 8 two orbital crossings are observed, and the odd electron goes to the lower LUMO of the system. Because we do not know the exact shape of the curves, we do not know whether the less stable radical is formed before the thermodynamically more stable anion is formed, and is then converted into the more stable radical by intramolecular electron transfer.

If the aromatic system has a lower LUMO than the carbonyl group, such as the naphthalene ring, the energy of the incipient σ^* MO equals that of the π^* MO of the aromatic ring earlier on the reaction coordinate, and the orbital crossing takes place before the energy of the incipient σ^* can reach the value of the π^* MO of the carbonyl moiety (Figure 9). All the other MOs involved are almost the same, because the energy of the radicals are similar in phenyl or 1-naphthyl radicals (*15*).

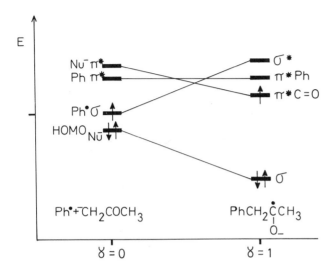

Figure 8. Changes in energy during the coupling of a phenyl radical and an acetone enolate anion.

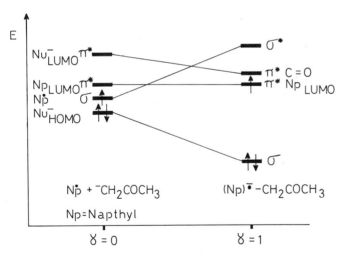

Figure 9. Changes in energy during the coupling of a naphthyl radical and an acetone enolate anion.

There is experimental evidence that the radical anion intermediate formed when naphthyl or phenyl radicals react with acetone enolate anions are those where the odd electron is in the LUMO of lowest energy of the molecule, as for example, the naphthyl moiety in naphthyl coupling and the carbonyl group in phenyl coupling.

Systems with Other Low Lying Antibonding MOs

The behavior of systems that have π^* MOs of low energy which stabilize the radical anion intermediate, decreasing the energy of the coupling process by orbital crossing from an incipient σ^* MO to a π^* MO of lower energy, has been described previously. The π^* MO is not always the one of lower energy in the system. It has been shown that σ^* radical anions are formed even when there are π^* MOs available (16).

Consider the reaction coordinate for the reverse reaction, the decomposition of a radical anion into an aryl radical and the anion of the leaving group; for instance, the decomposition of chlorobenzene radical anion (Reaction 18).

$$(PhCl)^{\overline{\cdot}} \longrightarrow Ph^{\cdot} + Cl^- \tag{18}$$

The principle of microscopic reversibility requires that the reaction coordinate for Reaction 18 be the same as that for the coupling of chloride ion with a phenyl radical (Figure 10).

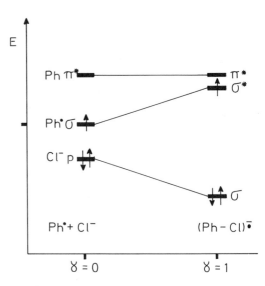

Figure 10. Changes in energy during the coupling of a phenyl radical and a chloride ion.

It can be seen from Figure 10 that the odd electron is not stabilized because the σ^* MO has a lower energy than the π^* MO of the system. Calculations carried out by a CNDO/2 method (17) indicate that in chlorobenzene radical anion, the lowest lying antibonding MO corresponds to the σ^* MO of the C–Cl bond.

The radical anion of 2-chlorothiophene was detected by ESR after gamma radiolysis of crystals of 2-chlorothiophene. It was suggested that this radical anion is a σ^* radical anion on the basis of its spin densities, and this interpretation was supported by CNDO calculations (18).

A CNDO/2 study of the reaction coordinate of the coupling of chloride ion with phenyl radical (or the decomposition of the radical ion) revealed that the lowest antibonding MO is always the σ^*; but, the same method applied to the coupling of fluoride ion with phenyl radical revealed that the fluorobenzene radical anion is a π^* radical anion (Reaction 19) (11).

$$Ph^{\cdot} + F^- \longrightarrow (PhF)^{\bar{\cdot}} \qquad (19)$$

It is known that fluorobenzene radical anion decomposes at a rate $\gg 10^6/\text{s}$ (19). Toluene radical anion does not decompose, and fluorobenzene anion does, although both are π^* radical anions. The difference in behavior may be explained by the stabilities of the methyl carbanion and fluoride ion that are formed. Because the latter is a much more stable anion, the change in energy from $\gamma = 1$ to $\gamma = 0$ occurs with a greater slope than in toluene radical anion, where the p orbital of the methyl carbanion is of much higher energy. Consequently, orbital crossing from $\pi^* \rightarrow \sigma^*$ takes place at shorter distances from the equilibrium in fluorobenzene than in toluene radical anions.

MO calculations indicate that the differences between the equilibrium distance and the distance at which the orbital crossing occurs are about 0.2 and 0.4 Å in fluorobenzene and toluene radical anion respectively. The likelihood that the vibration of the bond in the radical anion can reach this distance depends on the bond stretching energy and on the thermal distribution of molecules. Obviously, the larger the difference between the orbital crossing and equilibrium distances, the less likely its radical anions will dissociate.

It has been postulated that iodobenzene gives a σ^* radical anion, and probably also a π^* radical anion, in low yield, which in protic media become protonated. A further postulate is that dissociative electron capture to give a phenyl radical must proceed via the σ^* radical anion, and hence, if π^* addition occurs, Path $b \rightarrow c \rightarrow d$ (Reaction 20) must be followed (16).

It has been shown that γ-irradiated or photolyzed solutions containing 5-halouracils form radical anion in glasses. These radical anions are

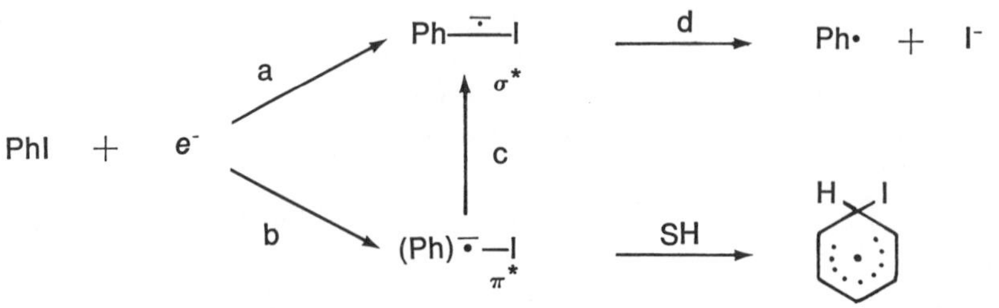

$$\text{PhI} \ + \ e^- \tag{20}$$

of π^* nature in fluoro or chloro derivatives **6**. However, with bromo or iodo derivatives the σ^* radical anion **7** is formed. In both cases, upon warming, an "uracil-yl" radical **8** is formed with halide elimination. The rate of halide detachment was found to increase in the order $I > Br > Cl$ (*20*). When a π^* radical anion is formed it has been suggested that subsequent dissociation is preceded by an electron migration from the π^* to the σ^* orbital (*20*).

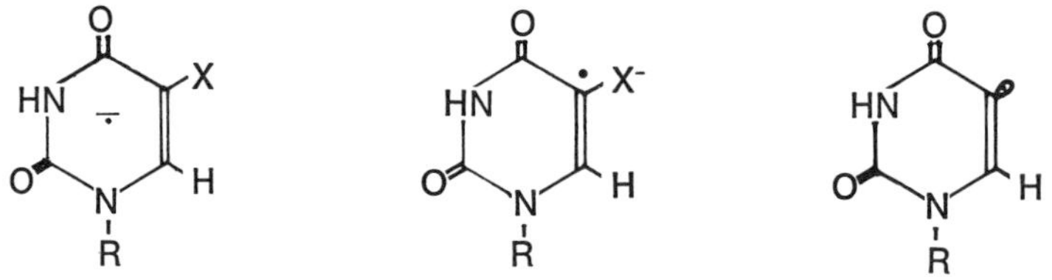

The potential curve for the dissociation of methyl chloride radical anion under vacuum, and solvated by two water molecules, has been calculated, and some interesting conclusions were extracted from the results. Namely, the dissociation barrier is decreased strongly by solvation, and as the barrier decreases the maximum is reached at distances closer to the equilibrium. The stability of the radical anions, and whether or not they can be observed as intermediates with a finite lifetime, depends crucially on the height of the dissociation barrier, and hence on the σ^* C–X orbital, and the position of the intersection with the orbital containing the odd electron (*21*).

The σ^* MOs of the carbon–halogen bonds are not the only low lying MOs because bond fragmentation is observed in other systems too. For instance, it has been shown that the coupling of aryl radicals with phenyl telluride ions gave not only the asymmetric substituted product, but also the symmetric ones. It was suggested that the σ^* MO of the C–Te bond has energy similar to the π^* system, and the orbital crossing can occur as in Reaction 21 (Figure 11).

$$Ar^{\cdot} + {}^{-}Te\text{-}Ph \rightleftharpoons Ar^{\underline{\cdot}}Te\text{-}Ph \longleftrightarrow Ar\text{-}Te^{\underline{\cdot}}Ph \rightleftharpoons ArTe^{-} + Ph^{\cdot} \quad (21)$$

Not only phenyl telluride ion gave this scrambling of aryl rings. The same behavior is observed with diphenyl arsenide ion. However, when this nucleophile reacts with an aryl radical having low-lying π^* MOs, such as 4-benzoylphenyl or 2-quinolyl radicals, scrambling of aryl rings is not observed, and only the straightforward substitution product is formed. In these cases the π^* MO is of lower energy than the σ^* MO of the C–As bond.

When the C–M bond has an even lower σ^* MO, such as C–Sb, even the p-benzoylphenyl moiety cannot protect the C–Sb bond from breaking, and scrambling of the aromatic moieties is observed.

Another example of a system with low lying σ^* MOs is the radical anion formed in the coupling of a phenyl radical with cyanomethyl anion. The main reaction is shown in Reaction 22.

$$Ph^{\cdot} + {}^{-}CH_2CN \longrightarrow (PhCH_2CN)^{\underline{\cdot}} \longrightarrow PhCH_2^{\cdot} + CN^{-} \quad (22)$$

However, when the aryl radical has a lower π^* MO than the benzene ring, such as naphthalene or phenanthrene, there is no bond breaking process, and only the substitution product is observed.

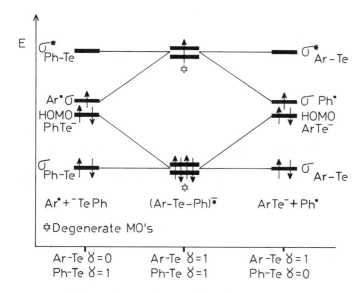

Figure 11. Changes in energy during formation and decomposition of an aryl phenyl telluride radical anion.

Another system where the reaction pathway depends strongly on the values of the π^* and σ^* MOs is the reaction of aryl radicals with alkyl thiolate ions. The two processes observed are the fragmentation of the radical anion intermediate and electron transfer of the odd electron.

The π^* MO depends on the aryl moiety, while the σ^* MO depends on the nature of the alkyl moiety, and is expected to decrease when the stability of the alkyl radical increases, or when the energy of the C–S bond decreases.

Because fragmentation depends on the probability of the odd electron reaching the σ^* orbital, it will increase as the σ^* MO energy decreases, just as observed.

Figure 12 is a schematic representation of all the combinations of the energies of the MOs involved in the process of fragmentation. We can distinguish three zones: a σ^*, a π^*, and a σ^*–π^* zone. In the σ^* zone the σ^* MO energy is higher than the π^* MO (π^* zone), thus the odd electron never reaches that zone; the radical anion intermediate is of π^* nature, and only electron transfer is observed. In the σ^*–π^* zone the odd electron can be in any of these orbitals, resulting in variable amounts of fragmentation competing with electron transfer reactions, which in turn depend on the relative differences of their energies.

The PMO method applied to the $S_{RN}1$ mechanism allows for the prediction of the coupling position of an aryl radical with a nucleophile, the structure of the radical anion intermediate and the products to be expected.

There are other possible approaches and other important factors (such as solvation, ion pairs, etc.) besides MO considerations that must be taken into account to deal with this problem in a more general way. There are also several nucleophiles that do not react with aryl radicals, such as oxygen nucleophiles (alkoxides, phenoxides, etc.) and the monoenolate ions of β-dicarbonyl compounds. With these nucleophiles there is an ample difference in energy between the SOMO of the aryl radical and the HOMO of the nucleophile, but still there should be an attractive interaction between the nucleophile and the aryl radical, if this simple approach is applied. As was demonstrated with more sophisticated methods (22–24), and with inclusion of overlap in the interaction of a SOMO with an HOMO with very large differences in energy, the attractive interaction may convert into a repulsive interaction. This is one explanation why these nucleophiles do not react with aryl radicals. Another is that the coupling of the aryl radical with the nucleophile first must form a radical anion with the odd electron in the σ^* MO of the bond being formed; this anion then slowly changes to a more stable radical anion. Thus, the lack of reactivity of oxygen nucleophiles can also be attributed to the exceptionally high energy level of the σ^* MO of the C–O bond (25).

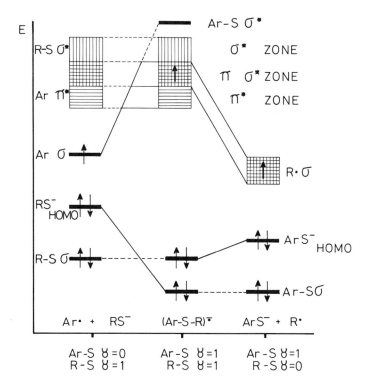

Figure 12. Schematic representation of the MOs involved when an aryl radical couples with an alkanethiolate ion forming a radical anion which fragments.

Literature Cited

1. Woodward, R. B.; Hoffman, R. *Angew. Chem.* **1969,** *8,* 781.
2. Dewar, M. J. S. "The Molecular Orbital Theory of Organic Chemistry"; Mc-Graw-Hill: New York, 1969; Chapter 6.
3. Dewar, M. J. S.; Dougherty, R. C. "The PMO Theory of Organic Chemistry"; Plenum: New York, 1975.
4. Fukui, K. *Acc. Chem. Res.* **1971,** *4,* 57.
5. Fleming, I. "Frontier Orbitals and Organic Chemical Reactions"; Wiley: New York, 1975; p. 28.
6. Klopman, G. *J. Am. Chem. Soc.,* **1968,** *90,* 223.
7. Salem, L. *J. Am. Chem. Soc.* **1968,** *90,* 543.
8. Rossi, R. A. *J. Chem. Ed.,* **1982,** *59,* 310.
9. Bernardi, F.; Guerra, M.; Pedulli, G. F. *Tetrahedron* **1978,** *34,* 2141.
10. Bigelow, R. W. *J. Chem. Phys.* **1979,** *70,* 2315.
11. Villar, H.; Castro, E. A.; Rossi, R. A. *Can J. Chem.,* in press.
12. Rossi, R. A.; de Rossi, R. H.; López, A. F. *J. Org. Chem.* **1976,** *41,* 3367.
13. Tolbert, L. M. *J. Am. Chem. Soc.* **1980,** *102,* 3531.
14. Zimmerman, H. E. *Acc. Chem. Res.* **1972,** *5,* 3939.

15. Kasai, P. H.; Clark, P. A.; Whipple, E. B. *J. Am. Chem. Soc.* **1970,** *92,* 2640.
16. Symons, M. C. R. *Pure and Appl. Chem.* **1981,** *53,* 223.
17. Beland, F. A.; Farwell, S. O.; Callis, P. R.; Geer, R. D. *J. Electroanal. Chem.* **1977,** *78,* 145.
18. Nagai, S.; Tomas, G. *J. Phys. Chem.* **1977,** *81,* 1793.
19. Andrieux, C. P.; Blocman, C.; Saveant, J. M. *J. Electroanal. Chem.* **1979,** *105,* 413.
20. Symons, M. C. R. *J. Chem. Soc., Chem. Comm.* **1978,** 313.
21. Canadell, E.; Karafiloglou, P.; Salem, L. *J. Am. Chem. Soc.* **1980,** *102,* 855.
22. Bernardi, F.; Epiotis, N. D.; Cherry, W.; Schelegel, H. B.; Whangbo, M. H.; Wolfe, S. *J. Am. Chem. Soc.* **1976,** *98,* 469.
23. Bernardi, F.; Cherry, W.; Shaik, S.; Epiotis, N. D. *J. Am. Chem. Soc.* **1978,** *100,* 1352.
24. Arnaud, R.; Court, J.; Bonnier, J. M.; Fossey, J. *Nouveau J. Chim.* **1980,** *4,* 299.
25. Galli, C.; Bunnett, J. F. *J. Am. Chem. Soc.,* **1981,** *103,* 7140.

7

The Initiation Step

A reaction that produces a reactive intermediate that can start a chain reaction becomes an initiation step. Although the initiation steps, together with the termination steps, may represent only a few reactions compared to the whole, they can decide if the reaction occurs or not.

Spontaneous Reactions

In several cases, the $S_{RN}1$ mechanism occurs without stimulation of any type other than temperature. These types of reactions are called spontaneous reactions.

Amide ion reacts with 1-iodo-2,4,5- and 1-iodo-2,3,5-trimethylbenzenes in liquid ammonia in the dark (1). Related nucleophiles, whose first atoms belong to the same group of elements, such as diphenyl phosphide (2), diphenyl arsenide (3), and diphenyl stibide (4) ions, also react with certain aromatic substrates. There are also examples of ketone enolate ions that react in the dark in liquid ammonia or DMSO (5).

Carbanions derived from diketones or diesters do not react with halobenzenes by the $S_{RN}1$ mechanism, by spontaneous, or by photo-stimulated reactions. However, arylation of these nucleophiles, brought about by diphenyliodonium ion, is a reaction that has been known for some time (6, 7, 8).

The proposed mechanism (6, 7, 8) is similar to the $S_{RN}1$ mechanism because it involves the formation of aryl radicals brought about by an electron transfer from the carbanion nucleophile to the substrate, yet differs because it is not a radical chain mechanism (Reaction 1). On the

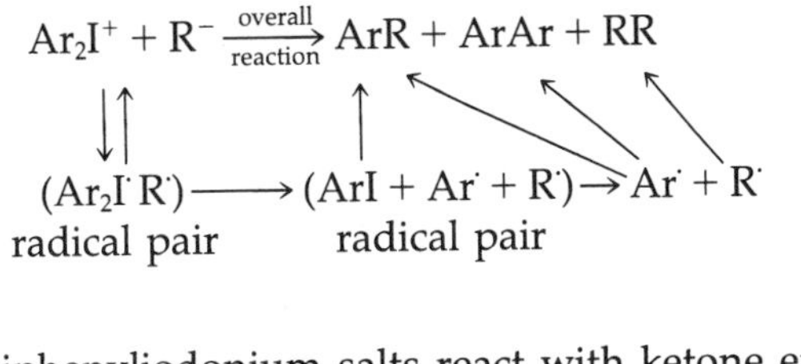

$$Ar_2I^+ + R^- \xrightarrow[\text{reaction}]{\text{overall}} ArR + ArAr + RR \tag{1}$$

$(Ar_2I^{\cdot} R^{\cdot}) \longrightarrow (ArI + Ar^{\cdot} + R^{\cdot}) \rightarrow Ar^{\cdot} + R^{\cdot}$
radical pair   radical pair

other hand, diphenyliodonium salts react with ketone enolate ions by the $S_{RN}1$ mechanism in the dark in liquid ammonia (9).

0065-7719-83/0178-0161$07.50/1

Direct electron transfer was also the mechanism proposed for some reactions of diazonium salts. It has been suggested that the uncatalyzed reaction of diazonium ions (1) with iodide ion or other easily oxidizable anions might proceed by an electron transfer process (Reaction 2) (10, 11).

$$\underset{\mathbf{1}}{ArN_2^+} + I^- \longrightarrow (Ar^. + N_2 + I^.) \longrightarrow ArI + N_2 \tag{2}$$

The uncatalyzed reaction of aryldiazonium tetrafluoroborate with potassium iodide in methanol at 0°C was found to form aryl iodide along with redox products such as arenes and formaldehyde. It was suggested that the initiation step involves the transfer of an electron to the aryl-diazonium ion (Reaction 3) followed by Reactions 4–6, which are analogous to the $S_{RN}1$ propagation steps (12, 13).

$$\mathbf{1} + I^- \longrightarrow (ArN_2^+)^{\overline{.}} + I^. \tag{3}$$

$$(ArN_2^+)^{\overline{.}} \longrightarrow Ar^. + N_2 \tag{4}$$

$$Ar^. + I^- \longrightarrow (ArI)^{\overline{.}} \tag{5}$$

$$(ArI)^{\overline{.}} + ArN_2^+ \longrightarrow ArI + (ArN_2^+)^{\overline{.}} \tag{6}$$

Based on kinetic studies a similar mechanism was proposed for the reaction of aryldiazonium salts with nitrite ion in water solution. In this case the spontaneous initiation step involves the formation of adduct 2 (Reaction 7) that fragments into Species 3 and nitrogen dioxide (Reaction 8); 3 then fragments with the formation of aryl radical and nitrogen (Reaction 9). The aryl radicals formed are trapped by nitrite ion (Reaction 10), and the radical anion formed transfers its odd electron to 2 (Reaction 11). Reactions 9–12 constitute the chain propagation circle (14).

$$\underset{\mathbf{2}}{\mathbf{1} + NO_2^- \longrightarrow Ar-N=N-NO_2} \tag{7}$$

$$\underset{\mathbf{3}}{\mathbf{2} \longrightarrow ArN_2^. + NO_2} \tag{8}$$

$$\mathbf{3} \longrightarrow Ar^. + N_2 \tag{9}$$

$$Ar^. + NO_2^- \longrightarrow (ArNO_2)^{\overline{.}} \tag{10}$$

$$\underset{\mathbf{4}}{(ArNO_2)^{\overline{.}} + \mathbf{2} \longrightarrow (Ar-N=N-NO_2)^{\overline{.}} + ArNO_2} \tag{11}$$

$$\mathbf{4} \longrightarrow \mathbf{3} + NO_2^- \tag{12}$$

Direct electron transfer from amide ions to the iodobenzenes (Reaction 13) in the dark reaction is believed unlikely because it is thermodynamically unfavorable by several kcal/mol (1).

$$ArI + NH_2^- \longrightarrow (ArI)^{\bar{\cdot}} + NH_2^{\cdot} \tag{13}$$

The ionization potential of the amide ion (25 kcal/mol) (15) is greater than the electron affinity of iodobenzene (13 kcal/mol) (16) in the gas phase. However, solvation must be taken into account. Although anilide ion is conceptually a more attractive electron donor, the reaction of iodotrimethylbenzenes with amide ion in the presence of 2,4-dimethyl anilide ion gives less products derived from a $S_{RN}1$ reaction (1).

Spontaneous electron transfer reactions from anions to several acceptors are well documented. Carbanions derived from weak acids are the best donors (17). On the other hand, with a particular donor, the transfer of an electron becomes more favorable with substrates that are easily reduced (less negative reduction potential). Consequently the order of reactivity of aryl halides is $Ph_2I^+ \rangle\rangle PhI \rangle PhBr \rangle PhCl$ with ketone enolate ions, and $PhI \rangle PhBr$ with diphenyl phosphide ions.

Chlorobenzene has never been found to react in the dark, even with diphenyl arsenide ion, but p-chlorobenzophenone, a more easily reduced compound, reacts with this nucleophile (3).

Because thermodynamic data are unavailable for all the substrates and nucleophiles that react in the dark, it is not possible to know if electron transfer is feasible for each case from a thermodynamic point of view. The formation of a charge-transfer complex between donor and acceptor prior to the actual electron-transfer reaction (Reaction 14) is a possibility, but these questions await further research.

$$Acceptor + Donor^- \rightleftharpoons (Acceptor\text{-}Donor)^- \rightarrow (Acceptor)^{\bar{\cdot}} + Donor \tag{14}$$

Reactions Stimulated by Solvated Electrons

When a substrate and a nucleophile do not have the appropriate combination of electron affinity–ionization potential for spontaneous transfer of one electron from the nucleophile to the substrate, reaction does not take place, unless it is promoted by other means.

Several substrates react with solvated electrons in liquid ammonia to give aryl radicals (Reactions 15–16), which are then trapped by acetone enolate ions to give substitution products (9).

$$ArX + e^- \longrightarrow (ArX)^{\bar{\cdot}} \tag{15}$$

$$(ArX)^{\bar{\cdot}} \longrightarrow Ar^{\cdot} + X^- \tag{16}$$

Some substrates are reduced totally to benzene, or suffer other types of reactions, and some are unreactive. The formation of benzene was suggested to arise from any of the alternatives shown in Reactions 17–18 (9).

$$ArY + e^- \longrightarrow (ArY)^{\cdot-} \longrightarrow Ar^- + Y^{\cdot} \xrightarrow{NH_3} ArH \qquad (17)$$

$$(ArY)^{\cdot-} \xrightarrow{e^-} (ArY)^{=} \longrightarrow Ar^- + Y^- \xrightarrow{NH_3} ArH \qquad (18)$$

$$Ar^- + NH_3 \longrightarrow ArH + NH_2^-$$

As discussed previously, the reaction of triphenylphosphine, arsine, stibine, and bismutine with solvated electrons in the presence of acetone enolate anion afforded only benzene. These results suggested that the bond fragmentation of the radical anion intermediate occurs as in Reactions 17–18. The scrambling of aryl rings found in the reaction of diphenyl arsenide ion (2) and diphenylstibide ion (2, 3) with aryl radicals, however, suggested that the radical anions of triphenylarsine and triphenylstibine decompose according to Reaction 16, and thus avert fragmentation according to Reaction 17. Reaction 18 is more difficult to reject under the strongly reducing conditions. An alternative explanation is based on the suggestion that the major reaction occurs during mixing.

3-Bromothiophene reacts with ketone enolate anions when treated with alkali metals in liquid ammonia giving good yields of substituted products. The 2-bromo and 2-chlorothiophenes are much less reactive and give mostly thiophene and unreacted material under similar reaction conditions (18). This behavior was attributed to differences in the decomposition rates of the radical anion intermediates 5 and 6, with 6 fragmenting faster than 5b.

5a, X = Cl   6
5b, X = Br

It is unlikely that 5b is more stable than 6 because, if there is any difference, 5b should eliminate bromide ion faster than 6, considering that the spin density in position two of the thiophene radical anion is higher than that in position three (19). It is also unlikely that 2-halothiophene radical anions 5 are more stable than the radical anion of 1-chloro or 1-iodonaphthalenes. The latter two compounds react with ketone enolate ions stimulated with solvated electrons giving

1-naphthylacetone (and derivatives) in good yields. Naphthalene is also formed in yields around 10%, although they vary somewhat from experiment to experiment (20).

The lower reactivity of 2-halothiophenes as compared with the 3-isomer can be attributed to the lower electrophilicity of 2-thienyl radicals as compared with 3-thienyl radicals. Therefore, Reaction 19 is slow, which makes propagation steps compete inefficiently with reduction by solvated electrons.

$$\text{(thiophene)} \cdot + \text{Nu}^- \longrightarrow \left[\text{(thiophene-Nu)} \right]^{\overline{\cdot}} \tag{19}$$

Similar behavior was found with 2-bromopyridine and acetone enolate ion, which gave mostly reduction products when stimulated by potassium metal (21).

Insight into the mechanism of $S_{RN}1$ reactions promoted by solvated electrons was obtained from studies of leaving-group effect on the yields of substitution versus reduction products in the reaction of acetone enolate ion with phenyl halides. Products are phenylacetone (7), the corresponding 1-phenyl-2-propanol (8), and the reduction product benzene (9) (Reaction 20) (22).

$$\text{PhX} + {}^-\text{CH}_2\text{COCH}_3 \xrightarrow{e^-} \underset{7}{\text{PhCH}_2\text{COCH}_3} + \underset{8}{\text{PhCH}_2\overset{\overset{\displaystyle \text{OH}}{|}}{\text{CHCH}_3}} + \underset{9}{\text{PhH}} \tag{20}$$

The ratio of **7:8** is strongly dependent on the halogen, and decreases in the order I〉Br〉Cl〉F. Minor changes in the aromatic moiety (i.e. *m*-anisyl), however, do not have much effect on the product distribution (Table I) (22).

When pairs of phenyl halides (for example iodobenzene and chlorobenzene) are treated with a deficiency of potassium metal, either as the solid or as a dilute solution in liquid ammonia, they show equal reactivity. This result indicates that the "macroscopic" reaction rate is controlled by the rate of mixing.

On the other hand, a mixture of iodobenzene and **10b** or chlorobenzene and **10a** gives, for each aryl group, a ketone:alcohol ratio characteristic of the halogen originally present (Reaction 21). Differences in product ratios are accentuated in experiments such as these. For instance, the ratio of **7:8** for chlorobenzene is 0.55 for a separated reaction (Entry 3, Table 1) and 0.47 for the reaction in presence of **10b** (Entry 9,

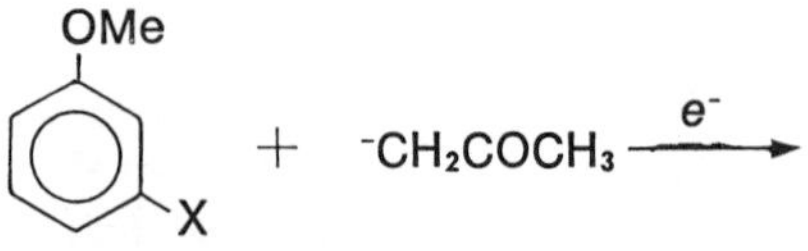

10a, X = I

10b, X = Cl

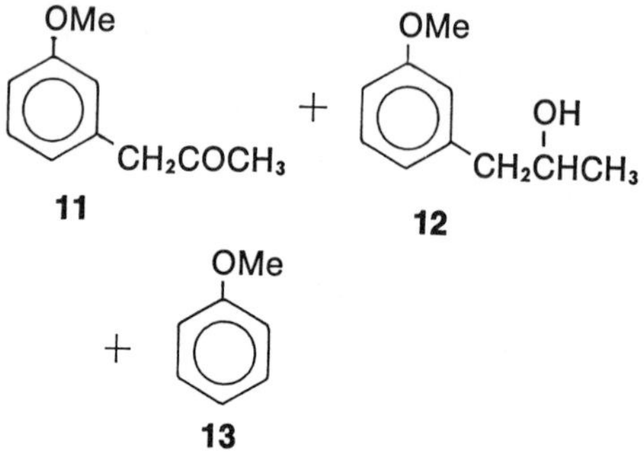

11   **12**

13

(21)

Table I. Ratio of Ketone : Alcohol Obtained in Reactions of Aryl Halides with Acetone Enolate Ion and Solvated Electrons in Ammonia

Entry	ArX	Ketone : Alcohol	Entry	(ArX + PhX)	Ketone : Alcohols
1	PhI	6.6	8 {	m-MeOC$_6$H$_4$Cl	0.64
2	PhBr	7.0		PhI	23.0
3	PhCl	0.55	9 {	m-MeOC$_6$H$_4$I	28
4	C$_6$D$_5$Cl	0.64		PhCl	0.47
5	PhF	0.28	10 {	C$_6$D$_5$Cl	0.38
6	m-MeOC$_6$H$_4$I	6.9		PhI	20
7	m-MeOC$_6$H$_4$Cl	0.51			

Source: Reproduced, from Ref. 22. Copyright, 1980, American Chemical Society.

Table I). The **11:12** ratios are 6.9 for *m*-iodoanisole in a separate reaction (Entry 6, Table I) and 28 (Entry 9, Table I) in the presence of chlorobenzene.

These reactions are interpreted in terms of the elaborated $S_{RN}1$ mechanism, sketched in Scheme I (22).

Step a is the initiation component of the radical chain reaction. Steps b–d are the propagation cycle, and Steps f and h effect termination. Steps e, g, and i are proton transfer reactions, and Step j involves electron capture of the substitution product.

Despite the observed leaving group effects on product composition, this mechanistic representation indicates that the leaving group is not present in the reacting species at the points where product selection occurs. These effects were postulated to arise from the microscopic effect occurring during the mixing process, with the rate of fragmentation of aryl radical anions (Step b) being identified as the major factor that determines product distribution.

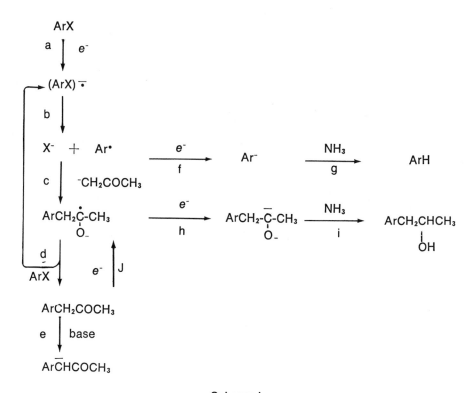

Scheme I

Generally, this mechanistic model suggests that reaction occurs during the mixing of a zone rich in solvated electrons with a zone free of solvated electrons, but that contains aryl halide as well as the nucleophilic enolate ion. When a portion of the solution containing the solvated electrons is swirled into the other zone, reaction with aryl halide molecules (Step a) occurs. What happens then depends on how fast these radical anions fragment in Step b. If they fragment rapidly, Steps b–e may be completed largely before massive mixing with further solvated electrons occurs (Figure I). However, if they fragment slowly, the system may only have progressed to the stage of ketyl radical anion or the aryl acetone by the time of massive mixing with solvated electrons. Once the electrons arrive in quantity, Steps h–j predominate and 1-aryl-2-propoxide ion is ultimately formed, together with the reduction product ArH.

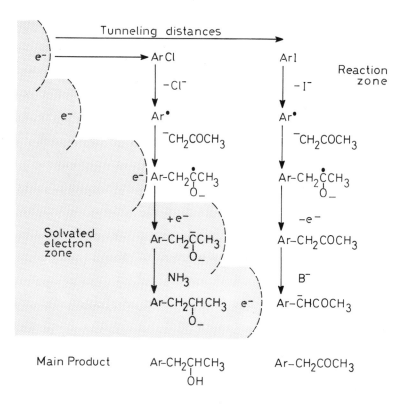

Figure 1. Schematic representation of the microscopic events that occur when aryl chloride and aryl iodide react with acetone enolate ions under solvated electron stimulation.

Fragmentation Step b is much faster for aryl iodide radical anions than for the corresponding aryl chloride radical anions. Therefore, the intermediates derived from aryl iodides have a better chance of completing Steps b–e before the main force of solvated electrons arrive.

It was suggested (22) that at least part of the electron advance is by tunneling. Because the tunneling distance is larger for aryl iodides (more positive reduction potential) than for aryl chlorides, the intermediate radical anions derived from them are spatially separated, with aryl chloride-derived species closer to the front of advancing solvated electrons. After most of the aryl iodide-derived intermediates have reacted, aryl chloride and intermediates derived from it combine sacrificially with an ensuing surge of solvated electrons so as to protect the aryl iodide derived intermediate from assault (22).

The mechanism suggested in Scheme I postulates that the relative reactivity of two competing nucleophiles should be insensitive to the identity of the halogen, despite the major dependence of other product ratios on this factor. Thus it was shown that in the reaction of the four 1-halo-2,4,6-trimethylbenzenes 14 as substrates in potassium metal stimulated reactions with a mixture of acetone enolate and amide ions (Reaction 22), the ratio of 1,3,5-trimethylbenzene (18) to combined substi-

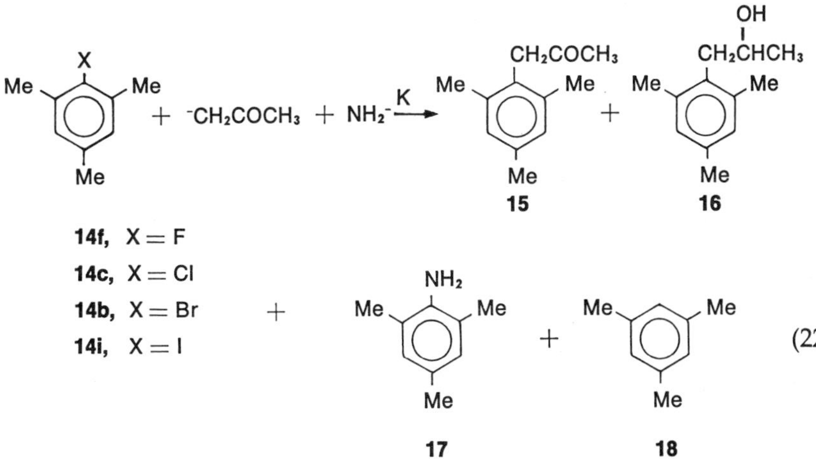

$$\text{14f, } X = F$$
$$\text{14c, } X = Cl$$
$$\text{14b, } X = Br$$
$$\text{14i, } X = I$$

tution products varies from 4:1 with 14f to as low as 1:4 with 14i. On the other hand the relative reactivity of the enolate ion with respect to the amide ion, as reckoned from the relative amounts of 15, 16, and 17 formed ($k_{enolate} / k_{amide}$), is the same (within experimental error) for all four 1-halo-2,4,6-trimethylbenzenes with value of 0.5 (23).

This mechanism offers an alternative explanation for the lack of reactivity of Ph_3M (M being P, As, or Sb) with ketone enolate ions in solvated electron stimulated reactions. Triphenylphosphine, arsine, and stibine have highly negative reduction potentials ($E_{1/2} = -3.5$, -3.4 and -3.5 V vs Ag/Ag$^+$ in glyme respectively (24), and triphenylphosphine $E_{1/2} = -2.9$ V in DMF or HMPT vs Ag/Ag$^+$ (25), and the tunneling distance for their reaction with the solvated electrons is probably very short. Besides, triphenylphosphine radical anions are probably more stable than halobenzene radical anions. For instance, cyclic voltammetry of triphenylphosphine gives a one-electron reversible wave (25), while that of chlorobenzene gives two-electron irreversible hydrogenolysis (26), indicating that the decomposition of chlorobenzene radical anion is faster on the time scale of cyclic voltammetry. Thus the decomposition of triphenylphosphine radical anions into aryl radicals occurs once the main force of electrons arrive, with subsequent reduction of aryl radicals to aryl anions to give benzene. Alternatively, triphenylphosphine radical anion takes another electron and fragments as in Reaction 18.

Photostimulated Reactions

Many examples show that when a substrate and a nucleophile do not react spontaneously, they can react if they are stimulated by near UV light. Until now, the nature of the initiation step and the actual mechanism of this photostimulated process were not clear. Furthermore, many photostimulated reactions may involve chain and nonchain processes that occur in a competitive manner.

Reactivity depends on the substrate, the solvent, and the nucleophile. Some conceivable mechanisms are: homolytic bond dissociation of the substrate, photoejection of electrons, photoassisted electron transfer, and electron transfer to excited states.

Probably no unique mechanism operates in every system. We will next analyze some of the factors that may favor one or several of these possibilities.

Homolytic Bond Dissociation of the Substrate. In homolytic bond dissociation of the substrate, the substrate must absorb a photon in a dissociative process to produce an aryl radical, that initiates the chain propagation steps.

$$ArX + h\nu \longrightarrow Ar^{\cdot} + X^{\cdot} \tag{23}$$

Considering the wavelength employed in these reactions (above 290 nm with pyrex-filtered light), it is unlikely that this process can operate in the reaction of halobenzenes, such as fluorobenzene, chlorobenzene, or bromobenzene.

The photochemistry of aryl halides has been extensively investigated. Aryl chlorides afford mainly products explicable in terms of intermediate aryl radicals, probably derived from the triplet excited state of the halide (27, 28). However it seems unlikely that simple homolytic C–Cl fission (Reaction 23) occurs, because there is usually a large energy gap between the triplet excitation energy (for chloronaphthalene for instance ~ 58 kcal/mol) (29) and the bond dissociation energy DC–Cl ~ 85 kcal/mol). The possibility of electron transfer has therefore been suggested (Reaction 24) (30–31).

$$\text{donor} + {}^3\text{ArCl} \longrightarrow D^{+\cdot} + Ar^\cdot + Cl^- \qquad (24)$$

Triplet energies are largely independent of the identity of the halogen so that the energy gap decreases in the order of bond dissociation energy, Cl > Br > I. In the case of aryl iodides, fragmentation may well occur by the simple fission corresponding to Reaction 23. However, the photochemistry of bromobiphenyls seems better explained by bond fission involving an electron transfer mechanism, as in Reaction 24 (32).

The mechanism of photodissociation was suggested to involve the promotion of a π electron to produce a singlet excited state that in principle could revert to an excited vibrational level of the ground state, and either dissociate from this state (S^0) or be transferred to a σ* molecular orbital.

Theoretical considerations indicate that if, during the internal conversion, most excitation energy is transformed into the high frequency C–H stretching modes, it would take longer than 10^{-12} s (the lifetime of the excited state) to redistribute the vibrational energy and to gather more than 57 kcal/mol in the C–I bond. The intersystem crossing to a σ* orbital mainly localized in the C–I bond is proposed as a better alternative (33). Moreover, in a model proposed for the photodissociation of aryl halides, the final state is a dissociative πσ* state localized on the C–X bond (34).

Although photochemical studies indicate that aryl iodides are photolyzed very slowly, the photodissociation according to Reaction 23 can still serve as an initiation step since $S_{RN}1$ reactions are chain reactions that, in favorable cases, may be very long.

Photoejection of Electrons. Photoejection from excited state anions is a well documented process. Not only can open shell anions photoeject electrons (Reaction 25), but closed shell anions do it as well (35, 36).

$$(\text{Arene-})^{\bar{\cdot}} + h\nu \longrightarrow \text{Arene} + e^- \qquad (25)$$

The absorption maximum of anions is shifted considerably to the red as compared with the parent neutral hydrocarbon, and certainly less energy is required to form the excited state (37).

Anionic photoionization is thought to proceed via charge-transfer states in which solvent orientation, and hence charge dispersal, is particularly important (*38*). This factor may be very important in liquid ammonia, where the solvated electron is a stable entity.

Thus, conceivably the nucleophile photoejects an electron that is then recaptured by the substrate, forming the first reactive intermediate to initiate the chain propagation of the $S_{RN}1$ mechanism (Reactions 26–27). The electron affinity of ketone enolate radicals in the gas phase is on the order of 35–40 kcal/mol (*39*), hence visible light has enough energy to provoke the photodetachment of an electron from ketone enolate anions.

$$Nu^- \xrightleftharpoons{h\nu} Nu^\cdot + e^- \tag{26}$$

$$ArX + e^- \xrightarrow{h\nu} (ArX)^{\bar{\cdot}} \tag{27}$$

Photoassisted Electron Transfer. As substrates are good electron acceptors, and nucleophiles good electron donors, they may form a charge-transfer complex. As was mentioned before, in the spontaneous reaction a charge-transfer complex could be an intermediate before the actual electron-transfer reaction, and the photostimulated reaction may be produced from the same intermediate (Reaction 28).

$$ArX + Nu^- \rightleftharpoons (ArX-Nu)^- \xrightarrow{h\nu} (ArX)^{\bar{\cdot}} + Nu^\cdot \tag{28}$$

The charge-transfer complex may be formed from the substrate and the nucleophile in their ground states, or from the excited state of one of the reactants (exciplex-like interactions).

The pairing of an anionic excited state donor with an electron acceptor allows the transfer of electrons without removing the electron to a free solvated state. Photocurrent can be observed, for example, when the cyclooctatetraene dianion is excited at the surface of an appropriate semiconductor (acceptor electrode) (*40*) despite the fact that normal photoejection occurs in solution only at shorter wavelengths.

Electron Transfer to Excited States. There are two possible situations for electron transfer to excited states. One is if the substrate absorbs at the wavelength of the radiation used, the substrate will be excited. This excited species then reacts with the nucleophile to give aryl radicals that initiate the reaction (Reactions 29–30).

$$ArX \xrightarrow{h\nu} (ArX)^* \tag{29}$$

$$(ArX)^* + Nu^- \longrightarrow Ar^\cdot + X^- + Nu^\cdot \tag{30}$$

The photoinduced dehalogenation of haloaromatic compounds is accelerated by addition of triethylamine. The mechanism of acceleration is suggested to be due to the amine reacting with the excited singlet state of the haloaromatic compound to give a radical anion that breaks down to give an aryl radical and halide anion (*41*).

The other possibility is that the nucleophile absorbs a photon to give the excited state, and then it transfers the electron to the substrate (Reactions 31–32). Oxidation and reduction potentials are enhanced in the excited state (*40*).

$$Nu^- \xrightarrow{h\nu} (Nu^-)^* \tag{31}$$

$$ArX + (Nu^-)^* \longrightarrow Ar^{\cdot} + X^- + Nu^{\cdot} \tag{32}$$

Experimental Observations. Most of the $S_{RN}1$ reactions studied have been carried out in liquid ammonia as solvent. Liquid ammonia is a very good solvent for this mechanism, but, due to its low boiling point, it is not a convenient solvent for quantitative studies of the reaction, such as the quantum yield determinations that are required in order to prove the nature of the initiation steps.

One quantitative study (*42*) has been reported on the photo-stimulated $S_{RN}1$ reaction of iodobenzene with potassium diethyl phosphite in DMSO. Results from this study indicate that, in all cases, the raw quantum yield exceeds unity, by as much as 20–50, and is independent of the substrate concentration, and the reaction rate depends on a power of light intensity slightly less than one, i.e. 0.84. The fact that the quantum yield exceeds unity indicates a chain mechanism with initiation by the action of photons, and propagation steps that do not require light.

Based on these results, two models were proposed for the photo-stimulated initiation step (Reactions 33–35 and 36–38).

In the first step of Model 1, iodobenzene absorbs a photon to give excited iodobenzene **19**, to which we can assign no definite structure (caged radical pair, π-PhI* (*42*). Step k_{-1} represents the reverse reaction to ground state iodobenzene. Iodobenzene absorbs at the wavelength of the irradiation (313 nm) employed.

Step 34 is the reaction of the excited species **19** with diethyl phosphite ion to give iodine atoms and the radical anion of the substitution product **20**, that can then enter into the propagation steps. Step 35 is an electron-transfer reaction to **19** from the nucleophile to give iodide, phenyl radical, and diethoxyphosphinyl radical **21**.

$$PhI \underset{k_{-1}}{\overset{h\nu}{\rightleftharpoons}} (PhI)^* \tag{33}$$
$$\mathbf{19}$$

$$19 + (EtO)_2PO^- \Big\langle \begin{array}{l} \nearrow (PhP(O)(OEt)_2)^{\cdot -} + I \qquad (34) \\ \mathbf{20} \\ \searrow Ph^{\cdot} + I^- + (EtO)_2PO^{\cdot} \qquad (35) \\ \mathbf{21} \end{array}$$

Model 1

In this model diethyl phosphite ion may act as an interceptor of the solvent-caged radical pair symbolized (PhI)* or a π-iodobenzene (*43*) (Reaction 34), which otherwise reverts to starting material (*44*).

The alternative route is the transfer of an electron to the excited iodobenzene from the nucleophile forming phenyl radicals, iodide ions, and the radical **21**. Phenyl radicals then enter into the propagation cycle.

Model 2 (Reactions 36–39) considers the possibility that the light is absorbed by a charge-transfer complex. Charge-transfer complex formation is indicated by spectroscopic studies of solutions of iodobenzene and potassium diethyl phosphite. The nucleophile is transparent in DMSO in the region of 300–270 nm. Iodobenzene absorbs weakly in this wavelength region (Figure 2). However, mixtures of potassium diethyl phosphite and iodobenzene absorb more strongly, showing that there is an interaction (charge-transfer complex) between both reactants (*42*).

$$PhI + (EtO)_2PO^- \rightleftharpoons (PhI-(EtO)_2PO)^- \qquad (36)$$
$$\mathbf{22}$$

$$\mathbf{22} \underset{k_{-37}}{\overset{h\nu}{\rightleftharpoons}} (Ph^{\cdot} \; I^- \; (EtO)_2PO^{\cdot}) \qquad (37)$$
$$\mathbf{23}$$

$$\mathbf{23} \longrightarrow PhP(O)(OEt)_2 + I^- \qquad (38)$$

$$\mathbf{23} + (EtO)_2PO^- \longrightarrow \mathbf{20} + \mathbf{21} + I^- \qquad (39)$$

Model 2

In Model 2 an equilibrium to form a charge-transfer complex **22** (Reaction 36), which absorbs a photon to give a solvent-caged-melange **23**, is formed by phenyl radical, iodide ion, and diethoxyphosphinyl radical, is suggested. Species **23** can collapse to give the substitution product and iodide ion (Reaction 38), but this is probably a very unimportant reaction since a quantum yield higher than unity was found. A more likely reaction is the attack of **23** by diethyl phosphite ion to form the radical anion **20**, which can enter into the propagation steps (Reaction 39).

The experimental results are in accordance with both models of the reaction, and the data probably are not precise enough to allow a choice of one of them as the most adequate.

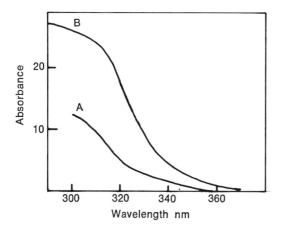

Figure 2. Spectra of 0.149 M iodobenzene in DMSO (A), and 0.149 M iodobenzene + 0.302 M potassium diethyl phosphite (B). Reproduced, from Ref. 42. Copyright 1977, American Chemical Society.

The formation of a free radical derived from the nucleophile was suggested in the photostimulated reaction of 1-iodonaphthalene with phenylmethanethiolate ion **24**. Because the radical anion intermediate fragments into unreactive species, the chain length is not very long. Therefore, products derived from initiation as well as termination steps could be isolated (45).

Products obtained from these reactions were the substitution product 1-naphthyl benzyl sulfide (2–3%), and 1-naphthalenethiolate ion **29** and dibenzyl sulfide **32,** both in about 15–20% yield. The formation of these products is explained by Reactions 40–43.

$$1\text{–}I\text{–}Np + PhCH_2S^{-} \xrightarrow{h\nu} (1\text{–}I\text{–}Np)^{-} + PhCH_2S^{\cdot} \qquad (40)$$
$$\phantom{1\text{–}I\text{–}Np + }\mathbf{24}\phantom{PhCH_2S^{-} \xrightarrow{h\nu} }\mathbf{25}\phantom{(1\text{–}I\text{–}Np)^{-} + }\mathbf{26}$$

$$\mathbf{25} \longrightarrow 1\text{–}N\overset{\cdot}{p} + I^{-} \qquad (41)$$
$$\phantom{\mathbf{25} \longrightarrow }\mathbf{27}$$

$$\mathbf{27} + \mathbf{24} \longrightarrow (1\text{–}Np\text{–}S\text{–}CH_2Ph)^{-} \qquad (42)$$
$$\phantom{\mathbf{27} + \mathbf{24} \longrightarrow }\mathbf{28}$$

$$\mathbf{28} \longrightarrow 1\text{–}Np\text{–}S^{-} + PhCH_2^{\cdot} \qquad (43)$$
$$\phantom{\mathbf{28} \longrightarrow }\mathbf{29}\phantom{1\text{–}Np\text{–}S^{-} + }\mathbf{30}$$

Np = naphthyl

Reaction 43 resembles the decomposition of the radical anion intermediate that occurs in the reaction of cyanomethyl anion with phenyl radicals (Reaction 44) (**46**).

$$Ph^{\cdot} + {}^{-}CH_2CN \longrightarrow (PhCH_2CN)^{\overline{\cdot}} \longrightarrow CN^{-} + PhCH_2^{\cdot} \qquad (44)$$
$$\mathbf{30}$$

In this reaction, 1,2-diphenylethane, presumably arising from the dimerization of benzyl radicals **30**, is formed in relatively high yield. However, in the reaction of 1-iodonaphthalene with **24**, the formation of 1,2-diphenylethane could not be detected.

The formation of **32** could be ascribed to the reaction of **30** with the nucleophile **24** (Reaction 45). This seems to be unlikely because the radical anion **31** is able to carry along the chain, and the overall reactivity should not be depressed, contrary to what is found.

$$\mathbf{30} + \mathbf{24} \longrightarrow (PhCH_2SCH_2Ph)^{\overline{\cdot}} \xrightarrow{1\text{-}I\text{-}Np} \mathbf{25} + PhCH_2SCH_2Ph \qquad (45)$$
$$\mathbf{31} \qquad\qquad\qquad\qquad \mathbf{32}$$

In addition, it was also found that **30** was unreactive toward cyanomethyl anion in liquid ammonia (**46**).

A more likely source of **32** could be the reaction of **30** with the nucleophile derived radical **26** formed in the Initiation Step 40 (Reaction 46).

$$\mathbf{30} + \mathbf{26} \longrightarrow \mathbf{32} \qquad (46)$$

The overall stoichiometry of the reaction is obtained by summing up Reactions 40–43 and 46, which gives Reaction 47.

$$1\text{-}I\text{-}Np + 2 \cdot \mathbf{24} \longrightarrow \mathbf{29} + \mathbf{32} + I^{-} \qquad (47)$$

The amount of **29** and **32** obtained is about the same and, therefore, is in agreement with the mechanism proposed.

Reactions Stimulated by Electrodes

The electrochemical reduction of haloaromatic compounds has been the subject of numerous investigations, and it has been determined that, in the first step, a radical anion is formed (Reaction 48). Different be-

$$ArX + e- \longrightarrow (ArX)^{\overline{\cdot}} \; (E_1^{\circ}) \qquad (48)$$

havior is observed, depending on the aromatic moiety, the halide and its position in the ring, and the nature of the other substituents.

The stability of the radical anion is strengthened by increasing delocalization of the odd electron over the ring and by the presence of electron withdrawing groups. When the radical anions have life times on the order of minutes or more, their ESR spectra can be recorded and the standard potential of the couple ArX /(ArX)⁻ can be determined by techniques such as conventional polarography and slow sweep cyclic voltammetry.

For instance, 2-fluorofluorenones, 2-chlorofluorenones, 2-bromofluorenones, (47, 48), 6-chloroquinoxaline, 1- and 2-chlorophenazine and 2-bromophenazine (49) and chloro and bromonitrobenzenes (50–52) form radical anions stable enough to record their ESR spectra. In other cases, the radical anion decomposes very fast on the time scale of conventional electrochemical techniques leading to the hydrogenolysis product ArH with the total consumption of two electrons per molecule (53, 54). Chlorobenzene and bromobenzene (26) and chlorobiphenyls behave this way (55).

The reduction of these four halobenzenes at the electrode is a stepwise process with the formation of halobenzene radical anion as an intermediate (56). These results are in agreement with pulse radiolysis studies regarding chlorobenzene and bromobenzene, but are contrary to those of iodobenzene (57). Pulse radiolysis studies indicate that the attachment of the electron to iodobenzene is a dissociative process without formation of a radical anion (57).

When radical anions are intermediates, the simplest reduction mechanism proposed involves Reactions 49–52 (58).

$$(ArX)^{\overline{\cdot}} \xrightarrow{k_f} Ar^{\cdot} + X^- \tag{49}$$

$$Ar^{\cdot} + e^- \longrightarrow Ar^- \tag{50}$$

$$(ArX)^{\overline{\cdot}} + Ar^{\cdot} \longrightarrow Ar^- + ArX \tag{51}$$

$$Ar^- + SH \longrightarrow ArH + S^- \tag{52}$$

The aryl radical formed in Reaction 49 takes another electron, either at the electrode (Reaction 50) or in solution (Reaction 51). Since the aryl anion is a strong base, it is protonated quickly by the solvent (usually DMSO, DMF, or acetonitrile), or by residual water in the solvent. Reaction 51 implies that aryl radical is easier to reduce than the starting material. Although data are not precise as to the reduction potential of aryl radicals they are known to have a more positive reduction potential than the aryl halides (59, 60).

Other reactions, such as Reaction 53 have been postulated in the reduction of several aryl halides in order to explain the experimental observations that the aryl halide reduction wave sometimes involves the exchange of less than two electrons per molecule (50–53, 58, 61).

$$Ar^{.} + SH \longrightarrow S^{.} + ArH \qquad (53)$$

The factors affecting the competition between hydrogen-atom abstraction from the solvent (Reaction 53) and electron-transfer reaction at the electrode (Reaction 50), or in the bulk of the solution (Reaction 51), have been analyzed in an electrochemical study of the reduction of aryl halides (62). The competition between these possibilities not only depends on the rates of Reactions 50 and 51 compared with 53, but also on the rate of Reaction 49 (k_f).

When k_f is large, the aryl radical is formed in a zone very close to the electrode and is reduced further at the electrode (Reaction 51) faster than it can encounter an aryl halide radical anion molecule. In this case Reaction 51 can be neglected, and competition between electron transfer and hydrogen-atom transfer depends only on the rate of Reaction 53 (thus, on the hydrogen-atom donor ability of the solvent).

When k_f is small, the aryl halide radical anion has time to diffuse to a zone relatively far from the electrode surface before it decomposes. Thus, the reduction of aryl radicals occurs mainly through Reaction 51. The competition between reduction by electron transfer and hydrogen atom transfer now depends on the relative rates of Reactions 51 and 53.

Because Reaction 51 is bimolecular, while Reaction 53 is pseudo-unimolecular (SH is the solvent), the competition between them is dependent not only on their relative rate constants, but also on the concentration of aryl halide radical anion.

The stationary concentration of aryl halide radical anions certainly depends on k_f. As k_f decreases, the radical anion concentration increases and the rate of Reaction 51 becomes more important compared to the rate of Reaction 53.

The knowledge of all these rate constants is very important regarding the S$_{RN}$1 mechanism and some valuable information was obtained by electrochemical methods (62). Thus it was found that the reduction of 1-halonaphthalenes in DMSO, and p-halobenzonitriles in acetonitrile, occurs mainly through Reaction 50 due to the high rate of decomposition of the derived radical anion. With 9-haloanthracenes, the behavior depends on the halogen; the chloro derivative is reduced mainly through Reaction 51, and the iodo derivative through Reaction 50. In all cases Reaction 53 is the competing reduction.

The rates of hydrogen atom abstraction are not very dependent on the aromatic moiety, whereas the rate of radical anion fragmentation depends strongly on the type of aromatic ring. Compare for instance the rates in Reactions 54–57 (62). where An = anthracenyl and Np = naphthyl.

$$(9\text{--I--An})^{\overline{\cdot}} \xrightarrow{\text{DMSO}} 9\text{--An}^{\cdot} + I^- \quad k_{An} = 7 \times 10^5 \text{ s}^{-1} \tag{54}$$

$$(1\text{--I--Np})^{\overline{\cdot}} \xrightarrow{\text{DMSO}} 1\text{--Np}^{\cdot} + I^- \quad k_{Np} = 6 \times 10^8 \text{ s}^{-1} \tag{55}$$

$$k_{Np}/k_{An} = 860$$

$$9\text{--An}^{\cdot} \xrightarrow[\text{SH}]{\text{(DMSO)}} AnH + S^{\cdot} \quad k_{An} = 8.5 \times 10^6 \text{ s}^{-1} \tag{56}$$

$$1\text{--Np}^{\cdot} \xrightarrow[\text{SH}]{\text{(DMSO)}} NpH + S^{\cdot} \quad k_{Np} = 1 \times 10^8 \text{ s}^{-1} \tag{57}$$

$$k_{Np}/k_{An} = 12$$

When there are nucleophiles in the solution that are able to react with the aryl radical formed as in Reaction 49, Reactions 58–62 may follow.

$$Ar^{\cdot} + Nu^- \longrightarrow (ArNu)^{\overline{\cdot}} \tag{58}$$

$$(ArNu)^{\overline{\cdot}} - e^- \rightleftharpoons ArNu \, (-E_2^{\circ}) \tag{59}$$

$$(ArNu)^{\overline{\cdot}} + ArX \rightleftharpoons ArNu + (ArX)^{\overline{\cdot}} \tag{60}$$

$$K_{60} = \exp F/RT \, (E_1^{\circ} - E_2^{\circ}) \tag{61}$$

$$(ArNu)^{\overline{\cdot}} + Ar^{\cdot} \longrightarrow ArNu + Ar^- \tag{62}$$

The coupling of an aryl radical with the nucleophile (Reaction 58) forms a radical anion whose parent compound may have a standard reduction potential (E_2°) more positive or more negative than the standard reduction potential of the aryl halide (E_1°, Reaction 48). Consequently, the equilibrium of Reaction 60 (with an equilibrium constant defined as in Equation 61) will lie on the right or left of Reaction 60.

When the substrate is easier to reduce than the substitution product $(E_1^{\circ} \rangle E_2^{\circ})$, Reaction 60 is thermodynamically favored and electrochemical injection of electrons catalyzes the nucleophilic reaction with a small consumption of electricity. However, if $E_1^{\circ} \langle E_2^{\circ}$, Reaction 60 is endergonic, but coupled with Reaction 49 it may, nevertheless, lead to the substitution product. The oxidation of the substitution product radical anion must, however, be carried out independently.

The efficiency of electron catalysis from the electrode depends on how well Reaction 58 can compete with electron-transfer Reactions 50,

51, and 62, or hydrogen-atom abstraction Reaction 53. Thus, in a preparative scale electrolysis, the amounts of ArH and substitution product ArNu obtained reflect the competition between Reactions 50, 51, 53, and 62, and 58, 59, and 60. The number of Faradays per mole tends toward zero as the yield of ArNu approaches 100%.

As discussed previously, the decomposition rate of aryl halide radical anions plays a major role in determining the importance of the reduction reaction, either at the electrode (Reaction 50) or in solution, in competition with other reactions.

The events that follow Reaction 48 depend on how far from the electrode the fragmentation of the aryl halide radical anion into aryl radical and halide ion occurs. In this regard, three zones can be distinguished (*see* Figure 3). Zone I, also known as the ECE zone (*63, 64*), is a narrow region close to the electrode where the aryl radical is immediately reduced to aryl anion and thus, has no time to react with the nucleophile. Reduction of aryl halides to arenes is the major reaction, and it takes place mainly through Reaction 50. Zone II follows spatially Zone I, and is a region where the main pathway for reduction of aryl radicals is by taking up electrons from the electrode (ECE mechanism),

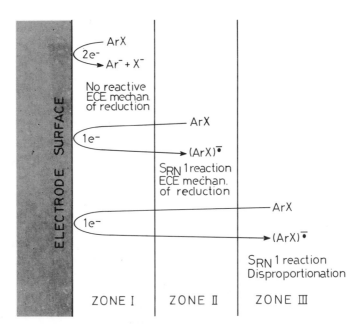

Figure 3. Hypothetical zones in electrochemically induced S$_{RN}$1 reactions determined by the stability of (ArX)$^{-}$.

but some competition with other chemical reactions, such as Reactions 51, 53, and 58 also takes place. Zone III, also known as the Disproportionation zone (65, 66) belongs to the bulk of the solution. Here the competition between nucleophilic reaction and reduction reaction depends mainly on the effectiveness of Reaction 58 compared to Reactions 51, 62, and certainly Reaction 53.

It can be concluded that, from the point of view of a $S_{RN}1$ reaction, those aryl halide radical anions that decompose very quickly, such as in Zone I, will not react under electrochemical stimulation. (There are, however, some possible ways to overcome this problem, as discussed later.) Halobenzenes are probably examples of this type of compound.

The decomposition of p-iodobenzonitrile is six times as fast as the decomposition of p-bromobenzonitrile (62) and, in the presence of benzenethiolate ion as nucleophile, the substitution to reduction ratio is 0.25 for the former and 8 for the latter (67).

The electrochemical reduction of 4-chlorobenzonitrile in the presence of tetraethylammonium cyanide forms benzonitrile as the major product (Reaction 63), whereas p-chloronitrobenzene under similar reaction conditions leads quantitatively to p-cyanonitrobenzene (Reaction 65).

These differences were attributed to differences in the rates of hydrogen-atom abstraction of the p-cyanophenyl and p-nitrophenyl radicals (59). However, these rates are not very dependent on the aromatic moiety, as mentioned previously, and a better explanation is probably the higher rate of decomposition of the p-chlorobenzonitrile radical anion, with formation of p-cyanophenyl radical in a zone close to the electrode, rendering the nucleophilic reaction inefficient.

The reaction of 2-iodoquinoline ($k_f = 3 \times 10^6$ s^{-1}) (68) and 2-chloroquinoline ($k_f = 1.7 \times 10^4$ s^{-1}) (68) with benzenethiolate ion in liquid ammonia are examples of substrates reacting in Zones II and III respectively.

Substrates that are very difficult to reduce or whose radical anion decomposes very close to the electrode (Zone I) with little possibility of giving further reaction besides reduction, might still be suitable for electrode-promoted radical nucleophilic aromatic substitution provided that a redox catalytic process can take place.

The method (69) consists of introducing the oxidized form of the redox catalyst P/Q into the solution. The standard reduction potential, $E°$, of P/Q is positive compared to the reduction potential of the aryl halide. The reduced form Q is then generated at the electrode and can transfer one electron to the aryl halide, giving the reaction sequence 66–69. Although Reaction 67 is thermodynamically unfavorable, the irreversibility of Reaction 68 drives the equilibrium to the right.

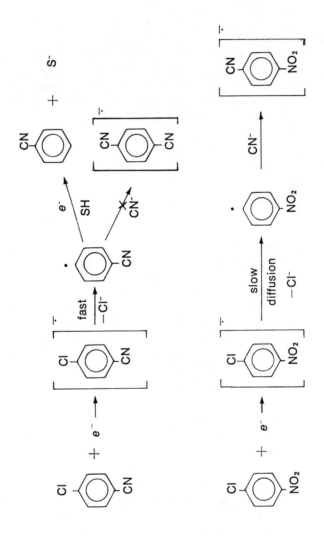

(63)

(64)

(65)

$$P + e^- \rightleftharpoons Q \tag{66}$$

$$Q + ArX \rightleftharpoons P + (ArX)^{\overline{\cdot}} \tag{67}$$

$$(ArX)^{\overline{\cdot}} \longrightarrow Ar^{\cdot} + X^- \tag{68}$$

$$Q + Ar^{\cdot} \longrightarrow P + Ar^- \tag{69}$$

This method was shown to allow the reduction of chlorobenzene at a potential more positive than required for its reduction. The catalysts used were naphthalene and phenanthrene (70).

The homogeneous electron transfer process can be considered as an electrolysis with an extremely small diffusion layer thickness, on the order of a few Angstrom units as opposed to electrolysis in the context of the usual electrochemical techniques where the diffusion layer thickness is at best one thousand times larger. This will favor reactions that aryl radicals can achieve out of the diffusion layer.

Theoretical treatment of cyclic voltammetry (56) and polarography data (71) for redox catalyzed processes allow for the determination of important parameters in regard to the $S_{RN}1$ mechanism, such as electron transfer rates and redox potentials of ArX.

This redox-catalyzed radical nucleophilic substitution was in fact observed in the reaction of diethyl phosphite ion with p-chlorobenzonitrile (reduction potential $E_1 = -1.66$ V) by scanning the reaction at a potential in the region of ArNu/(ArNu)$^{\overline{\cdot}}$ wave (reduction potential $E_2 = -1.29$ V) (72).

The results were interpreted by considering that the small amount of (ArX)$^{\overline{\cdot}}$ formed during the first cathodic scan decomposes into aryl radicals and halide ions. Then, the aryl radicals give rise to a small amount of (ArNu)$^{\overline{\cdot}}$ by reaction with the nucleophile, which triggers the autocatalytic production of ArNu through Reaction 67. Thus, ArNu builds up during the first scan. A reduction current is thus observed during the first part of the anodic scan. In the second part of the anodic scan the (ArNu)$^{\overline{\cdot}}$ thus formed is reoxidized, leading to a trace-crossing phenomenon (Figure 4). Further cyclic voltammometric scanning results in an increase of the ArNu/(ArNu)$^{\overline{\cdot}}$ couple wave until it corresponds to the total starting concentration.

The comparison of Figures 4a and 4b clearly shows the essential role of Reaction 67. Raising the starting concentration from 1.08×10^{-3} M(a) to 1.55×10^{-2} M(b) results in a dramatic increase in the efficiency of ArNu production, corresponding to an acceleration of Reaction 67 owing to its second-order character.

In addition, a preparative scale reaction with the electrode potential set at -1.35 V gave 100% of substitution product with a total of 0.01 electron/molecule consumption (72).

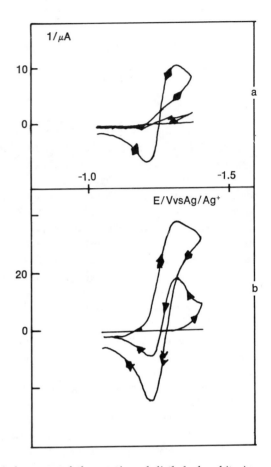

Figure 4. Electrochemical inducement of the reaction of diethyl phosphite ion (0.663 M) and 4-chlorobenzonitrile by scanning the potential in the region of the ArNu/(ArNu)⁻ wave; $v = 0.115$ V/s. Key: a, (4-chlorobenzonitrile)$_0$=1.08 × 10^{-3} M; b, (4-chlorobenzonitrile)$_0$=1.55 × 10^{-2} M. (Reproduced, with permission, from Ref. 72. Copyright 1980, Elsevier Sequoia.)

This method seems promising for promoting $S_{RN}1$ reactions with ArX substrates which otherwise give two electron reduction, although at present it has been used only in the reaction cited. It is uncommon in $S_{RN}1$ reactions that the reduction potential of ArX/(ArX)⁻ couple is negative to that of the ArNu/(ArNu)⁻ couple. When the reduction potential of the couples involved are in the reverse direction, an external redox couple P/Q, whose reduction potential is more positive than that of the ArX/(ArX)⁻ couple, could be added in order to use it as a catalyst.

The efficiency of these redox-catalyzed processes will depend on how Reaction 67 can compete with other electron-transfer processes, and this in turn will depend on the equilibrium constant for the reaction. Because Reaction 67 thermodynamically favors the left, k_{-67} approaches a diffusion controlled rate, and $k_{67} = K_{67} k_{diff}$ will be lower the lower K_{67} is. Thus the possibility of using redox catalysis depends on a delicate balance of the rate constants involved, but it looks to be a very interesting means of catalyzing radical nucleophilic aromatic substitution with a very small consumption of energy (73).

Literature Cited

1. Kim, J. K.; Bunnett, J. F. J. Am. Chem. Soc. **1970,** *92,* 7463, 7464.
2. Swarts, J. E.; Bunnett, J. F. J. Org. Chem. **1979,** *44,* 340.
3. Rossi, R. A.; Alonso, R. A.; Palacios, S. M. J. Org. Chem. **1981,** *46,* 2498.
4. Alonso, R. A.; Rossi, R. A. J. Org. Chem. **1982,** *47,* 77.
5. Scamehorn, R. G.; Bunnett, J. F. J. Org. Chem. **1977,** *42,* 1449.
6. Beringer, F. M.; Galton, S. A.; Huang, S. J. J. Am. Chem. Soc. **1962,** *84,* 2819.
7. Beringer, F. M.; Galton, S.A. J. Org. Chem. **1963,** *28,* 3417.
8. Beringer, F. M.; Daniel, W. J.; Galton, S. A.; Rubin, G. J. Org. Chem. **1966,** *31,* 4315.
9. Rossi, R. A.; Bunnett, J. F. J. Am. Chem. Soc. **1974,** *96,* 112.
10. Waters, W. A. J. Chem. Soc. **1942,** 266.
11. Nonhebel, D. C.; Waters, W. A. Proc. R. Soc. London Ser. A, **1957,** *242,* 16.
12. Kumar, R.; Sing, P. R. Tetrahedron Lett. **1972,** 613.
13. Sing, P. R.; Kumar, R. Aust. J. Chem. **1972,** *25,* 2133.
14. Opgenorth, H. J.; Ruchardt, C. Justus Liebigs Ann. Chem. **1974,** 1333.
15. Cuthrell, R. E.; Lagowski, J. J. J. Phys. Chem. **1967,** *71,* 1298.
16. Briegleb, G. Angew. Chem. **1964,** *76,* 326.
17. Russell, G. A.; Jansen, E. G.; Strom, E. T. J. Am. Chem. Soc. **1964,** *86,* 1807.
18. Bunnett, J. F.; Gloor, B. F. Heterocycles, **1976,** *5,* 377.
19. Rossi, R. A. unpublished data.
20. Rossi, R. A.; de Rossi, R. H.; Lopez, A. F. J. Am. Chem. Soc. **1976,** *98,* 1252.
21. Komin, A. P.; Wolfe, J. F. J. Org. Chem. **1977,** *42,* 2481.
22. Bard, R. R.; Bunnett, J. F.; Creary, X.; Tremelling, M. J. J. Am. Chem. Soc. **1980,** *102,* 2852.
23. Tremelling, M. J.; Bunnett, J. F. J. Am. Chem. Soc. **1980,** *102,* 7375.
24. Dessy, R. E.; Chivers, T.; Kitching, W. J. Am. Chem. Soc. **1966,** *88,* 467.
25. Saveant, J. M.; Binh, S. K. J. Electroanal. Chem. **1978,** *88,* 27.
26. Farwell, S. O.; Beland, F. A.; Geer, R. D. J. Electroanal. Chem. **1975,** *61,* 303.
27. Majer, R. J.; Simmons, J. P. Adv. Photochem. **1964,** *2,* 137.
28. Pinhey, J. T.; Rigby, R. D. G. Tetrahedron Lett. **1969,** 1267.
29. Thompson, L. G.; Webber, S. E. J. Phys. Chem. **1972,** *76,* 221.
30. Ruzo, L. O.; Bunce, N. J.; Safe, S. Can. J. Chem. **1975,** *53,* 688.
31. Bunce, N. J.; Pilon, P.; Ruzo, L. O.; Sturch, D. J. J. Org. Chem. **1976,** *41,* 3023.
32. Bunce, N. J.; Safe, S.; Ruzo, L. O. J. Chem. Soc. Perkin Trans. I **1975,** 1607.
33. Dzvonik, M.; Yang, S.; Bersohn, R. J. Chem. Phys. **1974,** *61,* 4408.
34. Marconi, G. J. Photochem. **1979,** *11,* 385.
35. Saeva, F. D.; Olin, G. R. J. Am. Chem. Soc. **1975,** *97,* 9631.
36. Fisher, M.; Rämme, G.; Claesson, S.; Szwarc, M. Proc. R. Soc. London Ser. A **1972,** *327,* 481.
37. Fox, M. A. Chem. Rev. **1979,** *79,* 253.
38. Cook, J. A. H.; Logan, S. R. J. Photochem. **1974,** *3,* 89.
39. Zimmerman, A. H.; Reed, K. J.; Brauman, J. I. J. Am. Chem. Soc. **1977,** *99,* 7203.
40. Fox, M. A.; Kabir-ud-Din J. Phys. Chem. **1979,** *83,* 1800.
41. Davidson, R. S.; Goodin, J. W. Tetrahedron Lett. **1981,** *22,* 163.

42. Hoz, S.; Bunnett, J. F. *J. Am. Chem. Soc.* **1977,** *99,* 4690.
43. Fox, M. A.; Nichols, Jr., W. C.; Lemal, D. M. *J. Am. Chem. Soc.* **1973,** *95,* 8164.
44. Levy, A.; Meyerstein, D.; Ottolenghi, M. *J. Phys. Chem.* **1971,** *75,* 3350.
45. Rossi, R. A.; Palacios, S. M. *J. Org. Chem.* **1981,** *46,* 5300.
46. Bunnett, J. F.; Gloor, B. F. *J. Org. Chem.* **1973,** *38,* 4156.
47. Nadjo, L.; Saveant, J. M. *J. Electroanal. Chem.* **1971,** *30,* 41.
48. Grimshaw, J.; Trocha-Grimshaw, J. *J. Electroanal. Chem.* **1974,** *56,* 443.
49. Alwair, K.; Grimshaw, J. *J. Chem. Soc., Perkin Trans 2* **1973,** 1811.
50. Lawless, J. G.; Hawley, M.D. *J. Electroanal. Chem.* **1969,** *21,* 365.
51. van Duyne, R. P.; Reilley, C. N. *Anal Chem.* **1972,** *44,* 158.
52. Nelson, R. F.; Carpenter, A. K.; Seo, E. T. *J. Electrochem. Soc.* **1973,** *120,* 206.
53. Houser, K. J.; Bartak, D. E.; Hawley, M. D. *J. Am. Chem. Soc.* **1973,** *95,* 6033.
54. Renaud, R. N. *Can. J. Chem.* **1974,** *52,* 376.
55. Farwell, S. O.; Beland, F. A.; Geer, R. D. *J. Electroanal. Chem.* **1975,** *61,* 315.
56. Andrieux, C. P.; Blocman, C.; Dumas-Bouchiat, J. M.; Saveant, J. M. *J. Am. Chem. Soc.* **1979,** *101,* 3431.
57. Steelhammer, J. C.; Wentworth, W. E. *J. Chem. Phys.* **1969,** *51,* 1802.
58. M'Halla, F.; Pinson, J.; Saveant, J. M. *J. Electroanal. Chem.,* **1978,** *89,* 347.
59. Bartak, D. E.; Houser, K. J.; Rudy, B. C.; Hawley, M. D. *J. Am. Chem. Soc.* **1972,** *94,* 7526.
60. Jaun, B.; Schwarz, J.; Breslow, R. *J. Am. Chem. Soc.* **1980,** *102,* 5741.
61. Gores, G. J.; Koeppe, C. E.; Bartak, D. E. *J. Org. Chem.* **1979,** *44,* 380.
62. M'Halla, F.; Pinson, J.; Saveant, J. M. *J. Am. Chem. Soc.* **1980,** *102,* 4120.
63. Testa, A. C.; Reinmuth, W. H. *J. Am. Chem. Soc.* **1961,** *83,* 781.
64. Testa, A. C.; Reinmuth, W. H. *Anal. Chem.* **1961,** *33,* 1320.
65. Mastragostino, M.; Nadjo, L.; Saveant, J. M. *Electroanal. Acta,* **1968,** *13,* 721.
66. Feldberg, S. H. *J. Phys. Chem.* **1969,** *73,* 1238.
67. Pinson, J.; Saveant, J. M. *J. Am. Chem. Soc.* **1978,** *100,* 1506.
68. Amatore, C.; Chaussard, J.; Pinson, J.; Saveant, J. M.; Thiebault, A. *J. Am. Chem. Soc.* **1979,** *101,* 6012.
69. Lund, H.; Simmonet, J. *J. Electroanal. Chem.* **1975,** *65,* 205.
70. Sease, J. W.; Reed, R. C. *Tetrahedron Lett.* **1975,** 393.
71. Andrieux, C. P.; Dumas-Bouchiat, J. M.; Saveant, J. M. *J. Electroanal. Chem.* **1978,** *87,* 39.
72. Amatore, C.; Pinson, J.; Saveant, J. M. Thiebault, A. *J. Electroanal. Chem.* **1980,** *107,* 59.
73. Saveant, J. M. *Acc. Chem. Res.* **1980,** *13,* 323.

Chain Propagation Steps

Once the initiation step occurs, a reactive intermediate is formed that enters into the propagation steps. The main propagation steps are:

a. decomposition of the radical anion into an aryl radical and nucleofugal group (Reaction 1).
b. coupling of the aryl radical and the nucleophile (Reaction 2).
c. electron transfer reaction (Reaction 3).

Although this sequence of reactions is followed in the majority of the $S_{RN}1$ reactions, the actual sequence depends on the initiation step. For example, if the initiation step produces an aryl radical rather than a radical anion, the sequence would be:

$$b \longrightarrow c \longrightarrow a$$

On the other hand, if the initiation step produces the radical anion of the substitution product, the sequence would be:

$$c \longrightarrow a \longrightarrow b$$

The effectiveness of a chain reaction depends on the competition between the chain propagation steps and the termination steps. Also, the dominant termination steps will depend on the rates of the propagation steps. Let us look at the propagation cycle (Reactions 1–3).

$$(ArX)^{\bar{\cdot}} \xrightarrow{k_1} Ar^{\cdot} + X^- \tag{1}$$

$$Ar^{\cdot} + Nu^- \xrightarrow{k_2} (ArNu)^{\bar{\cdot}} \tag{2}$$

$$(ArNu)^{\bar{\cdot}} + ArX \xrightarrow{k_3} ArNu + (ArX)^{\bar{\cdot}} \tag{3}$$

If $(ArX)^{\bar{\cdot}}$ is a radical anion that decomposes very slowly, the overall propagation rate will depend on Reaction 1, and competing termination

0065-7719/83/0178-0187$15.00/1
© 1983 American Chemical Society

steps involving $(ArX)^{\overline{\cdot}}$ are likely to be important. However if the coupling Reaction 2 is slow compared with the other two, coupling will be the rate determining step and termination steps involving the consumption of aryl radicals to form nonreactive intermediates (from the point of view of the $S_{RN}1$ reaction) will be important. When electron transfer Step 3 is inefficient, it becomes the rate determining step of the entire propagation cycle, and termination reactions involving $(ArNu)^{\overline{\cdot}}$ are likely to be important.

Decomposition of a Radical Anion

The most common initiation step is the formation of a radical anion of the substrate with an appropriate leaving group. Entering into the $S_{RN}1$ propagation steps, the next reaction is decomposition of the radical anion to a free aryl radical and the nucleofugal group (Reaction 4).

$$(ArX)^{\overline{\cdot}} \longrightarrow Ar^{\cdot} + X^{-} \tag{4}$$

In order to eliminate alternative pathways, such as the reaction of the nucleophile with the radical anion in a bimolecular step, the reactions of several monosubstituted benzenes competing for *two* nucleophiles (e.g. diethyl phosphite and pinacolone enolate ions) have been studied (1). If the mechanism involves a free phenyl radical (Reaction 4), the relative reactivity should not depend on the leaving group, whereas, if there is a bimolecular step, such as in Reaction 5, the relative reactivity should depend on the nature of the leaving group.

$$(ArX)^{\overline{\cdot}} + Nu^{-} \longrightarrow (ArNu)^{\overline{\cdot}} + X^{-} \tag{5}$$

Results have been obtained showing that, with both nucleophiles present, the relative reactivity of the four halobenzenes towards diphenyl sulfide and phenyltrimethylammonium ion is the same. That is, the product distribution is independent of the substrate used, despite large steric differences between nucleofugal groups, especially fluoride and trimethylammonium groups (1).

Evidence suggests that Reaction 4 is followed when X represents halogens, trimethylammonium ion, and others. However, not every aromatic substrate having these substituents will decompose as in Reaction 4 at a rate suitable to maintain the chain propagation steps. For instance, the radical anions of m- and p-nitrohalobenzenes decompose very slowly or not at all; thus, they do not react under the usual $S_{RN}1$ conditions to give substitution products. Products derived from other types of reactions are obtained instead.

The decomposition rate of the radical anion depends not only on the substituent on the aromatic ring, but also on its position. For instance, it has been shown that the electrochemical reduction of chloro, bromo, and iodoacetophenones in DMF is a one-electron reduction that forms an unstable radical anion which decomposes to an aryl radical and halide ion. The decomposition rates of o- and p-chloroacetophenones are greater than 10^4 s^{-1}, whereas that of the m-isomer is only 5 s^{-1} (2).

The rates of C–X bond fragmentation of haloaromatic compounds have been found to correlate with the free electron density, as calculated by HMO methods (Table I).

Molecular orbital studies by the CNDO/2 method of chlorinated benzenes indicate that the LUMO is a σ^* MO. Electron density in the radical anions is therefore localized in the C–Cl bond that undergoes cleavage to the reaction products. The product ratio correlates with the relative electron density of the different C–Cl bonds when several products are formed (3).

Moreover, it has been shown that substrates such as 1 and 2 can be selectively monodehalogenated if the spin density at the carbon bonded to the halogen is different. For instance, 1a and 2a give the correspond-

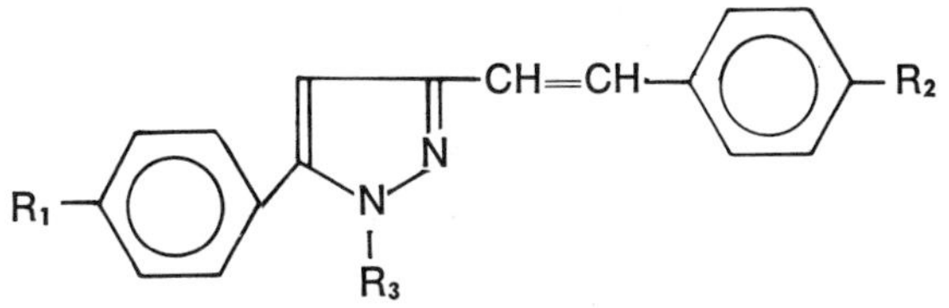

1a: $R_1 = R_2 = $ Cl or Br; $R_3 = $ H
1b: $R_1 = $ Cl or Br; $R_2 = $ H; $R_3 = $ H
1c: $R_1 = R_2 = $ Cl or Br; $R_3 = $ Ph

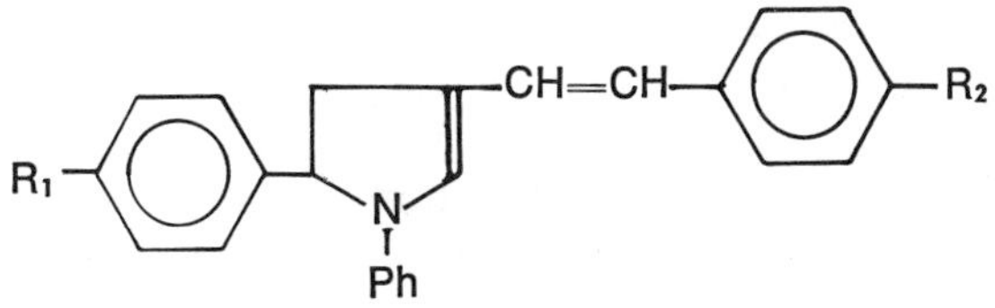

2a: $R_1 = R_2 = $ Cl or Br
2b: $R_1 = $ Cl or Br; $R_2 = $ H

Table I

Rates of Fragmentation of Radical Anions as a Function of Free Electron Density

Radical Anion Derived from	Position of X	HMO Free Electron Density	k (s^{-1}) Br	k (s^{-1}) Cl	Ref.
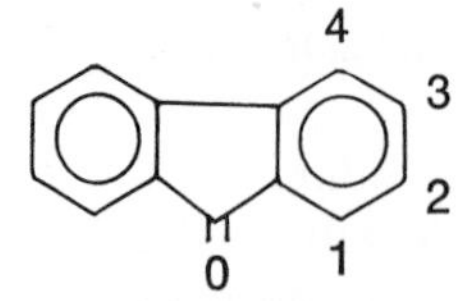	1	0.196	0.15	110	
	2	0.007	0.07	110	
	3	0.209	0.24	110	
	4	0.026	0.07	110	
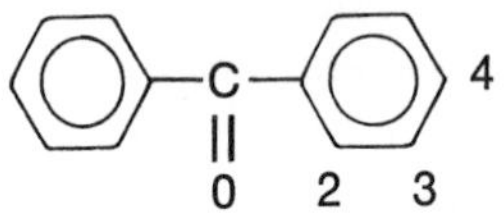	2	0.076			
	3	0.006	7.4×10^2	very slow	111, 42
	4	0.097	800×10^2	10	111, 42
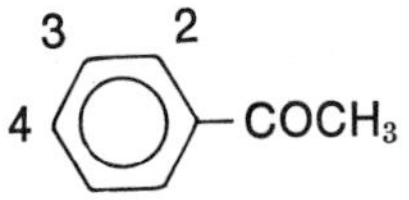	2	0.070		$> 10^4$	2
	3	0.019		5	2
	4	0.193		$> 10^4$	2

ing derivatives, **1b** and **2b,** by electrochemical reduction in DMF at the potential of the first polarographic wave. By HMO calculations it has been shown that the spin density of the p-carbon in the styryl group (which is dehalogenated) is substantially different from that of the p-position of the other phenyl group (which is not dehalogenated) (4).

In contrast, **1c** cannot be selectively dehalogenated. It was suggested that the spin densities of the carbons bonded to the halogens have almost the same value (4).

Another factor that seems to be important regarding the rate of Reaction 4 is the steric effect. This effect has been demonstrated for halonitrobenzenes.

The four o-haloderivatives show a higher decomposition rate than the p-isomers, and the ratio of k_o to k_p decreases in the order $I > Br > Cl > F$ (5) and is also the decreasing order for the van der Waals radii. Several methyl-substituted halonitrobenzenes, with the methyl group ortho to either the halogen or the nitro group, also give enhanced rates of decomposition (6). This important steric effect may be attributed to an increase in the energy of the LUMO π^* due to lack of planarity of the phenyl ring with the nitro group when the latter is flanked by one or two $ortho$ substituents. An increase in the LUMO π^* energy decreases the gap between this orbital and the LUMO σ^* of the C–X bond, thus making the intramolecular electron transfer easier.

Solvent and temperature effects on the stability of halobenzophenones have been investigated by electrochemical techniques (7, 8). Representative data are reported in Table II, from which it can be seen that the stability of the radical anions is dependent on the solvent, and decreases in the order acetonitrile ≫ DMF ≫ ammonia. Similar solvent dependence was reported for halonitrobenzenes for which the stability decreases in the order DMSO > acetonitrile > DMF (9). The Arrhenius parameters indicate that the difference in stability due to the solvent is caused by changes in activation enthalpy, which is somewhat compensated for by the difference in the pre-exponential factor.

The effect of solvents on the stability of radical anions has been attributed to solvation of the radical anion (7). The better the ion is solvated, the more delocalized the negative charge; thus, less charge

Table II
Stability of Halobenzophenone Radical Anions as a Function of Solvent and Temperature

Halobenzophenone	Solvent	T, °C	k^a	E_a^b units
m-F	all[c]	—	stable	
m-Cl	all[c]	—	stable	
p-Cl	DMF	20	10 ± 3	
	DMF	−10	stable	
	CH_3CN	20	stable	
	NH_3	−40	stable	
m-Br	DMF	20	$(7.4 \pm 2) \times 10^2$	20.8
	DMF	−10	13 ± 3	
	CH_3CN	20	9 ± 3	
	NH_3	−40	stable	
p-Br	DMF	20	$(8 \pm 2) \times 10^4$	20.8
	DMF	−10	$(1.4 \pm 0.4) \times 10^3$	
	CH_3CN	20	$(2.4 \pm 0.7) \times 10^3$	
	NH_3	−60	40 ± 10	13.4
	NH_3	−50	160 ± 10	
	NH_3	−40	590 ± 35	
	NH_3	20^d	2×10^5	

[a] Rate constant in s^{-1} from Ref. 7. Stability in the time range of cyclic voltammetry implies a life time > 5 s.
[b] Arrhenius activation energy in real mol.
[c] DMF, 20°C; acetonitrile, 20°C; and ammonia, −40°C.
[d] Extrapolated from an Arrhenius plot.

resides at the carbon bearing the halogen, and stability is favored. Therefore, acetonitrile appears to have better solvating ability than DMF.

If the transition state theory is applied to the decomposition of the
radical anion (Reaction 6), we can see that the charge initially delocalized
in the molecule becomes localized in the C–X bond, and especially in the
leaving group of the activated complex.

$$(ArX)^{\cdot-} \Longrightarrow \left[\overset{\delta-}{Ar} \ldots \overset{\delta-}{X} \right]^{\ddagger} \longrightarrow Ar^{\cdot} + X^{-} \qquad (6)$$

Dipolar aprotic solvents are known to stabilize large delocalized
anions, thus giving better solvated radical anions than the transition
state (10). In contrast, protic solvents like ammonia stabilize the transition state better than the radical anion provided that there is considerable
bond rupture in the transition state. In fact, the solvent effect on the
decomposition of these radical anions indicates that bond rupture in the
transition state is quite advanced. The activation parameters are in agreement with this interpretation. The lower activation energy in ammonia
can be attributed to better stabilization of the incipient anion in the
transition state, which in turn causes a decrease in the pre-exponential
factor (which reflects a more negative entropy of activation).

Coupling of an Aryl Radical with a Nucleophile

Once an aryl radical is formed, it should react with a nucleophile to
form a new radical anion intermediate if it is to produce substitution
products.

Figure 1 depicts three possible electronic ground states of an aryl
radical, depending on the relative energies of the three molecular orbitals
(MOs) involved.

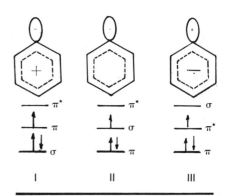

Figure 1. Relative MO energies of the phenyl radical. (Reproduced, from Ref. 14. Copyright, 1970, American Chemical Society.)

It has been shown by several methods, such as the electronic spectrum (11) and ESR spectroscopy in matrices at 77 K and 4 K (12, 13) that phenyl radicals are σ radicals (structure II, Figure 1). An ESR study was conducted to determine the electronic nature of several large aryl radicals, and indicated that 1-naphthyl, 2-naphthyl, 1- and 9- anthracyl, and 1-pyrenyl radicals are also σ radicals (14). Molecular orbital calculations by the INDO method (15) indicate that phenyl, as well as 1- and 2-naphthyl radicals, are σ radicals. However, in the case of the naphthyl radicals, the energy of the σ orbital bearing the odd electron is lower than that of the π MO bearing two electrons. It was suggested that the σ orbital remains singly occupied, even though it has lower energy, because of the greater repulsion that would be experienced if two electrons were to occupy this highly localized MO (14). Moreover, ESR spectra of a number of phenyl and 2-pyridyl radicals show very low g values (near or below the free electron value) reflecting their σ nature (16).

Because of the σ character of all aryl radicals, MO calculations predict that the electron affinity of aryl radicals does not depend on the type of substituents on the ring. For instance, the electron affinity of 2,5-dichlorophenyl, benzoquinoly, and naphthyl radicals is similar to that of phenyl radicals (45 kcal/mol) (17). It follows that the electrophilicity of aryl radicals should not be greatly affected by the substituents on the ring as long as the transition state for the coupling reaction resembles the starting materials more closely than the radical anion intermediate. If, on the other hand, it resembles the radical anion structure, the substituent that stabilizes the odd electron will favor the coupling.

Questions arise as to what types of nucleophiles are able to react with aryl radicals. Because aryl radicals are soft electrophiles, they should react with soft nucleophiles, and indeed that is observed in most cases. An exception is the amide ion which, although being considered a hard nucleophile, reacts with aryl radicals.

Soft nucleophiles, such as nitrite or cyanide ions, do not react in photostimulated or solvated-electron-stimulated reactions. But in this case, the lack of reactivity is not due to the inability of aryl radicals to couple with these nucleophiles, but rather to the slowness of the reaction that follows, namely, the electron transfer reaction to the substrate (Reaction 3). The radical anion formed in Reaction 2 is not able to continue the chain propagation steps because it has a higher electron affinity than the substrate, making Reaction 3 thermodynamically disfavored. In electrochemically induced reactions, both nucleophiles were shown to be very efficient scavengers of aryl radicals.

With bidentate nucleophiles, reaction always occurs with the softer part of the nucleophile. Examples include:

$$Ar\text{-}O\text{-}C\overset{\displaystyle CH_2}{\underset{\displaystyle R}{\Bigg<}} \longleftarrow\!\times\!\longleftarrow {}^-CH_2COR \longrightarrow ArCH_2COR$$

$$Ar\text{-}O\text{-}P(OEt)_2 \longleftarrow\!\times\!\longleftarrow (EtO)_2PO^- \longrightarrow ArP(O)(OEt)_2$$

$$ArNC \longleftarrow\!\times\!\longleftarrow CN^- \longrightarrow ArCN$$

$$ArONO \longleftarrow\!\times\!\longleftarrow NO_2^- \longrightarrow ArNO_2$$

The rates of coupling of 2-quinolyl and 1-naphthyl radicals with several nucleophiles in liquid ammonia have been calculated using electrochemical techniques (*see* Table III) (*18*).

The rate of reaction of 1-naphthyl radical and benzenethiolate ion was also determined in DMSO by an independent method, and is reported to be $1.7 \times 10^8 \ M^{-1} \ s^{-1}$ (*19*). It is remarkable that a 60°C increase in temperature and a change from a protic solvent, such as ammonia, to a dipolar aprotic solvent, such as DMSO, changes the rate by a factor of only 7.

Electron Transfer Reactions

Since the coupling of the aryl radical and the nucleophile produces a radical anion, the radical anion must lose the odd electron in order to form the substitution product. We will consider two possibilities: electron transfer to the substrate and electron ejection to the solvent.

Electron Transfer to the Substrate. The mechanisms of electron transfer reactions have been studied extensively and their dependence on different factors, such as reduction potential of the donor–acceptor pair, solvent, etc., have been treated by many authors. Several theories, which allow the prediction of the rate of an electron transfer reaction, have been developed. Among these, Marcus' theory seems to give best concordance between experimental and theoretical values. An extensive discussion of these theories is beyond the coverage of this monograph, and we refer the readers to more specialized literature (*20–22*).

The mechanism of an electron transfer reaction may be written as in Reaction 7.

$$D^{\bar{\cdot}} + A \underset{k_{-d}}{\overset{k_d}{\rightleftharpoons}} (D^{\bar{\cdot}}A) \underset{k_{-act}}{\overset{k_{act}}{\rightleftharpoons}} (DA^{\bar{\cdot}}) \underset{k_d}{\overset{k_{-d}}{\rightleftharpoons}} D + A^{\bar{\cdot}} \tag{7}$$

Here k_d represents the rate of diffusion of reactants (or products) to form the encounter pair, k_{-d} is the diffusional separation rate of the pair, and k_{act} is the rate of the electron transfer reaction where the actual chemical event occurs.

Table III

Rate Constants of 2-Quinolyl Radical with Nucleophiles in Liquid Ammonia

Nucleophile	10^{-6} k, M^{-1} s^{-1}
PhS^-	$14(23)^a$
$p-ClC_6H_4S^-$	6
$(EtO)_2PO^-$	18
$CH_3COCH_2^-$	7.5
$PhCOCH_2^-$	45

[a]Values in parentheses correspond to the reaction with 1-naphthyl radical.
[b]Source: Reproduced, from Ref. 18. Copyright 1979, American Chemical Society.

The observed rate constant for an electron transfer reaction is given in Equation 8.

$$k_{obs} = \frac{k_d k_{act}}{k_{-d} + k_{act}} \tag{8}$$

Calculations of k_d are possible from Equation 9, where in D is the diffusion coefficient of the species participating, a is the collision diameter (distance of closest approach which leads to reaction), and N is Avogadro's number (23).

$$k_d = \frac{4\pi a N D}{1000} \tag{9}$$

According to Marcus theory the rate for the actual electron transfer step, k_{act}, can be calculated by Equation 10, where Z represents the rate

$$k_{act} = Z e^{-\Delta G^*/RT} \tag{10}$$

of electron transfer for a reaction having no free energy of activation, and it is $\sim 10^{11}$ M^{-1} s^{-1}; and ΔG^* is the free energy of activation which, according to the theory, is given by Equation 11. Wr and Wp are coulom-

$$\Delta G^* = \frac{Wr + Wp}{2} + \frac{\lambda}{4} + \frac{\Delta G^\circ}{2} + \frac{(\Delta G^\circ + Wp - Wr)^2}{4\lambda} \tag{11}$$

bic work terms for bringing reactants or products together, respectively, and are considered to be very small for the type of reaction that we are considering; λ is a solvent reorganization parameter which depends on the radius of the activated complex and reactants, the dielectric properties of the solvent, and the number of electrons transferred. It also includes the energy required to equalize the energies of the orbitals participating in the reaction (24). ΔG° is the standard free energy of the reaction, and can be taken as the difference in standard reduction potentials of the donor–acceptor pair. For a family of related compounds in the same solvent there should be a direct relationship between the standard reduction potential and the rate of the electron transfer process.

Figure 2 shows the dependence of the rate of electron transfer as a function of the difference of reduction potentials. The data come from different sources and are not expected to fall on one single curve (different solvents, temperature, etc.). However, it is felt that the correlation is useful for predicting orders of magnitude of $S_{RN}1$ reaction rates.

The reactions of most interest are Reactions 12–14 where E°_s is the standard reduction potential of the couple $ArX/(ArX)^{\overline{\cdot}}$, and E°_p is the standard reduction potential of the substitution product $ArNu/(ArNu)^{\overline{\cdot}}$.

$$ArX + e^- \rightleftharpoons (ArX)^{\overline{\cdot}} \qquad E^\circ_s \qquad\qquad (12)$$

$$ArNu + e^- \rightleftharpoons (ArNu)^{\overline{\cdot}} \qquad E^\circ_p \qquad\qquad (13)$$

$$(ArNu)^{\overline{\cdot}} + ArX \rightleftharpoons ArNu + (ArX)^{\overline{\cdot}} \qquad K_{14} \qquad\qquad (14)$$

If $E^\circ_s > E^\circ_p$, Reaction 14 is favored thermodynamically toward the right (Equation 15), and the electron transfer reaction will be fast. If the difference in reduction potential is large enough (see Figure 2), electron transfer will be diffusion-controlled. However, if $E^\circ_s < E^\circ_p$, electron trans-

$$(E^\circ_s - E^\circ_p) = RT/F \ln k_{14} \qquad\qquad (15)$$

fer Reaction 14 is thermodynamically unfavorable, and may be quite slow compared to other competing reactions (i.e., disproportionation, protonation, or electron transfer to aryl radicals). Because $(ArX)^{\overline{\cdot}}$ decomposes in an irreversible reaction, the equilibrium may be driven to the right. In electrochemically induced reactions, the whole system is driven toward the products through the reoxidation of $(ArNu)^{\overline{\cdot}}$ at the electrode.

The behavior of systems where both situations hold has been analyzed by electrochemical techniques (18). The cyclic voltammetry of 2-chloroquinoline in liquid ammonia exhibits an irreversible wave followed by a reversible wave. The first one, called wave H (18), corresponds to the two-electron hydrogenolysis of the C–X bond (Reactions 16–18). The second wave is the reversible formation of quinoline radical anion, and is called wave D (Reaction 19) (18).

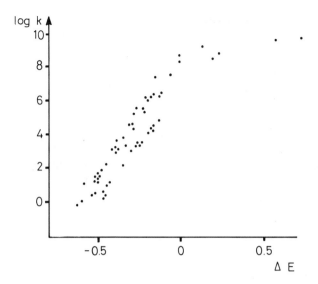

Figure 2. Log k *versus* ΔE *(difference in reduction potentials of donor and acceptor) for the reaction* $D^{\cdot -} + A \xrightarrow{k} D + A^{\cdot -}$. *Data taken from Ref. 112–117.*

$$2\text{–Cl–Q} + e^- \rightleftharpoons (2\text{-Cl–Q})^{\cdot -} \tag{16}$$

$$(2\text{–Cl–Q})^{\cdot -} \longrightarrow 2\text{–Q}^{\cdot} + \text{Cl}^- \tag{17}$$

$$2\text{–Q}^{\cdot} + e^- \longrightarrow \text{Q}^- \xrightarrow{\text{SH}} \text{QH} \tag{18}$$

$$\text{QH} + e^- \rightleftharpoons (\text{QH})^{\cdot -} \tag{19}$$

Q = quinolyl

Addition of benzenethiolate as nucleophile to 2-chloroquinoline decreases Waves H and D in the cyclic voltammogram and, at the same time, a new wave appears, called Wave S, between Waves H and D. Wave S is attributed to the reduction of the couple 2-quinolyl phenyl sulfide/2-quinolyl phenyl sulfide radical anion, formed as in Reaction 20.

$$2\text{–Q}^{\cdot} + {}^-\text{SPh} \longrightarrow (2\text{–QSPh})^{\cdot -} \rightleftharpoons 2\text{-QSPh} + e^- \tag{20}$$

Q = quinolyl

Increasing the amount of benzenethiolate ion decreases Waves H and D and increases Wave S. The cyclic voltammogram of 2-quinolyl phenyl sulfide shows a single peak at the same potential of wave S (Figure 3).

The situation is quite different with diethyl phosphite ion as nucleophile because $E°_p > E°_s$; that is, the substitution product is easier to reduce than the substrate. Wave S indeed appears only in the anodic scan and in the second cycle of the cathodic scan (Figure 4).

As discussed by Saveant (25), electrochemical techniques provide a useful tool, not only to stimulate radical nucleophilic aromatic substitution, but to determine also the rates of coupling of aryl radicals and nucleophiles and the decomposition rates of radical anions.

The fact that the electron transfer reaction (Reaction 14) in the $S_{RN}1$ mechanism is a second order reaction has been established several times (26, 27). For instance, in the reaction of m-bromoiodobenzene 1 with diethyl phosphite ion in liquid ammonia (26), two types of products are formed, the mono-substituted product 3 and the disubstituted product 4 (Reaction 21). These two products are formed according to Reactions 22 and 23 (28).

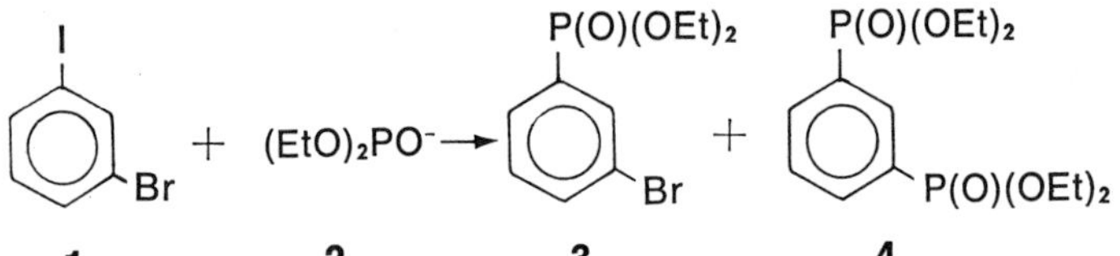

$$ \text{(21)} $$

The ratio of 3 : 4 was found to increase linearly with the concentration of 1, as required by Reactions 22 and 23 (Figure 5).

Electron Ejection to the Solvent. There are examples in the literature where the expected relationship between products coming from fragmentation (Reaction 23) and products coming from electron transfer reactions (Reaction 22) are not observed. A case in point is the reaction

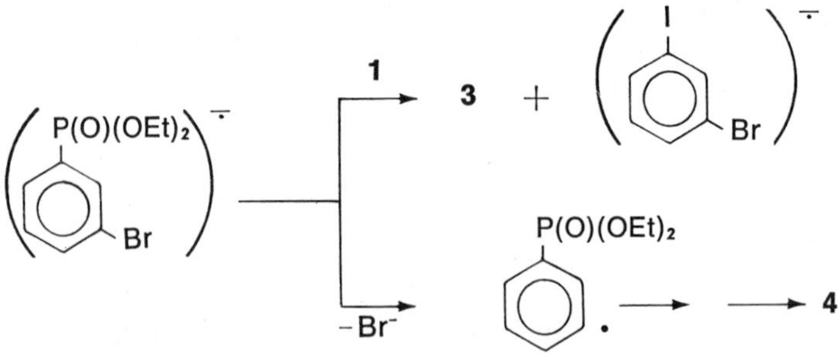

$$ \text{(22)} $$

$$ \text{(23)} $$

of cyanomethyl anion with halobenzenes in liquid ammonia. As was discussed previously, phenylacetonitrile radical anion 5 gives products derived from fragmentation and from electron transfer reactions (Reactions 24–25).

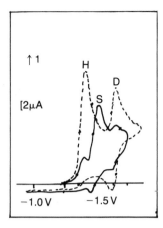

Figure 3. Cyclic voltammetry of 2-chloroquinoline 2.5×10^{-3} M without nucleophile added (- - -), or in the presence of benzenethiolate ion, 5.6×10^{-2} M (—). Scan rate 0.2 V/s. (Reproduced, from Ref. 18. Copyright 1979, American Chemical Society.)

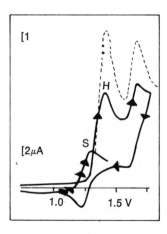

Figure 4. Cyclic voltammetry of 2-chloroquinoline, 3.3×10^{-3} M, without nucleophile added (- - -), or in the presence of diethyl phosphite ion, 9.4×10^{-2} M (——→) first cycle, and (——→—) tenth cycles. Scan rate 0.2 V/s. (Reproduced, from Ref. 18. Copyright 1979, American Chemical Society.)

$$(PhCH_2CN)^{\cdot -} \overset{PhX}{\underset{5}{\longrightarrow}} \begin{cases} \longrightarrow PhCH_2CN + (PhX)^{\cdot -} & (24) \\ \qquad\qquad 6 \\ \longrightarrow PhCH_2^{\cdot} + CN^- & (25) \\ \qquad\qquad 7 \end{cases}$$

If the electron transfer reaction were bimolecular, the ratio of products should depend on the nature and concentration of the substrate. Because the reduction potentials of halobenzenes become progressively more negative in the order PhI > PhBr > PhCl > PhF, their rate of reac-

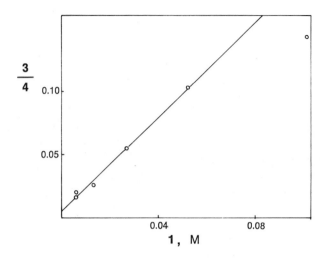

*Figure 5. The product ratio **3:4** as a function of m-bromoiodobenzene **1**, concentration for reactions of **1** with diethyl phosphite ion. (Reproduced from Ref. 28. Copyright 1978, American Chemical Society.)*

tion with the hydrated electron should decrease in the same order (*29*), and the ratio of products coming from Reactions 24 and 25 should decrease accordingly. In Table IV it can be seen that these expectations are not borne out by the results because, if anything, the ratio increases in the reverse order (*30*). Moreover, a five-fold increase in the concentrations of reactants does not change the product distribution (*30*).

Similar results were obtained in the reaction of diphenyl arsenide ion with *p*-halotoluenes and *p*-haloanisoles (X = Cl, Br, I). This nucleophile also gives rise to products coming from electron transfer reactions and fragmentation of the radical anion intermediate, yet a ten-fold increase in the concentration of *p*-bromoanisole does not change the product distribution (*31*).

Two mechanistic models·were proposed to account for the results of the photostimulated and solvated electron stimulated reactions.

MECHANISTIC MODEL I. Mechanistic model I suggests that phenylacetonitrile radical anion **5** forms an adduct with the substrate, or with any other electron acceptor present in the reaction medium. This adduct would then collapse into the radical anion of the electron acceptor and the substitution product phenylacetonitrile **6** (Reaction 26), or decompose giving benzyl radical, cyanide ion, and the electron acceptor (Reac-

Table IV

Ratio of Products Derived from Reactions 25/24 in the Photostimulated Reactions of Halobenzenes with Cyanomethyl Anion

PhX	Reaction 25/24
PhF	1.5
PhCl	2.0
PhBr	3.8 ± 1.4
PhI	4.5

Source: Reproduced from Ref. 30. Copyright 1979, American Chemical Society.

tion 27). Both processes should have the same concentration dependence.

$$5 + EA \longrightarrow (PhCH_2CN-EA)^{\overline{\cdot}} \longrightarrow \begin{cases} (EA)^{\overline{\cdot}} + 6 & (26) \\ PhCH_2^{\cdot} + CN^- + EA & (27) \end{cases}$$

(EA = any electron acceptor, including the substrate.)

There is precedent for adduct formation prior to the actual electron transfer reaction (32). The relative rate of Reactions **26** and **27** need not be dependent on the electron-acceptor ability of the substrate. This mechanistic model however, requires that **5** have a life time long enough to survive until it encounters an electron-acceptor molecule.

MECHANISTIC MODEL II. Mechanistic model II involves the release of the odd electron to the solvent, as a solvated electron (Reaction 28), in competition with radical anion fragmentation to give **7** and cyanide ion (Reaction 29). Both reactions have the same concentration dependence (first order reactions) and are independent of the reduction potential of the substrate.

$$5 \longrightarrow \begin{cases} 6 + e^-_{\text{solvated}} & (28) \\ 7 + CN^- & (29) \end{cases}$$

The rate of reaction of acetonitrile with the solvated electron in liquid ammonia has been measured, and the fact that it is very slow ($k_{obs} = 2\ M^{-1}\ s^{-1}$) compared with the rate of the reaction in water ($k_{obs} = 3 \times 10^7\ M^{-1}\ s^{-1}$) led to the suggestion that the overall rate is depressed in liquid ammonia because k_2 in Reaction 30 is preceded by an unfavorable equilibrium (33).

$$CH_3CN + e^-_{\text{solvated}} \underset{k_{-1}}{\overset{k_1}{\rightleftharpoons}} (CH_3CN)^{\overline{\cdot}} \xrightarrow{k_2} \text{products} \qquad (30)$$

The low value of k_1 in liquid ammonia is attributed to the high stability of the ammoniated electron which increases k_{-1} relative to k_2 and k_1. Similar results have been found for the reactions of benzene (34), biphenyl (35), and fluorobenzene (36) with the solvated electron in liquid ammonia. In these cases, reversible attachment of the solvated electron was suggested.

Phenylacetonitrile radical anion is expected to be about as stable as acetonitrile radical anion; thus, it may lose the odd electron at a rate similar to k_{-1} (Equation 30), and the electron may be transferred to the solvent in a first order process instead of by second order electron transfer to the substrate.

The alternative to electron transfer to the solvent is the transfer to an electron-acceptor molecule, and may be represented as in Reaction 31.

$$(RA)^{\overline{\cdot}} + EA \underset{k_{-1}}{\overset{k_1}{\rightleftharpoons}} (EA)^{\overline{\cdot}} + {}^-RA$$
$$k_{-2} \diagdown \quad k_2 \qquad\qquad k_3 \diagup\!\!\diagup k_{-3} \qquad\qquad (31)$$
$$RA + EA + e^-$$

where RA = any radical anion; EA = any electron acceptor.

$$E°_{EA} - E°_{RA} = (RT/F) \ln K_1 \qquad (32)$$

$$E°_{e^-} - E°_{RA} = (RT/F) \ln K_2 \qquad (33)$$

$$E°_{EA} - E°_{e^-} = (RT/F) \ln K_3 \qquad (34)$$

Microscopic reversibility requires that:

$$k_1 = k_2 \, k_3 \qquad (35)$$

As discussed previously, there is considerable evidence that when the difference in reduction potential of the donor–acceptor pair surpasses a certain value in the thermodynamically favored direction, the rate of the electron transfer reaction is diffusion controlled. As in the analogous situation of proton transfer between an acid–base pair, the overall rate of electron transfer can be expressed by Equation 36 (37).

$$\vec{k} = k_2 + k_1 \, (EA) \qquad (36)$$

If $E°_{EA} > E°_{RA}$, but $E°_{e^-} < E°_{RA}$, and $k_1 > 1$ but $k_2 < 1$, then the main pathway for electron transfer will be direct electron transfer between the donor–acceptor pair, RA and EA.

If both k_1 and k_2 are diffusion controlled ($E°_{EA} > E°_{RA}$ and $E°_{e^-} > E°_{RA}$), it is seen that, under the usual conditions of solution reactions and with a moderate concentration of EA, about 0.01–0.10 M, the electron will be transferred to the solvent. That is, K will be given by k_2. In other words, if the reduction potential $E°_{RA}$ is more negative than the reduction potential of the electron in solution, the radical anion in solution will function as a cathode ejecting the odd electron into the solvent.

In a potential-sweep voltammogram of 0.5 M alkali metal iodide in ammonia at $-40°C$, a sharp rise of current occurs at about -2 V (vs. Pb/PbSO$_4$ electrode, 0.05 M), corresponding to the dissolution of electrons from the cathode according to Reaction 37, and is indicated by blue coloring around the electrode surface (38).

$$e_M^- \rightleftharpoons e_S^- \tag{37}$$

The metal electrode electron is e_M^-, and e_S^- is the electron solvated in solution. This is called the electron–electrode (38).

The standard reduction potential of the electron–electrode in ammonia, as determined by voltammetry and potentiometry is -2.14 V (vs. Pb/Pb(NO$_3$)$_2$ electrode, 0.05 M), -1.86 V versus the standard hydrogen electrode (38, 39), or -2.69 V (vs. Ag/AgNO$_3$ electrode, 0.1 M) (40). Therefore any RA/(RA)$^-$ couple with a standard reduction potential more negative than these values will function as an electrode in liquid ammonia, releasing electrons into the solvent.

It follows from these arguments that whenever an aryl radical reacts with a nucleophile to give a radical anion, whose parent molecule has a standard reduction potential more negative than the standard potential of the electron–electrode in liquid ammonia, it will act as a cathode ejecting the odd electron to the solvent (Reactions 38–40).

$$Ar\cdot + Nu^- \longrightarrow (ArNu)^- \tag{38}$$

$$(ArNu)^- \longrightarrow ArNu + e_S^- \quad E°_{ArNu} < E°_{e^-} \tag{39}$$

$$ArX + e^- \longrightarrow (ArX)^- \tag{40}$$

The sequence of reactions under these circumstances will be $38 \rightarrow 39 \rightarrow 40$. The possibility that Reactions 38 and 39 are not separate steps and that the (ArNu)$^-$ is not in fact formed as an intermediate cannot be rejected. However, no data are available to allow the distinction between the two possible situations and it will not be considered further.

On the other hand, if $E°_{ArNu} > E°_{e^-}$, the odd electron will be transferred in a bimolecular step, and the reaction sequence will be $38 \rightarrow 41$.

$$(ArNu)^{\cdot -} + ArX \longrightarrow ArNu + (ArX)^{\cdot -} \tag{41}$$

The summation of Reactions 39 and 40 equals 41. Both yield a molecule of the radical anion of the substrate and one of the substitution product, but the mechanism of electron transfer is different. Reactions 39–40 involve the ejection of electrons into the solvent as a major reaction path. Thus, the observed reaction rate is independent of the concentration and the nature of the aryl halide. Because Reaction 41 represents direct electron transfer to the aromatic substrate, the observed rate will depend on the concentration of the aryl halide.

This mechanistic model allows the interpretation of some results. As mentioned before, in the reaction of m-iodobromobenzene with diethyl phosphite ion, the dependence of the product ratios on the concentration of substrate indicates that the electron transfer step is bimolecular. On the other hand, the reactions of p-halotoluenes and p-haloanisoles with diphenyl arsenide ion, and the reaction of halobenzenes with cyanomethyl anion, give product ratios coming from fragmentation and electron-transfer reactions that are independent of concentration and the nature of the substrate. The substitution products formed in these reactions are, respectively, 5, 6, and 7.

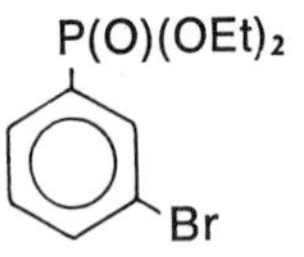

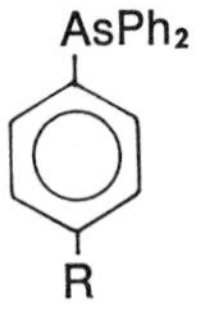

R = Me, OMe

6

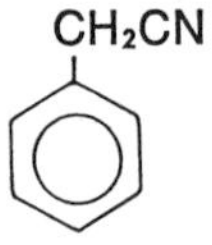

The reduction potentials of Compounds 5, 6, and 7 are not known in liquid ammonia, but it is likely that 5 has a reduction potential more positive than that of the electron in liquid ammonia, and thus its radical anion reacts following Reaction 41, giving the observed dependence on the concentration of the substrate. In comparison, 6 and 7, being more difficult to reduce, possibly have a reduction potential more negative than the electron and react following Reactions 39–40.

Mechanistic models I and II consider only processes that take place in the dark; yet, most $S_{RN}1$ reactions occur under light. Thus, as was pointed out, (30) there is a possibility that electron ejection from the radical anion is a photoassisted process. The ionization potential of organic radical anions, or the electron affinity of the parent compound, is usually in the range of 1–2 V (41); thus, visible light could be effective in promoting photoejection from these species.

The intimate nature of the mechanism is not yet known, and more work is required to distinguish between these and other mechanistic possibilities.

Competing Reactions

Once a radical anion is formed by the coupling of an aryl radical with a nucleophile, several competing reactions besides electron transfer can follow, and the more frequently encountered reaction involves fragmentation to give a radical and an anion (Reactions 42–44).

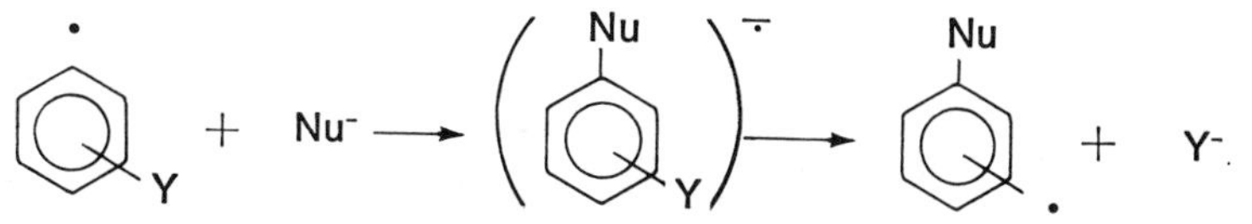

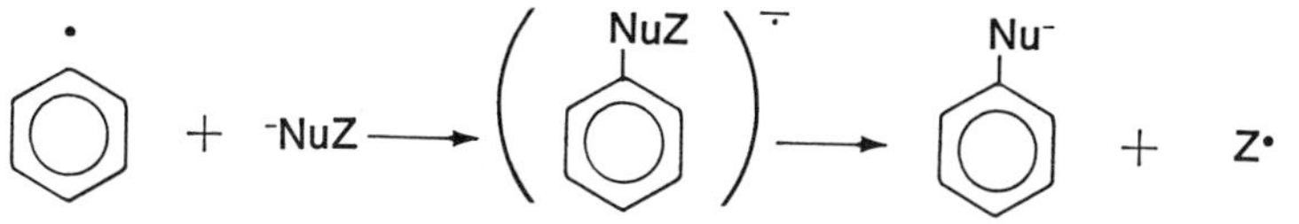

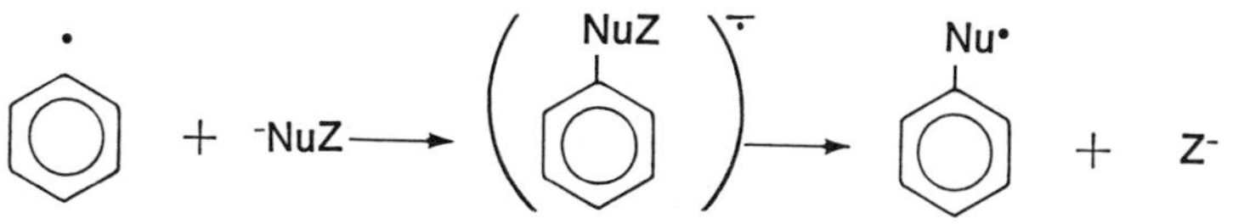

Reaction 42 is characteristic of substrates bearing two leaving groups, and the intermediate aryl radical formed is a reactive species, thus the overall reaction rate is not greatly depressed.

The radicals formed in Reactions 43 and 44 must be reactive toward the nucleophile in order to be effective as chain carriers such as aryl radicals, but if they are alkyl radicals they become products of a pretermination step.

Fragmentation of Radical Anions into Reactive Intermediates. SUB-
STRATES WITH TWO LEAVING GROUPS. In the photostimulated reaction of substrates with two leaving groups (for example, dihalobenzenes) with a suitable nucleophile, a radical anion intermediate is formed which can either transfer its odd electron (Reaction 45) or lose its second leaving group (Reaction 46).

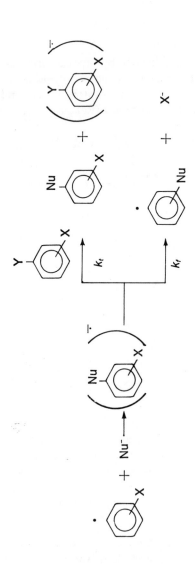

The factors that influence the electron transfer rate were discussed previously. The fragmentation reaction is discussed here.

The rate of decomposition of haloarene radical anions has been found to depend not only on solvent and temperature but also on: the reduction potential of the parent haloarene (the more negative the reduction potential, the faster is its decomposition); the nature of the group bonded to the same aryl moiety (the nucleofugality increases as the C–X bond strength decreases, and, for the halogen series, the order of leaving group ability is $I > Br > Cl > F$); and the spin density of the carbon bonded to the leaving group.

The reduction potential is a measure of the energy available for bond fragmentation, and it has been suggested that the reduction potentials above which the haloarene radical anions can be considered stable $(t_{1/2} > 5 s)$ are the following: Cl, $E° > -1.6$ V; Br, $E° > -1.2$ to -1.6; and I, $E° > -1.1$ V. (in DMF, vs. SCE) (42). That is to say, a radical anion whose reduction potential is more negative than these limits will have a life time of less than 5 s.

Molecular orbital calculations (3), and ESR spectroscopy (43), indicate that the haloarenes can form π^* or σ^* radical anions, but dissociation takes place only when the odd electron is located in the σ^* orbital, which is a repulsive state (44). The probability that an electron is located in either of these MOs depends on their relative energies. If the energy of the σ^* orbital is lower than that of the π^* orbital, the radical anion will be a σ^* radical anion, and vice versa.

It follows that leaving group expulsion in a radical anion depends on the probability that the odd electron is located in the C–X σ^* orbital. The likelihood that an electron located in a π^* orbital will be transferred intramolecularly to a particular σ bond depends on the energy gap between the two orbitals and on the spin density at the carbon atom bonded to the leaving group, because spin density is a measure of the probability of finding the electron on that carbon.

The energy of an aryl–halogen bond, and hence the energy of the σ^* orbital, is not expected to be greatly affected by the nature of the aromatic moiety, whereas the π^* MO is very much dependent on the aromatic ring system. In a series of compounds with the same leaving group, it is expected that the difference in energy between σ^* and π^* MOs will decrease as the reduction potential becomes more negative. The rate of decomposition will then increase. On the other hand, the σ^* MO energy is related to the C–X bond energy, and, in the series of halobenzenes, is expected to decrease in the order $F > Cl > Br > I$. These effects are apparent in the data of Table V.

In the reaction of several dihalobenzenes with nucleophiles, disubstitution products are formed, and in two cases it has been proven that the monosubstitution product is not an intermediate in the main pathway for the formation of the disubstitution product. This result

Table V
Reduction Potential and Rate of Fragmentation of Haloarenes

Parent	Substrates	$-E_{1/2}{}^a$	$k_f(s^{-1})$	Footnote	Ref.
Benzonitrile		2.32			118
	3–F	2.19	10^3	b	46
	4–F	2.33	11	c	46
	3–Cl	1.99	10^{10}	d	118
	4–Cl	2.40	10^{10}	d	118
Acetophenone		2.03			2
	3–Cl	1.84	10		2
	4–Cl		10^5		2
4-Styrylpyridine		1.88			119
	4'–F	1.86	stable		119
	4'–Cl	1.83	1.5		119
	4'–Br	1.86	fast		119
	3'–Cl	1.78	stable		119
Benzophenone		1.61			111
	4–F	1.71	stable		111
	4–Cl	1.56	10		111
	3–Cl	1.56	stable		111
	4–Br	1.48	8×10^4		111
	3–Br	1.50	7.4×10^2		111
Fluorenone		1.295			110
	1–Br	1.195	0.15		110
	3–Br	1.192	0.24		110
	4–Br	1.135	0.007		110
Nitrobenzene		1.06			6
	o–I	0.95	8×10^4		6
	m–I	0.94	0.31		9

indicates that fragmentation is faster than electron transfer ($k_f \gg k_t[\text{ArX}]$) as in Reactions 45–46.

In reactions of fluorohalo compounds, fluorine is always retained in the product. For instance, m-fluoroiodobenzene reacts with acetone enolate ion, diethyl phosphite and benzenethiolate ions to give the corresponding m-fluorophenyl derivatives. Diethyl phosphite ion reacts with p- and o- fluoroiodobenzene under photostimulation to give the fluoro-substitution product.

The high stability of aryl fluorides is evident because the radical anion of 2-fluoropyridine has a life time long enough in liquid ammonia so that its ESR spectrum can be recorded (45). All the other halopyridine radical anions decompose by loss of halide ion.

Moreover, the electrochemical reduction of m-fluorobenzonitrile (46) and p-dicyanoperfluorobenzene (47) yields the products of C–CN bond rupture, not C–F bond rupture (Reactions 47–48).

Table V (Continued)
Reduction Potential and Rate of Fragmentation of Haloarenes

Parent	Substrates	$-E_{1/2}{}^a$	k_f (s^{-1})	Footnote	Ref.
	p–I	1.06	0.9		9
	o–Br	1.03	1.1×10^2		9
	p–Br	0.98	4×10^{-3}		6
	o–Cl	1.05	10^{-2}		6
Nitrobenzene		1.14^e			5
	o–Cl		9×10^{-3}	e	5
	p–Cl		6×10^{-3}	e	5
	m–Cl		stable	e	5
			(3×10^{-3})	e	5
	o–Br		20^e		5
	p–Br		12×10^{-3}	e	5
	m–Br		5×10^{-3}	e	5
	o–I		10^2	e	5
	m–I		57×10^{-3}	e	5
	p–I		640×10^{-3}	e	5
	o–F		stable	e	5
			(4×10^{-3})	f	5
	m–F		stable	e	5
			(2×10^{-3})	f	5
	p–F		stable	e	5
			(3×10^{-3})	f	5

^{a} Values in DMF vs. SCE.
^{b} Cyanide ion is the leaving group.
^{c} Fluoride ion is the leaving group.
^{d} Lower limit.
^{e} Values in acetonitrile vs SCE.
^{f} Values in polyethylene carbonate.

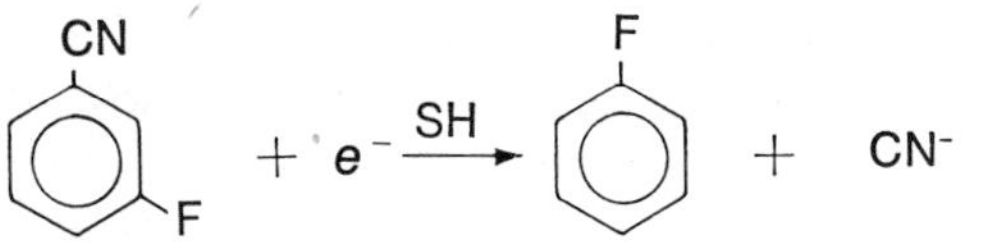

$$(47)$$

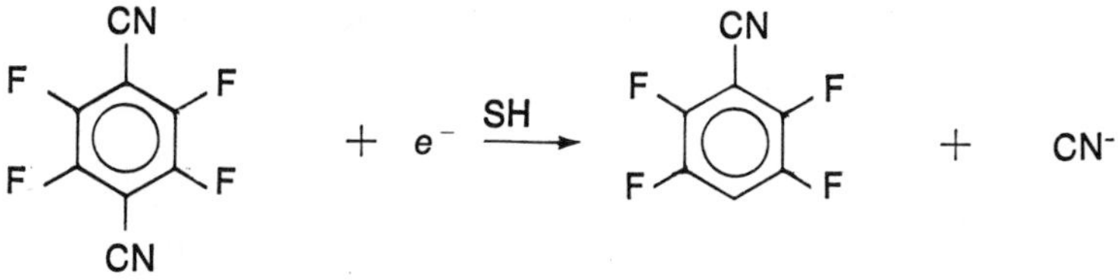

$$(48)$$

For dihalobenzenes bearing a chlorine atom as one of the leaving groups (*see* Reactions 49–51), the ratio of $k_f : k_t$ (ArX) depends on the nature of the nucleophile and its position in the molecule. *p*-Chloro-iodobenzene reacts with diethyl phosphite or benzenethiolate ions to give the disubstitution product. *m*-Chloroiodobenzene gives the disubstitution product with benzenethiolate ion, but the monosubstitution product with diethyl phosphite ion.

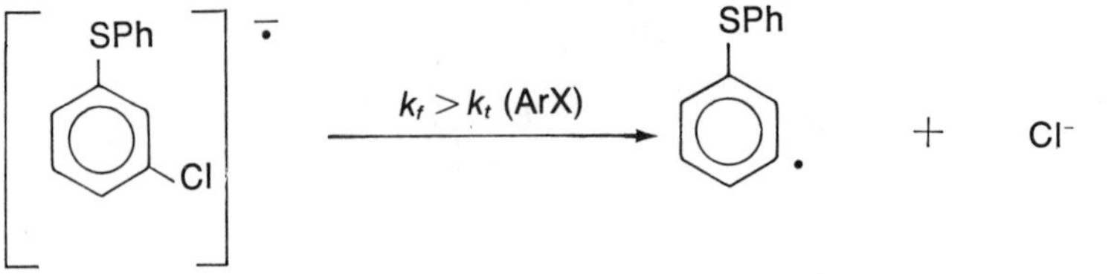

$$k_f > k_t \text{ (ArX)} \qquad\qquad\qquad + \quad Cl^- \qquad (49)$$

$$k_f < k_t \text{ (ArX)} \qquad\qquad\qquad + \qquad\qquad (50)$$
$$m\text{-ClC}_6\text{H}_4\text{I}$$

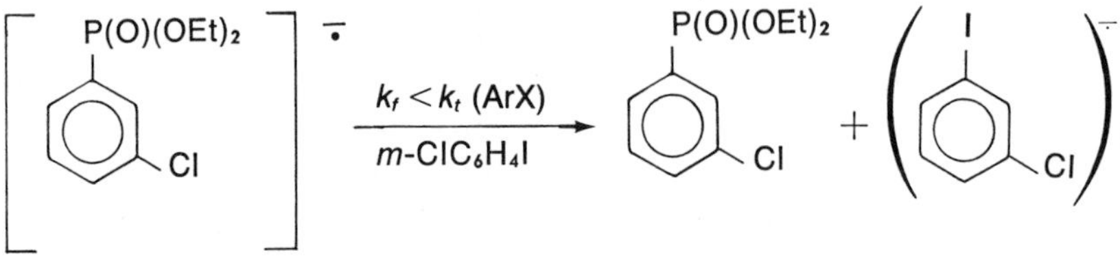

$$k_f > k_t \text{ (ArX)} \qquad\qquad\qquad + \quad Cl^- \qquad (51)$$

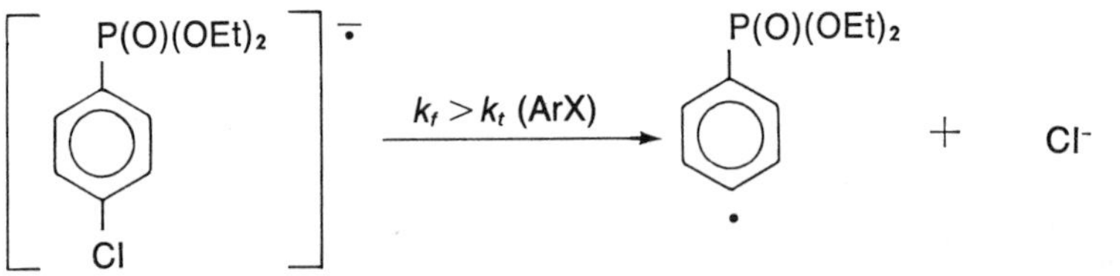

The electron-transfer rates of Reactions 49–51 are not expected to be very different because in all cases, they are probably close to diffusion. It follows that the product ratio must be controlled by the rate of fragmentation, with k_f for Reaction 49 higher than k_f for Reaction 50.

Based on these considerations it may be concluded that $\Delta E_{\pi^* \longrightarrow \sigma^*}$ is higher in the diethyl arylphosphonate radical anion (Reaction 50) than the phenyl aryl sulfide radical anion (Reaction 49). This difference is probably due to differences in π^* energies of both systems, because the σ^* energy of the C–Cl bond is not likely to be very affected by the presence of *meta*-substituents.

The energy of the π^* MO is probably lower for the diethyl arylphosphonate derivatives, as indicated by the relationship of the reduction potentials of the following compounds (*18*):

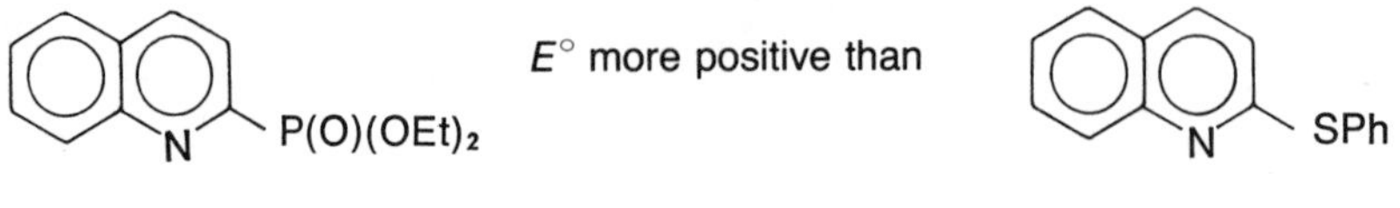

$E°$ more positive than

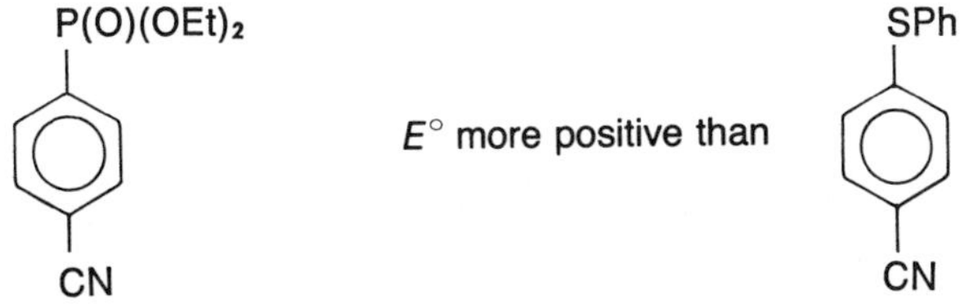

P(O)(OEt)2
CN
E° more positive than
SPh
CN

On the other hand the different rates of fragmentation of the *m*- and *p*-chlorophenyl phosphonate radical anions are probably due to different spin densities at *meta*- and *para*- carbon atoms.

A remarkable effect in regard to the partition of the radical anion between electron transfer and fragmentation products was reported in the photostimulated reaction of *p*-iodophenyltrimethylammonium ion **8** with benzenethiolate ion as nucleophile, which, in liquid ammonia gave only the *p*-disubstituted product **9** (*48*), but in water gave both products (*49*) (Reactions 52, 53).

I
+NMe3
8
+ PhS⁻
hν
NH3
SPh
SPh
9
(95%)
(52)
H2O
9 (12%) +
SPh
+NMe3
(38%)
(53)

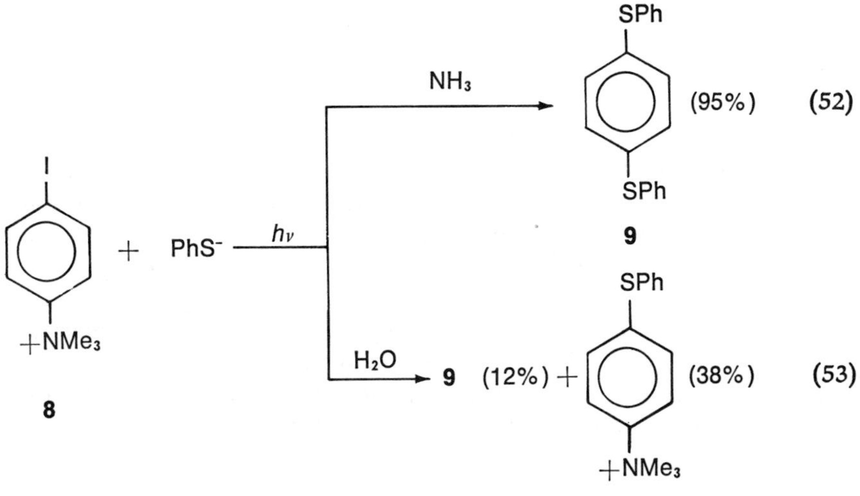

The fact that k_f/k_t(ArX) is greater than 100 in ammonia and less than one in water could be due to either a solvent effect or to a temperature change.

Taking the product ratio as the ratio of $k_f : k_t$ (ArX), and assuming that the ratio of pre-exponential rate factors in water and ammonia are about the same, it can be estimated that the activation energy of the electron transfer reaction should be about 14 kcal/mol higher than that of the fragmentation reaction to account for the differences in the product ratios as a result of a temperature effect only. Even if the fragmentation reaction has a very small activation energy (say 3–4 kcal/mol), the activation energy for the electron transfer reaction, which is probably thermodynamically favored, seems quite high.

The results are better rationalized by the solvent effect on electron transfer rates and bond fragmentation. The radical anion intermediate **10** formed in these reactions is uncharged, although, perhaps, polar. The electron transfer reaction leads to charge separation, and thus is expected to be favored by an increase in solvent polarity. On the other hand the transition state leading to bond fragmentation is expected to be less polar than the starting material, because two uncharged species are formed. An increase in solvent polarity is expected to decrease the rate of fragmentation. Thus, the increase in k_t and decrease in k_f caused the observed changes in the $k_f : k_t$ ratio.

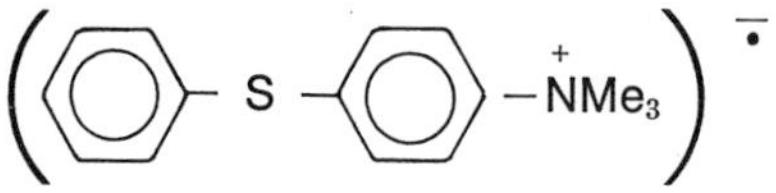

10

NUCLEOPHILES WITH LOW CARBON–NUCLEOPHILE BOND ENERGIES. Nucleophiles of the type $(Ph)_nZ^-$, where the Ph–Z bond energy has a low lying σ^* MO, couple with aryl radicals to form radical anion intermediates that fragment to form phenyl radicals. These phenyl radicals enter into the chain propagation steps leading ultimately to the scrambling of aryl rings.

For example, aryl scrambling occurs in the photostimulated reaction of p-iodoanisole with phenyl telluride ion in liquid ammonia. This result has been ascribed to the reversible coupling of the nucleophile with aryl radicals, as with the p-anisyl radical (Reaction 54) (50).

$$An\cdot + {}^-TePh \Longrightarrow (An–Te–Ph)^{\overline{\cdot}} \Longrightarrow AnTe^- + Ph\cdot \qquad (54)$$

$$\downarrow \text{electron transfer}$$

$$An–Te–Ph$$

An = anisyl

The two carbon–tellurium bonds of p-anisyl phenyl telluride are of similar energy, and the two σ^* MOs are probably of comparable or lower energy to that of the π^* MO. Thus, the radical anion intermediate formed when an aryl radical couples with phenyl telluride ion can suffer three competitive reactions: reversion to starting material, fragmentation into p-anisyl telluride ion and phenyl radical, or electron transfer to give the substitution product (Reaction 54). Either one of these competing reactions can give reactive intermediates that enter into the chain propagation steps and lead to the observed products.

Benzenethiolate ion gave the straightforward substitution products without scrambling of aryl rings. The difference in behavior is ascribed to differences in the σ^* MO energy in the group VIA series, or rather to decreasing $\Delta E_{\pi^* \to \sigma^*}$ going from S $\to$ Se $\to$ Te derivatives.

Similar differences in behavior were found in the VA group of elements. Whereas diphenyl phosphide ion reacts with p-iodoanisole giving the substitution product p-anisyldiphenylphosphine (51), diphenyl arsenide ion gave a mixture of four arsines. Their formation was explained as in the reaction with phenyl telluride ion (Reaction 55) (31).

$$ArX + Ph_2As^- \xrightarrow{\ h\nu\ } Ph_3As + Ph_2ArAs + PhAr_2As + Ar_3As \qquad (55)$$

ArX = p-X-toluene, p-X-anisole (X = Cl, Br, I), 1-chloronaphthalene, 9-bromophenanthrene.

Figure 6 represents the reactions involved and shows how the initial coupling of an aryl radical with diphenyl arsenide ion leads ultimately to the statistical distribution of products, because for all the radical anions involved, k_f/k_t ratios are about the same. (The concentration of ArX is not included because the ratio of fragmentation vs. electron transfer products is independent of [ArX].)

2-Chloroquinoline and 4-chlorobenzophenone on the other hand gave normal substitution products. The fact that these substrates do not lead to aryl ring scrambling indicates that, in these cases, the radical anion intermediates are π^* in nature, and that the $\Delta E_{\pi^* \to \sigma^*}$ is large enough so that the intramolecular electron exchange does not compete with other reactions, such as electron transfer to the substrate.

However, with diphenyl stibide ion as nucleophile, scrambling of aryl rings is observed even with 4-chlorobenzophenone. Even for this substrate, then, the σ^* MO is lower in energy than the π^* MO (52).

The reduction potentials of aromatic compounds are related to their π^* LUMO, that is, the more negative the reduction potential the higher the LUMO (53). On the other hand, the phenyl–M (M being P, As, Sb) bond strengths (54) give a measure of the σ MO energy, which in turn is related to the σ^* MO energy.

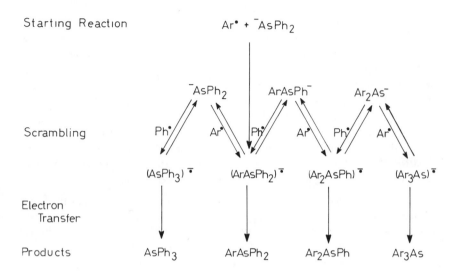

Figure 6. Schematic representation of the reactions leading to the scrambling of aryl rings during the coupling of an aryl radical with diphenyl arsenide ion.

In Figure 7 the energies involved in reactions of Ph_2M^- (M being P, As, and Sb) nucleophiles with aryl radicals are represented on a relative scale. On the right scale are the bond dissociation energies of the Ph–M bonds, and on the left scale are the reduction potentials of the aromatic moiety. The reduction potentials of benzene derivatives, such as toluene or anisole, are not known, but it is known that they are reduced at potentials more negative than the discharge of the supporting electrolyte (more negative than -3.0 V) (55).

The bond dissociation energy of the Ph–P bond is set above the limit of -3.0 V because, in the reaction of diphenyl phosphide ion with p-halotoluene, only normal substitution occurs. On the other hand, the bond dissociation energy of the Ph–Sb bond is set below the reduction potential of benzophenone for the reason that scrambling of aryl rings is observed in the reaction of 4-chlorobenzophenone with diphenyl stibide ion.

If the bond strength is above the reduction potential on this relative energy scale, the predominant radical anion will be a π^* radical **11**, and transference to the σ^* MO will be unfavorable and scrambling will not be observed. If the bond strength is below the reduction potential of the aryl moiety, the predominant radical anion will be a σ^* radical **12**, and fragmentation leading to scrambling of aryl rings will occur.

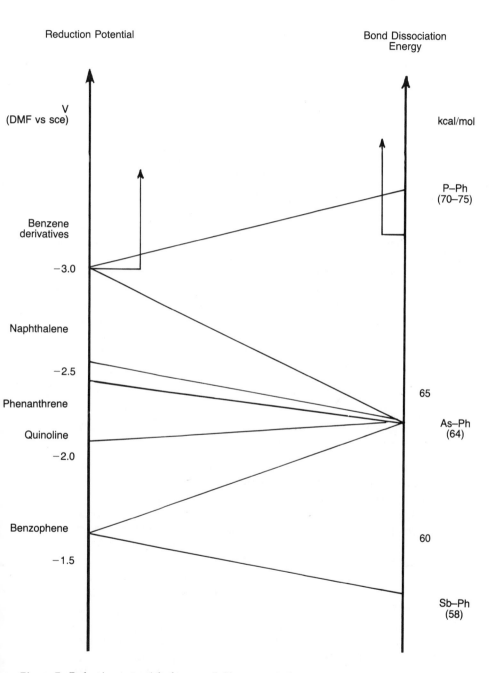

Figure 7. Reduction potential of arenes (left) represented on a relative scale with respect to the bond dissociation energy of Ph–M (right).

$$(Ar)^{\bar{\cdot}}\!\!-\!MPh_2$$

11

π^* nature

$$Ar\overset{\bar{\cdot}}{-}M\overset{Ph}{\underset{Ph}{\diagdown}}$$

$$Ar\!-\!M\overset{\bar{\cdot}Ph}{\underset{Ph}{\diagdown}}$$

$\updownarrow$

$$Ar\!-\!M\overset{\bar{\cdot}Ph}{\underset{Ph}{\diagdown}}$$

12

σ^* nature

The phenyl–As bond dissociation energy lies between the reduction potential of phenanthrene (−2.4 V) and quinoline (−2.1 V). Thus scrambling is found with the phenanthrene derivatives and all other derivatives with more negative reduction potentials, but is not observed with 2-chloroquinoline or 4-chlorobenzophenone.

It is concluded that, in photostimulated reactions in liquid ammonia, diphenylphosphide ion will react by normal substitution with every substrate having a reduction potential more positive than −3.0 V, whereas diphenylstibide ion will react with substrates having more negative reduction potentials than −1.6 V to scramble the aryl rings. The value of the reduction potential of the aromatic moiety required to avoid aryl scrambling with this nucleophile is not yet known.

Fragmentation of Radical Anions into Unreactive Intermediates. When the radical anion intermediate formed in the coupling of an aryl radical and a nucleophile decomposes into an unreactive radical, the reaction becomes a pretermination step, and the overall reactivity decreases.

CYANOMETHYL ANION. As discussed previously phenylacetonitrile radical anion suffers two competitive reactions. These are C–CN bond fragmentation to give a benzyl radical and cyanide ion, or electron transfer to give the substitution product. The latter reaction was discussed previously.

Because the benzyl radical is unable to react with nucleophiles, it dimerizes, abstracts hydrogen, or is reduced to benzyl anion, giving ultimately 1,2-diphenylethane or toluene, respectively (56). In this case, the fragmentation reaction is a termination step, and the overall reactivity is low. Similar behavior is observed in the reaction of 2-halothiophene. That is, the 2-thienyl-acetonitrile radical anion gives fragmentation and electron-transfer-derived products (57).

On the other hand, the radical anion intermediates formed when the substrates are halo derivatives of pyridine, biphenyl, or naphthalene give only the electron transfer reaction, with no indication of fragmentation (58).

As an explanation, it was suggested that, in the case of phenyl derivatives, the odd electron is located in the π^* MO of the –CN moiety 13, whereas, in all others (except thiophene), the odd electron occupies a π^* MO of the aryl ring 14 (59).

$$Ph-CH_2-(CN)^{\bar{\cdot}} \qquad\qquad (Ar)^{\bar{\cdot}}-CH_2-CN$$

$$\textbf{13} \qquad\qquad\qquad\qquad \textbf{14}$$

HMO calculations of the species involved agree with this interpretation because the LUMO of the –CN moiety has an energy of $\alpha + 0.82\beta$, which is lower than that of benzene ($\alpha + 1.00\beta$) or thiophene ($\alpha + 1.30\beta$), and higher than that of the other aromatic substrates used (59).

The different nature of the radical anion involved is further demonstrated in the reduction of phenylacetonitrile and 1-naphthylacetonitrile with solvated electrons in liquid ammonia. In the first case, the reaction leads to toluene (56), whereas, in the second case, a mixture of 1,4- and 1,2-dihydronaphthylacetonitriles is formed (60).

The decomposition rate of several 4-nitrobenzyl compounds has been found to parallel roughly the C–X bond energy (Table VI) (61), and it was suggested that the odd electron in the π^* MO "jumps" to the σ^* MO of the C–X bond, which then dissociates, (62). As implied earlier, passage of an electron from a π^* MO to a σ^* MO will certainly require more energy, and consequently a slower reaction rate the greater the difference in energy between these two MOs. Recently the fragmentation rate of o-, m-, and p-nitrobenzyl halides, which is believed to be a measure of the rate of intramolecular transformation of $\pi^* \to \sigma^*$, has been determined (Reaction 56) (63).

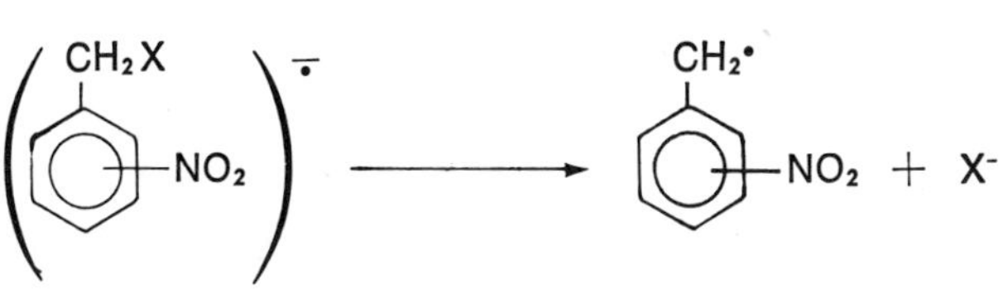

$$(56)$$

Table VI
Decomposition Rates of Radical Anions Derived from 4-Nitrobenzyl Compounds in Acetonitrile

$p-O_2-NC_6H_4CH_2X$ X	$k(s^{-1})$	Bond Strength[a] (kcal/mol)
Br	10^2	55
Cl	10–20	69
F	10^{-2}	90^b
CN	10^{-1}	
SCN	7–8	

Source: Reproduced, from Ref. 61. Copyright 1971, American Chemical Society.
[a]Bond strengths of $PhCh_2X$ taken from Ref. 120.
[b]Estimated value.

It was found that this decomposition rate not only depends on the halogen, but also on its position in the ring. The rate decreases in the order $I > Br > Cl$, and $o- > p- > m-$ (Table VII). These results agree with the fact that the σ^* MO energy of the C–X bond decreases as the bond energy decreases. The lower rates for the m-substituted compounds are attributed to the low spin density of the m-carbon atom. In the p- and o-substituted compounds the highest spin density is found in the nitro group, but a certain amount can be expected on the o- and p- positions, and this is channeled through the $-CH_2-$ to the halogen (63). Based on the previous results, the fast rate of decomposition of phenylacetonitrile or thienylacetonitrile radical anions can be attributed either to facile odd electron transfer from a relatively high energy π^* MO to a σ^* MO of the C–CN bond, or to the formation of a σ^* radical anion when the cyano-methyl anion couples with phenyl or thienyl radicals. When the radical anion intermediate has its odd electron in a low energy π^* MO, intra-molecular electron transfer to the σ^* MO of the C–CN bond is slow and cannot compete with the intermolecular electron transfer process which forms the normal substitution product.

ALKANETHIOLATE IONS. The photostimulated reaction of iodobenzene with benzenethiolate ions gives the substitution product in high yield. However, in the reaction of alkanethiolate ions with aryl radicals, the radical anion intermediate fragments according to Reaction 58, to give an alkyl radical that is unreactive as a chain carrier. Alternatively, the radical anion transfers the odd electron giving the substituted product (Reaction 57 (64).

$$
Ar^{\cdot} + {}^{-}SR \longrightarrow (Ar\text{–}S\text{–}R)^{\overline{\cdot}}
\begin{array}{l}
\xrightarrow{\quad k_t \quad} ArSR \qquad (57) \\
\text{electron transfer} \\
\xrightarrow{\quad k_f \quad} ArS^- + R^{\cdot} \qquad (58)
\end{array}
$$

Table VII
Decomposition Rates of Nitrobenzyl Halide Radical Anions in Water

| Isomer | Halogen, k (s^{-1}) | | |
	Cl	Br	I
$o-$	1×10^4	4×10^5	
$p-$	4×10^3	1.7×10^5	5.7×10^5
$m-$	5	60	3×10^3

The importance of each reaction pathway depends on the nature of the alkyl moiety, R, and the aryl radical.

Phenyl radicals give only substitution products with benzenethiolate ions (65), but almost equal amount of substitution and fragmentation with alkanethiolate ions (methyl, ethyl, n-butyl, and t-butyl). With benzenethiolate only fragmentation was found (Table VIII, Reaction 59) (66).

TABLE VIII
Photostimulated Reaction of Haloaromatic Compounds with Thiolate Ions in Liquid Ammonia

ArX^a	$-E_{1/2}$ ArH^b	RS^-	$D_{R-H}{}^c$	$D_{PhS-R}{}^d$	$ArSR$	ArS^-	$Ref.$
					\multicolumn{2}{c}{$Yield^e$}	
PhI	3.0	PhS^-	110	76	100	0	65
PhI	"	MeS^-	104	65	39	61	66
PhI	"	EtS^-	98	62	41	59	64
PhI	"	$n\text{-}BuS^-$	98	60	47	53	66
PhI	"	$t\text{-}BuS^-$	92	59	52	48	66
PhI	"	$PhCH_2S^-$	85	51	0	100	66
2–Cl–Pyr	2.7	$n\text{-}BuS^-$	98	60	84	16	66
2–Cl–Pyr	"	$PhCH_2S^-$	85	51	21	79	66
2–Cl–Pyr^f	"	$PhCH_2S^-$	85	51	32	68	66
1–I–Naph	2.5	PhS^-	110	76	100	0	65
1–Cl–Naph	"	$n\text{-}BuS^-$	98	60	100	0	60
1–I–Naph	"	$t\text{-}BuS^-$	92	59	100	0	66
1–Br–Naph	"	$PhCH_2S^-$	85	51	15	85	66
9–Br–Phen	2.4	$PhCH_2S^-$	85	51	12	88	66
2–Cl–Q	2.1	$PhCH_2S^-$	85	51	100	0	66

[a]Pyr = pyridine; Naph = naphthalene; Phen = phenanthrene; Q = quinoline.
[b]Reduction potential of the parent aromatic compound (DMF vs SCE).
[c]Bond dissociation energy of the R moiety (R–H $\longrightarrow$ R· + H·), taken from Ref. 124 (kcal/mol).
[d]Estimated values from Ref. 67.
[e]Relative amount of products.
[f]Ten fold excess of the substrate.

$$Ph· + {}^-S\text{–}R \xrightarrow{h\nu} PhSR + PhS^- \qquad (59)$$

	Relative amount	
	$PhSR$	PhS^-
Ph	100	0
Me, Et, n-Bu, t-Bu	50	50
$PhCH_2$	0	100

Because the fragmentation Reaction 58 is a pretermination step, the overall yield is low in all the reactions where Reaction 55 is important The relative amount of bond fragmentation increases as the C–H or C–S bond dissociation energy decreases (Table VIII) (67).

Thus, we see that with phenyl radicals whose Ph–H bond dissociation energy is 110 kcal/mol, only the electron transfer reaction occurs (Reaction 57). With alkyl radicals (CH$_3$–H bond dissociation energy of 104 kcal/mol, and t-Bu–H bond dissociation energy of 92 kcal/mol) almost the same amount of fragmentation and electron transfer products are formed. With benzyl radicals, only fragmentation occurs (Reaction 58).

Because electron transfer is a propagation step, whereas fragmentation is a termination step, overall substitution will decrease as fragmentation increases. Accordingly, benzenethiolate ion gives 94% of substitution product, alkanethiolate ions give 30–70%, and phenylmethanethiolate gives only about 4%.

The relative importance of these competing reactions also depends on the nature of the aryl moiety (Reaction 60).

$$Ar^{\cdot} + n\text{-BuS}^- \xrightarrow{h\nu} Ar\text{–S–}n\text{-Bu} + ArS^- \qquad (60)$$

Ar	Relative amount	
	Ar–S–Bu	ArS
Ph$^{\cdot}$	50	50
2-pyridyl	84	16
1-naphthyl	100	0

The fragmentation to electron transfer product ratio increases as the reduction potential of the parent aromatic moiety becomes more negative.

When the aryl is a phenyl group (benzene has a $E_{1/2}$ more negative than -3.0 V), the fragmentation to electron transfer ratio is 1:1. With the aryl being 2-pyridyl (pyridine $E_{1/2} = -2.7$ V) more electron transfer than fragmentation is observed, and, as a 1-naphthyl radical (naphthalene, $E_{1/2} = -2.5$ V), only the electron transfer product is observed (Reaction 60).

In the electrochemically induced reaction of 4-bromobenzophenone with benzenethiolate and alkanethiolate ions in DMSO or acetonitrile, only substitution and reduction products (benzophenone) were reported. No bond fragmentation was observed (68).

Phenylmethanethiolate ions as the nucleophile gave similar results. With phenyl radical, $k_f \gg k_t$ (ArX), only bond fragmentation is observed. With 2-pyridyl, 1-naphthyl, and 9-phenanthryl radicals, $k_f > k_t$ (ArX), some electron transfer is observed, but mainly bond fragmentation occurs. With 2-quinolyl radical (quinoline, $E_{1/2} = \sim 2.1$ V), $k_f \ll k_t$ (ArX), only electron transfer is observed (66).

The contrasting behavior of 1-naphthyl radicals and phenyl radicals in their reactions with alkanethiolate ions was attributed to the increased stability of 1-naphthyl alkyl sulfide radical anion, as compared with phenyl alkyl sulfide radical anion (44).

The stability of an aryl alkyl sulfide radical anion increases as the LUMO of the aryl moiety decreases (for instance, naphthalene compared to benzene). If the R–S bond energy, and hence the σ^* MO energy is constant, the energy gap between π^* and σ^* MOs will increase as the LUMO of the aromatic moiety decreases. Thus, the bond fragmentation rate will decrease.

On the other hand the reactions of a particular aryl halide with alkanethiolates of decreasing R–S bond dissociation energy and consequently *decreasing* σ^* MO energy, form radical anions where the energy of the σ^* MO becomes closer to the π^* MO, and even lower. Thus, bond fragmentation becomes easier.

Recently it has been shown that the nitrogen hyperfine splitting in the radical anion of p-alkylthionitrobenzenes, which is certainly a π^* radical anion, is smaller than expected due to electron transfer between the π^* MO of the nitrophenyl moiety containing the unpaired electron to the σ^* MO of the S–R bond. This charge transfer also increases with the bulkiness of the alkyl group (69). This interpretation is supported by INDO calculations (70). As the gap between σ^* and π^* MO energies decreases, the intramolecular transfer of the electron from a π^* MO to a σ^* MO, that ultimately leads to bond fragmentation, becomes more favorable (63).

The electrochemical reduction of **15** in DMF leads to bond fragmentation following Reactions 61, 62 with a rate of 9.5 s^{-1} at 19.5°C and an activation energy of 21 kcal/mol. Because ESR studies indicate that the radical anion **16** is of a π^* type, the high activation energy probably reflects the unfavorable electron transfer from the π^* MO of the p-nitrophenyl system to the σ^* MO of the C–S bond (71).

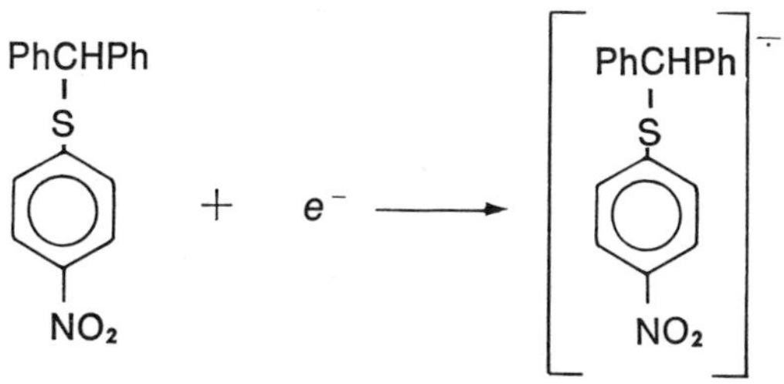

$$\text{(61)}$$

15   **16**

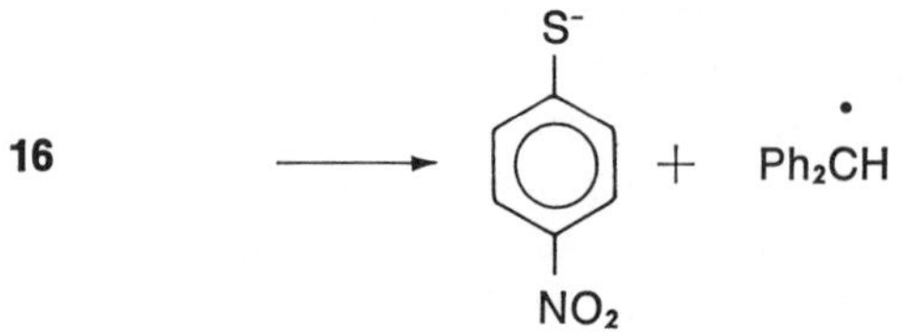

$$16 \longrightarrow \underset{NO_2}{\overset{S^-}{\bigcirc}} + Ph_2\overset{\bullet}{C}H \qquad (62)$$

It can be concluded that the changes in product ratio in Reactions 59 and 60 are due mainly to changes in bond fragmentation rates (Reaction 58).

MISCELLANEOUS FRAGMENTATION OF RADICAL ANIONS. Ketone enolate ions react with aryl halides under photostimulation to give α-arylketones in high yields. However, the corresponding reactions of methoxyacetone enolate ion **17** with bromobenzene gave less than 1% of 1-phenyl-1-methoxy-2-propanone **18**. The main product, obtained in 17% yield, was phenylacetone **19** (Reaction 63) (72).

$$PhBr + {}^-\underset{\underset{\textbf{17}}{OCH_3}}{CHCOCH_3} \xrightarrow{h\nu} Ph\underset{\underset{\textbf{18}}{OCH_3}}{CHCOCH_3} + PhCH_2COCH_3 \qquad (63)$$

$$\textbf{19}$$

Formation of **19** is suggested to come from the fragmentation of the radical anion formed in the coupling of phenyl radicals with **17** (Reactions 64–66).

$$Ph^{\cdot} + 17 \longrightarrow Ph\underset{\underset{\textbf{20}}{OCH_3}}{CH}-\overset{\overset{O^-}{|}}{\underset{}{C}}-\overset{\bullet}{C}H_3 \qquad (64)$$

$$20 \underset{-\ ^-OCH_3}{\overset{PhBr}{<}} \begin{array}{l} \longrightarrow 18 + (PhBr)^{\cdot-} \qquad (65) \\ \longrightarrow Ph\overset{\bullet}{C}HCOCH_3 \xrightarrow{SH} 19 \qquad (66) \\ \textbf{21} \end{array}$$

Relative amounts of **18** and **19** indicate that the rate of fragmentation to **21** is higher than the rate of electron transfer, with the result that termination (by way of **21**) dominates over chain propagation.

Electrochemical reduction of haloarenes to arene and halide ions is a two-electron reduction. However, quantitative yields of arene are not always obtained, and the consumption of electrons is below the theoretical value of two needed for the hydrogenolysis reaction. In a detailed study of the electrolysis of 4-bromobenzophenone in DMSO, several other products besides benzophenone, were formed. These products arise from the reaction of 4-benzoylphenyl radicals with the solvent (Reaction 67) (73).

$$ArBr + e^- \xrightarrow{DMSO} ArH + ArCH_3 + ArSCH_3 + ArSOCH_3 \qquad (67)$$

$$\begin{array}{cccc} \mathbf{22} & \mathbf{23} & \mathbf{24} & \mathbf{25} \\ (60\%) & (3\%) & (8\%) & (5\%) \end{array}$$

Ar = 4-benzoylphenyl

It was suggested that the 4-bromobenzophenone radical anion, formed by reaction of the substrate with an electron from the cathode, diffuses to the bulk solution and then decomposes into an aryl radical and bromide ion. The aryl radical abstracts hydrogen from the solvent (Reaction 68) and the dimsyl radical is further reduced to dimsyl anion (Reaction 69). The dimsyl anion acts as a nucleophile toward the aryl radical forming a radical anion intermediate that then fragments following Reaction 71.

$$Ar^. + CH_3SOCH_3 \longrightarrow ArH + {}^.CH_2SOCH_3 \qquad (68)$$

$$^.CH_2SOCH_3 + e^- \longrightarrow {}^-CH_2SOCH_3 \qquad (69)$$

$$Ar^. + {}^-CH_2SOCH_3 \longrightarrow (ArCH_2SOCH_3)^{\overline{.}} \overline{\begin{array}{l} \xrightarrow{-e^-} ArCH_2SOCH_3 \quad (70) \\ \qquad\qquad \mathbf{27} \\ \longrightarrow ArCH_2^. + {}^-SOCH_3 \quad (71) \end{array}}$$

$$\qquad\qquad\qquad\qquad\qquad \mathbf{26} \qquad\qquad\qquad\qquad \mathbf{28} \qquad\quad \mathbf{29}$$

Ar=4-benzoylphenyl

Radical **28** abstracts a hydrogen atom from DMSO to give product **23**. Ion **29**, formed in Reaction 71, also acts as a nucleophile and reacts with aryl radicals to give ultimately **25**, which, under the experimental conditions is reduced to **24**.

Decomposition of the radical anion **26** seems to be faster than electron transfer because **27** is not found among the products. Further confirmation that **26** decomposes according to Reaction 71 comes from the electrolysis of **27** under similar conditions. The electrolysis leads to the same products found in Reaction 67 (Reaction 72) (73).

$$27 + e^- \longrightarrow 26 \longrightarrow 28 + 29 \qquad (72)$$

These results contradict a report in which the photostimulated reaction (sunlight) of dimsyl anion with halobenzenes led to a high yield of the substitution product (benzyl methyl sulfoxide) by the $S_{RN}1$ mechanism (74). However, they agree with the well-known tendency of DMSO to produce methyl radicals upon dissociative electron capture (75, 76), and by the action of light on carbanions in DMSO (77).

Hydrogen Atom Abstraction with Formation of Reactive Intermediates. As is evident from the foregoing discussion, hydrogen atom abstraction is an important, competing reaction with aryl radicals and radical anions, especially with solvents that are good hydrogen donors.

The importance of this competing reaction in relation to $S_{RN}1$ substitution also depends on the hydrogen donor ability of the other species present (the substrate itself, the nucleophile, etc.), and on the reactivity of the aryl radical and the nucleophile.

If the radical formed after hydrogen atom abstraction is unreactive as a chain carrier, this competing reaction becomes a termination step (*see* Chapter 9). If it is a reactive intermediate it becomes part of the propagation steps.

In the photostimulated reaction of halobenzenes with acetone enolate ion in liquid ammonia, about 1% of benzene is formed. However, in the presence of equal concentrations of 2-propoxide ion, the yield of benzene increases to 32% at the expense of phenylacetone, which drops to 58% (the yield in the absence of 2-propoxide ion was 88%) (78). Contrary to other examples where hydrogen atom abstraction is significant (*see* Chapter 9), there is no decrease in the overall reactivity.

Alkoxides are known to be good hydrogen atom donors (79–81), and the resulting radical anion can act as a chain carrier in the $S_{RN}1$ mechanism by transferring the odd electron to the substrate (Reactions 73, 74).

$$\text{Ar}^{\cdot} + \text{H--}\overset{\overset{\displaystyle R}{|}}{\underset{\underset{\displaystyle R'}{|}}{\text{C}}}\text{--O}^{-} \longrightarrow \text{ArH} + \cdot\,\overset{\overset{\displaystyle R}{|}}{\underset{\underset{\displaystyle R'}{|}}{\text{C}}}\text{--O}^{-} \qquad (73)$$

$$\cdot\,\overset{\overset{\displaystyle R}{|}}{\underset{\underset{\displaystyle R}{|}}{\text{C}}}\text{--O}^{-} + \text{ArX} \longrightarrow (\text{ArX})^{\bar{\cdot}} + \text{R--}\overset{\overset{\displaystyle O}{||}}{\text{C}}\text{--R}' \qquad (74)$$

In fact, methoxide ion has been found to catalyze the reaction of 4-bromoisoquinoline with benzenethiolate ion in methanol (82).

The enolate ion of aldehydes gives substitution when irradiated in the presence of haloarenes in liquid ammonia, but substantial amounts of reduction also occur (83). Although the reaction was not fully investigated, the high yield of reduction products probably comes from hydrogen atom abstraction of the aldehyde function to produce a ketene type radical anion (Reactions 75, 76).

$$Ar^. + {}^-CHRCOH \begin{cases} (ArCHRCOH)^{\bar{}} & (75) \\ ArH + (CHR{=}C{=}O)^{\bar{}} & (76) \end{cases}$$

The importance of this reaction is evident in the reaction of o-iodoaniline with the enolate of acetaldehyde (84). As much as 25% of aniline was formed, although the substitution product (indole) was also obtained and the reactivity was unaffected. Similar results were obtained with o-iodoanisole (85)

Entrainment

Lack of reactivity in $S_{RN}1$ reactions is due occasionally to the failure to generate efficient initiation steps, even though the reactants may be quite reactive in the propagation steps. This circumstance was first recognized in the $S_{RN}1$ mechanism at aliphatic sites, where the addition of as little as 0.1 mol % of a reactive nucleophile to an otherwise unreactive mixture of a substrate and nucleophile, led to rapid formation of substitution products in high yield (86).

Similar effects have been observed in aromatic $S_{RN}1$ reactions. Diethyl phosphite ion is much less reactive than diphenylphosphide ion with aryl iodides in spontaneous initiation of the $S_{RN}1$ reaction (dark reaction). Diethyl phosphite reacts sluggishly with either p-iodotoluene (51) or iodobenzene (87), while diphenylphosphide is very reactive under the same conditions (51).

In experiments where potassium diethyl phosphite is allowed to react with p-iodotoluene in the dark with 10 mol % of potassium diphenyl phosphide, formation of diethyl p-tolylphosphonate was dramatically stimulated. A satisfactory explanation is that diphenyl phosphide ion is more reactive than diethyl phosphite at initiation, although both ions are about as reactive in propagation (relative reactivity ~ 1.7) (51).

Induction of a less reactive substrate has also been achieved by the addition of a reactive substrate. Thus, the thermal reaction of diphenyl phosphide ion with p-bromotoluene in DMSO is quite slow, but the addition of 33 mol % of p-iodotoluene sharply accelerates the release of bromide ion from p-bromotoluene (Figure 8) (*51*). In this case, addition of a more reactive substrate at initiation generates reactive intermediates, that thereby drive a species that is unreactive at initiation into the propagation steps (Reactions 77–80).

Initiation $\quad\quad\quad\quad\quad$ $p-\text{ITo} + \text{Ph}_2\text{P}^{-} \longrightarrow p-\text{To}^{\cdot} + \text{I}^{-} + \text{Ph}_2\text{P}^{\cdot}$ $\quad\quad$ (77)

Propagation $\quad\quad\quad\quad$ $p-\text{To}^{\cdot} + \text{Ph}_2\text{P}^{-} \longrightarrow (p-\text{ToPh}_2\text{P})^{\overline{\cdot}}$ $\quad\quad$ (78)

$$p - \text{BrTo}$$
$$(p-\text{ToPh}_2\text{P})^{\overline{\cdot}} \longrightarrow \begin{array}{l} \xrightarrow{} p-\text{To}^{\cdot} + \text{Br}^{-} + p-\text{ToPh}_2\text{P} \quad\quad (79) \\ \\ \xrightarrow{} p-\text{To}^{\cdot} + \text{I}^{-} + p-\text{ToPh}_2\text{P} \quad\quad (80) \end{array}$$
$$p-\text{ITo}$$

To = toluene (tolyl)

This effect has been called entrainment, and p-iodotoluene or potassium diphenylphosphide ion are, respectively, the entraining substrate and nucleophile, whereas p-bromotoluene and potassium diethyl phosphite are the entrained substrate and nucleophile, respectively.

Entrainment is also found in the reaction of acetone enolate ion with

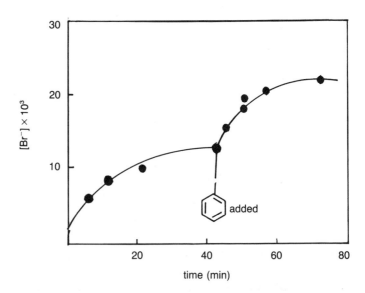

Figure 8. Reaction of p-bromotoluene with potassium diphenyl phosphide in DMSO in the dark. p-Iodotoluene in 33 mol % is added at 42.6 min. (Reproduced, from Ref. 51. Copyright 1979, American Chemical Society.)

2-chloroquinoline, which is very slow in the dark, but can be accelerated with the addition of 10 mol % of benzoylacetone dianion (*88*).

The enormous differences in the relative rates of substitution of aryl iodides with pinacolone enolate ion, depending on the method of measurement, can be similarly understood. For example, when the rate for each substrate is measured separately, the reactivity spread between the fastest and the slowest reaction is about four-hundred-fold; but when the relative reactivity is measured in competition experiments, the spread is less than two-fold (*89*). These differences were attributed to differences in the rates that are actually measured in the two sets of experiments.

When the two substrates react competitively, the relative reactivity is determined by the relative rates of electron transfer to the two competing substrates. When the two substrates react separately, overall reactivity is determined not only by the events within the propagation cycle, but by the relative rates of initiation and termination also. Examples of entrainment are also found in other instances where reactivity is determined (*90*).

Care should be taken in choosing the entraining substrate because a very reactive substrate does not always entrain the reaction of a less reactive one. This is because the overall reactivity of a substrate is determined by its reduction potential and the nucleofugality of the leaving group. For example, in the hypothetical reaction of 4-chlorobenzophenone, which is much more reactive than chlorobenzene with a particular nucleophile Nu^-, 4-chlorobenzophenone will not entrain the reaction of chlorobenzene (Reactions 81–83) because electron transfer, Step 83, is thermodynamically disfavored. In this situation inhibition of the reaction of chlorobenzene is expected.

Therefore, for entrainment to take place, the product of the entraining substrate with the nucleophile should have a similar or lower (more negative) reduction potential than the entrained substrate.

In the example of p-bromotoluene entrained by p-iodotoluene in reaction with diphenylphosphide ion, the electron transfer reaction (Reaction 79) is thermodynamically favored with both substrates, p-bromotoluene and p-iodotoluene. The triphenyl phosphine reduction potential is -2.68 V (*91*) and probably the reduction potential of p-tolyldiphenylphosphine is similar, whereas the reduction potential of bromobenzene (-2.23 V) and iodobenzene (-1.64 V) are more positive (*92*). Entrainment is thus very effective.

Inhibition of $S_{RN}1$ Reactions

The strong evidence for the radical character of the $S_{RN}1$ mechanism comes from the finding that radical scavengers inhibit reaction. Thus, addition of small amounts of certain compounds (less than 1 mol % and

(81)

(82)

(83)

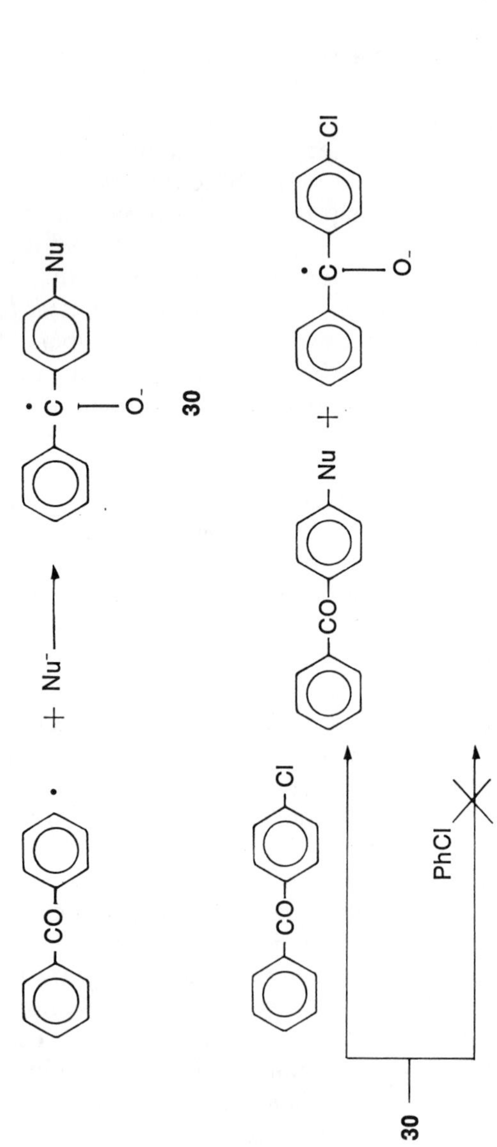

30

PhCl

30

up to 20 mol %) can slow or totally depress $S_{RN}1$ reactions. Conversely, in some cases, added substances increase rather than decrease the rate of reaction. In summary, added compounds can suppress or inhibit reaction by depressing the rate of the initiation step, by inhibiting the chain propagation steps, by reacting with reactive intermediates, radical or radical anions, to form lethargic radical intermediates that are unable to propagate the chain, and by increasing the rate of termination steps.

Depressing the rate of the initiation step applies mainly to photostimulated reactions and occurs with substances that either absorb the useful light or quench the excited state of the species that initiates the reaction.

Inhibiting the chain propagation steps and increasing the rate of termination steps are very important in all the $S_{RN}1$ reactions, whatever the initiation step. This inhibition is observed with compounds that are good hydrogen atom donors (*see* Chapter 9).

In Table IX the literature data available have been compiled regarding the effect of added substances on $S_{RN}1$ reactions.

When the reaction of 5- and 6-iodotrimethylbenzenes with amide ions is carried out in the presence of 2-methyl-2-nitrosopropane or tetraphenylhydrazine, $S_{RN}1$ reactions are inhibited because these compounds act as aryl trapping agents (*93*).

In some cases it is not entirely clear which inhibition mechanism is operative. Dioxygen, for instance, can react with aryl radicals (Reaction 84), or can accept an electron from a radical anion (Reaction 85).

$$Ar^{\cdot} + O_2 \longrightarrow ArOO^{\cdot} \qquad (84)$$

$$(RA)^{\overline{\cdot}} + O_2 \longrightarrow RA + O_2^{\overline{\cdot}} \qquad (85)$$

The couple $O_2/O_2^{\overline{\cdot}}$ has a relatively positive reduction potential (-0.77 V) (*96*). Thus, dioxygen radical anion, being unable to transfer its odd electron to the substrate, can stop chain propagation because Reaction 86 is thermodynamically unfavorable and slow.

$$ArX + O_2^{\overline{\cdot}} \longrightarrow (ArX)^{\overline{\cdot}} + O_2 \qquad (86)$$

Dioxygen was shown to be a strong inhibitor of the photostimulated reaction of bromobenzene with acetone enolate ion, which is very slow in liquid ammonia under air (Figure 9) (*78*). It also inhibits the dark reaction of 4-chlorobenzophenone and diphenylarsenide ion in ammonia (*31*). On the other hand, the photostimulated reaction of iodobenzene and diethyl phosphite ion in DMSO or in ammonia is not affected by dioxygen (*87*). In the dark reaction of iodobenzene with pinacolone enolate ion in DMSO, under dioxygen, there is a spectacular increase in

Table IX
Effect of Added Substances on S$_{RN}$1 Reactions

System	Solvent	Promoter	Other	Results	Ref.
Iodotrimethyl-benzene + NH_2^-	NH_3	dark	$(Me)_3CNO$	inhibition	93
			Ph_2NNPh_2	inhibition	93
$PhBr + {}^-CH_2COCH_3$	NH_3	$h\nu$	$(t-Bu)_2NO^.$	inhibition	78
			O_2	inhibition	78
$PhI + {}^-CH_2COC(CH_3)_3$	DMSO	dark	$p-(NO_2)_2C_6H_4$	inhibition	95
			$PhCOPh$	inhibition	95
			$(t-Bu)_2NO^.$	inhibition	95
			O_2	catalysis	95
			$PhNO_2$	inhibition and then catalysis	95
			$Fe(NO_3)_3$	catalysis	95
			$Fe_2(SO_4)_3$	slight inhibition	95
$PhI + (EtO)_2PO^-$	DMSO	$h\nu$	$(t-Bu)_2NO^.$	inhibition	87
			O_2	no effect	87
			Cu^{++}	no effect	87
			$FeCl_3{}^a$	inhibition	87
p-Iodotoluene + Ph_2P^-	DMSO	dark	$(t-Bu)_2NO^.$	inhibition	51
			$m-(NO_2)_2C_6H_4$	inhibition	51
			PhN_2Ph	inhibition	51
			$Ph_2C=CH_2$	slight inhibition	51
			$p-(NO_2)_2C_6H_4$	slight inhibition	51
	NH_3	dark	$p-(NO_2)_2C_6H_4$	inhibition	51
			$PhNO_2$	inhibition	51
			$(t-Bu)_2NO^.$	inhibition	51
			galvinoxyl	no effect	51
$PhBr + {}^-CH_2CN$	NH_3	$h\nu$	naphthalene	inhibition	30
			$PhCOPh$	inhibition	30
			anthracene^b	slight inhibition	30
$PhI + {}^-CH_2CN$	NH_3	$h\nu$	pyridine	inhibition	30
$4-ClC_6H_4COC_6H_5 + Ph_2As^-$	NH_3	dark	$m-(NO_2)_2C_6H_4$	inhibition	31
			O_2	inhibition	31

Table IX (Continued)
Effect of Added Substances on $S_{RN}1$ Reactions

System	Solvent	Promoter	Other	Results	Ref.
4-R-5-Halopyrimi- dines + $^-CH_2COR$	NH_3	dark	$(t-Bu)_2NO^{\cdot}$	inhibition	121
PhI + $(EtO)_2PO^-$	NH_3	$h\nu$	$p-(NO_2)_2C_6H_4$	inhibition	122
2-Cl–quinoline + $^-CH_2CO\bar{C}HCOPh$	NH_3	dark	Ph_2NNPh_2	inhibition	123
			$p-(NO_2)_2C_6H_4$	inhibition	123
			O_2	inhibition	123
2-Cl–quinoline + $^-CH_2COCH_3$	NH_3	$h\nu$	$p-(NO_2)_2C_6H_4$	inhibition	88, 124
2-Br–pyridine + $^-CH_2COCH_3$	NH_3	$h\nu$	$(t-Bu)_2NO^{\cdot}$	inhibition	125
4-Br–isoquinoline + PhS^-	CH_3OH	CH_3O^-	PhN_2Ph	inhibition	82
N-Acyl– o-chloroaniline	THF– hexane	$h\nu$	$(t-Bu)_2NO^{\cdot}$	inhibition	126

^{a}FeCl$_3$ absorbs at the wavelength used.
^{b}Slightly soluble in ammonia.

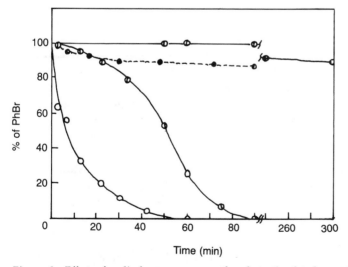

Figure 9. Effect of radical scavengers on the photostimulated reaction of bromobenzene with acetone enolate ion. Key: ○, without scavengers; ◐, with 0.68 mol % of di-t-butyl nitroxide; ⊕, with 4.3 mol % of di-t-butyl nitroxide; and ●, under air. (Reproduced, from Ref. 78. Copyright 1973, American Chemical Society.)

the release of iodide ion (95). The substitution product, 1-phenyl-3-methylbutanone could not be isolated in the latter reaction because it was shown to be oxidized under the reaction conditions with the formation of some benzoic acid. Conceivably some intermediates in the oxidation, such as peroxides, can act as initiators of the chain propagation sequence (95).

Lack of inhibition in the reaction of diethyl phosphite with iodobenzene was attributed to the low reactivity of dioxygen toward phenyl radicals in competition with diethyl phosphite (Reaction 87). This is consistent with the fact that diethyl phosphite ion behaves as a superb nucleophile.

$$Ph^{\cdot} + (EtO)_2PO^- \longrightarrow [PhP(O)(OEt)_2]^{\overline{\cdot}} \qquad (87)$$

Because the rate of reaction of phenyl radicals with dioxygen (Reaction 84) is $5 \times 10^9 \ M^{-1} \ s^{-1}$ in hydrocarbon solvents (96), and because dioxygen does not inhibit Reaction 87, it follows that the rates of Reactions 84 and 87 may be comparable.

Di-t-butyl nitroxide is also a very efficient inhibitor (Figure 9). So far, it has inhibited all reactions tested with it, not only in S$_{RN}$1 aromatic systems, but also in aliphatic systems (97). Inhibition may occur not only because it acts as a radical trap, but also because it is an effective excited-state quencher (98).

Substances that are very good electron acceptors act as inhibitors because they accept electrons more easily than the substrate, acting as a sink for the electrons, and decreasing the overall reactivity. The type of compound that inhibits a reaction by this mechanism depends on the reduction potential of the substrate and inhibitor.

Probably for this reason, the reactivity of some aromatic substrates toward nucleophiles cannot be determined by direct competition experiments, inasmuch as one of the substrates inhibits the reaction of the other. When 2-chloroquinoline and iodobenzene were allowed to react with acetone enolate ion in liquid ammonia under photostimulation, 2-chloroquinoline reacted, but iodobenzene did not react, in 1 h (99). However when iodobenzene reacted alone, 100% of reaction was obtained in less than 5 min (78).

$$PhI + {}^-CH_2COCH_3 \xrightarrow[5 \ min]{h\nu} PhCH_2COCH_3 \ (100\%) \qquad (88)$$

PhI + 2 Cl-quinoline

$$+ \ ^-CH_2COCH_3 \xrightarrow[1 \ h]{h\nu} \begin{cases} \longrightarrow PhCH_2COCH_3 \ (<1\%) & (89) \\ \longrightarrow 2\text{-quinolyl acetone } (65\%) & (90) \end{cases}$$

The selectivity of acetone enolate ion toward 2-chloroquinoline might be attributed to the relative ease of reduction of individual substrates to the corresponding radical anion in the initiation step, with 2-chloroquinoline being able to accept the electron more readily than iodobenzene (99).

Remarkably, the photochemical reactivity of bromobenzene with acetone enolate ion is severely depressed by admixture of cyanomethyl anion in liquid ammonia (56).

m-Dinitrobenzene and p-dinitrobenzene are strong electron acceptors and usually inhibit these reactions, although to various degrees. Their action is complicated by the fact that they may form σ-complexes with the nucleophile, or react irreversibly with it (Reactions 91, 92) (100).

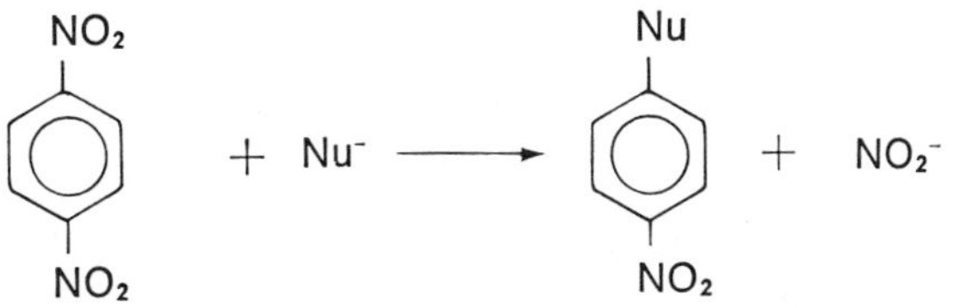

$$\text{p-(NO}_2)_2\text{C}_6\text{H}_4 + \text{Nu}^- \longrightarrow \text{p-NO}_2\text{-C}_6\text{H}_4\text{-Nu} + \text{NO}_2^- \qquad (91)$$

$$\text{m-(NO}_2)_2\text{C}_6\text{H}_4 + \text{Nu}^- \longrightarrow \text{(σ-complex)} \qquad (92)$$

A strange effect was found in the dark reaction of iodobenzene and pinacolone enolate ion in DMSO with nitrobenzene as inhibitor (95). The reaction is inhibited initially, but then it becomes faster than the reaction in its absence (Figure 10).

The nitrobenzene was completely consumed and good yields of products were obtained. The possibility that a σ-complex 31 is formed, and it, or a species derived from it, acts as initiator of the reaction has been suggested. Meisenheimer σ-complexes, such as 31, are good electron donors.

Comparative Reactivity

The propagation steps for $S_{RN}1$ reactions of one aromatic substrate (ArX) with two nucleophiles, Y^- and Z^-, are shown in Reactions 93–95.

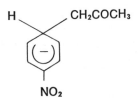

31

$$(ArX)^{\bar{\cdot}} \longrightarrow Ar^{\cdot} + X^- \tag{93}$$

$$Ar^{\cdot} + Y^- \longrightarrow (ArY)^{\bar{\cdot}} \tag{94y}$$

$$Ar^{\cdot} + Z^- \longrightarrow (ArZ)^{\bar{\cdot}} \tag{94z}$$

$$(ArY)^{\bar{\cdot}} + ArX \longrightarrow (ArX)^{\bar{\cdot}} + ArY \tag{95y}$$

$$(ArZ)^{\bar{\cdot}} + ArX \longrightarrow (ArX)^{\bar{\cdot}} + ArZ \tag{95z}$$

According to this mechanism, nucleophiles participate in steps leading to their incorporation into reaction products chiefly in Reactions 94y and 94z. If propagation events are much more frequent than termination events, the relative yields of ArY and ArZ should, when account is also taken of the nucleophile concentrations, be indicative of the relative rate constants for Reactions 94y and 94z.

Pairs of nucleophiles Y^- and Z^- were allowed to react in ammonia with single substrates, and relative nucleophilicities were evaluated from product yields (*101*).

Using iodobenzene as substrate, the order of nucleophilic reactivity versus phenyl radical is $Ph_2P^-(5.9) > Ph_2PO^-(2.7) > (EtO)_2PO^-(1.4) > Me_3CCOCH_2^-$ (1.00) $> PhS^-$ (0.08).

The small spread in nucleophilic reactivity is remarkable and is consistent with previous data. For instance, in reactions provoked by solvated electrons, amide ion (NH_2^-) is 2.0 times as reactive as acetone enolate ion versus the 2,4,6-trimethylphenyl radical (*102*), and 1.9 times as reactive versus phenyl radical (*103*). The enolate ion from 2,4-dimethyl-3-pentenone is 3.2 times as reactive as that of acetone versus 2-quinolyl radical (*99*) and 3.7 times as reactive versus 2-pyridyl radical (*104*).

All this information indicates that the reaction of aryl radicals with nucleophiles occurs virtually at encounter-controlled rates. To account for the lower reactivity of benzenethiolate ions relative to other nucleophiles, it was suggested that the encounter complex of phenyl radicals and benzenethiolate ions forms reversibly (Reaction 96) and dissociates faster than it undergoes C–S bond formation (Reaction 97) to form diphenyl sulfide radical anion (*101*).

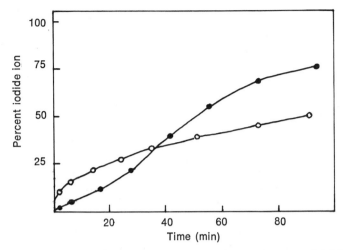

Figure 10. The effect of nitrobenzene on the reaction rate of iodobenzene and pinacolone enolate ion. Key: ○, without nitrobenzene; and ●, with 6×10^{-3} M nitrobenzene added. (Reproduced, from Ref. 95. Copyright 1977, American Chemical Society.)

$$Ph^{\cdot} + PhS^{-} \rightleftharpoons \left[Ph^{\cdot} PhS^{-} \right] \qquad (96)$$

encounter complex

$$\left[Ph^{\cdot} PhS^{-} \right] \longrightarrow (PhSPh)^{\overline{\cdot}} \qquad (97)$$

Among the nucleophiles of the PhZ^{-} family ($Z = O$, S, Se, and Te), the first member of the series, phenoxide ion, as well as alkoxide ion, is unreactive under most reaction conditions (*105–107*), and the reactivity order of the other members of the series toward 2-quinolyl radical is as follows (*108*):

PhO^{-}	unreactive	
PhS^{-}	(1.00)	
$PhSe^{-}$	5.8	vs. 2-Quinolyl
$PhTe^{-}$	39	

The relative reactivities as measured by electrochemical techniques are acetone enolate ion 1.00, benzenethiolate ion 1.9, and diethyl phosphite ion 2.4. If acetone enolate ion and pinacolone enolate ion have nearly the same reactivity as has been suggested (*109*), the relative reactivity previously estimated for diethyl phosphite ion (2.4) (*18*) is fairly close to the one measured by competition (*101*). However, the relative reactivity of benzenethiolate ion is about 24 times greater than indicated from previous data (101). The difference might be attributed to different selectivity of aryl and 2-quinolyl radicals, since the electrochemical data refer to reactions with this radical (101).

Literature Cited

1. Galli, C.; Bunnett, J. F. *J. Am. Chem. Soc.* **1979**, *101*, 6137.
2. Gores, G. J.; Koepped, C. E.; Bartak, D. E. *J. Org. Chem.* **1979**, *44*, 380.
3. Beland, F. A.; Farwell, S. O.; Callis, P. R.; Geer, R. D. *J. Electroanal. Chem.* **1977**, *78*, 145.
4. Grimshaw, J.; Trocha-Grimshaw, J. *J. Chem. Soc., Perkin 1*, **1974**, 1383.
5. Nelson, R. F.; Carpenter, A. K.; Seo, E. T. *J. Electrochem. Soc.* **1973**, *120*, 206.
6. Danen, W. C.; Kensler, T. T.; Lawless, J. G.; Marcus, M. F.; Hawley, M. D. *J. Phys. Chem.* **1969**, *73*, 4389.
7. Saveant, J. M.; Thiebault, T. *J. Electroanal. Chem.* **1978**, *89*, 335.
8. van Duyne, R. P.; Reilley, C. N. *Anal. Chem.* **1972**, *44*, 158.
9. Lawless, J. G.; Hawley, M. D. *J. Electroanal. Chem.* **1969**, *21*, 365.
10. Parker, A. J. *Chem. Rev.* **1969**, *69*, 1.
11. Porter, G.; Ward, B. *Proc. R. Soc. London Ser. A* **1965**, *287*, 457.
12. Bennett, J. E.; Mile, B.; Thomas, A. *Proc. R. Soc. London Ser. A* **1966**, *293*, 246.
13. Kasai, P. H.; Hedaya, E.; Whipple, E. B. *J. Am. Chem. Soc.* **1969**, *91*, 4364.
14. Kasai, P. H.; Clark, P. A.; Whipple, E. B. *J. Am. Chem. Soc.* **1970**, *92*, 2640.
15. Pople, J. A.; Beveridge, D. L.; Dobosh, P. A. *J. Chem. Phys.* **1963**, *39*, 1307.
16. Zemel, H.; Fessenden, R. W. *J. Chem. Phys.* **1975**, *79*, 1419.
17. Gaines, A. F.; Page, F. M. *Trans. Faraday Soc.* **1966**, *62*, 3086.
18. Amatore, C.; Chaussard, J.; Pinson, J.; Saveant, J. M.; Thiebault, A. *J. Am. Chem. Soc.* **1979**, *101*, 6012.'
19. Helgee, B.; Parker, V. D. *Acta Chem. Scand. Ser. B.* **1980**, *34*, 129.
20. Marcus, R. A. *J. Chem. Phys.* **1956**, *24*, 966.
21. *Ibid.*, **1965**, *43*, 679.
22. Marcus, R. A.; Sutin, N. *Inorg. Chem.* **1975**, *14*, 213.
23. Frost, A. A.; Pearson, R. G. "Kinetics and Mechanism"; 2nd ed.; Wiley: New York, 1961, p. 271.
24. Lewis, N. A. *J. Chem. Ed.* **1980**, *57*, 478.
25. Saveant, J. M. *Acc. Chem. Res.*, **1980**, *13*, 323.
26. Amatore, C.; Pinson, J.; Saveant, J. M. *J. Electroanal. Chem.* **1980**, *107*, 59.
27. van Tilborg, W. J. M.; Smit, C. J.; Scheele, J. J. *Tetrahedron Lett.* **1977**, 2113.
28. Bunnett, J. F.; Shafer, S. J. *J. Org. Chem.* **1978**, *43*, 1877.
29. Hart, E. J.; Anbar, M. "The Hydrated Electron"; Wiley: New York, 1970.
30. Rossi, R. A.; de Rossi, R. H.; Pierini, A. B. *J. Org. Chem.* **1979**, *44*, 2662.
31. Rossi, R. A.; Alonso, R. A.; Palacios, S. M. *J. Org. Chem.* **1981**, *46*, 2498.
32. Szwarc, M. *Acc. Chem. Res.* **1972**, *5*, 169.
33. Schindewolf, U.; Wünschal, P. *Can. J. Chem.* **1977**, *55*, 2159.
34. Schindewolf, U.; Neuman, B. *J. Phys. Chem.* **1979**, *83*, 423.
35. Farhataziz, Perkey, L. M. *J. Phys. Chem.* **1976**, *80*, 122.
36. San Roman, E.; Krebs, P.; Schindewolf, U. *Chem. Phys. Letters* **1977**, *49*, 98.
37. Eigen, M. *Angew. Chem. Int. Ed.* **1964**, *3*, 1.
38. Laitinen, H. A.; Nyman, C. J. *J. Am. Chem. Soc.* **1948**, *70*, 3002.
39. Harima, Y.; Aoyagui, S. *Isr. J. Chem.* **1979**, *18*, 81.
40. Teherani, T.; Itaya, K.; Bard, A. J. *Nouv. J. Chim.* **1978**, *2*, 481.
41. Szwarc, M. *Pure Appl. Chem.* **1979**, *51*, 1049.
42. Alwair, K.; Grimshaw, J. *J. Chem. Soc., Perkin 2*, **1973**, 1811.
43. Symons, M. C. R. *J. Chem. Soc., Chem. Commun.* **1978**, 313.
44. Symons, M. C. R. *Pure. Appl. Chem.* **1981**, *53*, 223.
45. Buick, A. R.; Kemp, T. J.; Neal, G. T.; Stone, T. J. *J. Chem. Soc. A* **1969**, 666.

46. Houser, K. J.; Bartak, D. E.; Hawley, M. D. *J. Am. Chem. Soc.* **1973**, *95*, 6033.
47. Volke, J.; Manousek, O.; Troyepolskaya, T. V. *J. Electroanal. Chem.* **1977**, *85*, 163.
48. Bunnett, J. F.; Creary, X. *J. Org. Chem.* **1974**, *39*, 3611.
49. Bunnett, J. F.; Scamehorn, R. G.; Traber, R. P. *J. Org. Chem.* **1976**, *41*, 3677.
50. Pierini, A. B.; Rossi, R. A. *J. Org. Chem.* **1979**, *44*, 4667.
51. Swartz, J. E.; Bunnett, J. F. *J. Org. Chem.* **1979**, *44*, 340.
52. Alonso, R. A.; Rossi, R. A. *J. Org. Chem.* **1982**, *47*, 77.
53. Streitwieser, A. "Molecular Orbital Theory for Organic Chemistry"; Wiley: New York, 1961.
54. Levason, W.; McAuliffe, C. A. *Acc. Chem. Res.* **1978**, *11*, 363.
55. Lehnkuhl, H. *Synthesis (Stuttgart)* **1973**, *7*, 1.
56. Bunnett, J. F.; Gloor, B. F. *J. Org. Chem.* **1973**, *38*, 4156.
57. Goldfarb, I. L.; Ikubov, A. P.; Belenki, L. I. *Zhur. Geter. Soedini* **1979**, 1044.
58. Rossi, R. A.; de Rossi, R. H.; Lopez, A. F. *J. Org. Chem.* **1976**, *41*, 3371.
59. Rossi, R. A.; de Rossi, R. H.; López, A. F. *J. Org. Chem.* **1976**, *41*, 3367.
60. Rossi, R. A.; de Rossi, R. H.; López, A. F. *J. Am. Chem. Soc.* **1976**, *98*, 1252.
61. Mohammad, M.; Hajdu, J.; Kosower, E. M. *J. Am. Chem. Soc.* **1971**, *93*, 1792.
62. Pearson, R. G. "Symmetry Rules for Chemical Reactions"; Wiley: New York, 1976; p. 440.
63. Neta, P.; Behar, D. *J. Am. Chem. Soc.* **1980**, *102*, 4798.
64. Bunnett, J. F.; Creary, X., *J. Org. Chem.* **1975**, *40*, 3740.
65. Ibid., **1974**, *39*, 3173.
66. Rossi, R. A.; Palacios, S. M. *J. Org. Chem.* **1981**, *46*, 5300.
67. Benson, S. W. *Chem. Rev.* **1978**, *78*, 23.
68. Pinson, J.; Saveant, J. M. *J. Am. Chem. Soc.* **1978**, *100*, 1506.
69. Alberti, A.; Martelli, G.; Pedulli, G. F. *J. Chem. Soc., Perkin 2* **1977**, 1252.
70. Alberti, A.; Guerra, M.; Bernardi, F.; Mangini, A.; Pedulli, G. F. *J. Am. Chem. Soc.* **1979**, *101*, 4627.
71. Farnia, G.; Severin, M. G.; Capobianco, G.; Vianello, E. *J. Chem. Soc., Perkin 2* **1978**, 1.
72. Bunnett, J. F.; Sundberg, J. E. *J. Org. Chem.* **1976**, *41*, 1707.
73. M'Halla, F.; Pinson, J.; Saveant, J. M. *J. Electroanal. Chem.* **1978**, *89*, 347.
74. Rajan, S.; Muralimohan, K. *Tetrahedron Lett.* **1978**, 483.
75. Chung, Y. J.; Nishikida, K.; Williams, F. *J. Phys. Chem.* **1974**, *78*, 1882.
76. Cooper, T. K.; Walker, D. C.; Gillis, H. A.; Klassen, N. V. *Can. J. Chem.* **1973**, *51*, 2195.
77. Tolbert, L. M. *J. Am. Chem. Soc.* **1980**, *102*, 3531, 6808.
78. Rossi, R. A.; Bunnett, J. F. *J. Org. Chem.* **1973**, *38*, 1407.
79. Bunnett, J. F.; Wamser, C. C. *J. Am. Chem. Soc.* **1967**, *89*, 6712.
80. Bunnett, J. F.; Takayama, H. *J. Am. Chem. Soc.* **1968**, *90*, 5173.
81. Zoltewicz, J. A.; Oestreich, T. M.; Sale, A. A. *J. Am. Chem. Soc.* **1975**, *97*, 5889.
82. Zoltewicz, J. A.; Oestreich, T. M. *J. Am. Chem. Soc.* **1973**, *95*, 6863.
83. López, A. F., Ph. D. Thesis, Universidad Nacional de Córdoba, 1979.
84. Beugelmans, R.; Roussi, G. *J. Chem. Soc., Chem. Commun.* **1979**, 950.
85. Beugelmans, R.; Ginsburg, H. *J. Chem. Soc., Chem. Commun.* **1980**, 508.
86. Kornblum, N.; Swiger, R. T.; Earl, G. W.; Pinnick, H. W.; Stuchal, F. W. *J. Am. Chem. Soc.* **1970**, *92*, 5513.
87. Hoz, S.; Bunnett, J. F. *J. Am. Chem. Soc.* **1977**, *99*, 4690.
88. Hay, J. V.; Hudlicky, T.; Wolfe, J. F. *J. Am. Chem. Soc.* **1975**, *97*, 374.
89. Scamehorn, R. G.; Bunnett, J. F. *J. Org. Chem.* **1979**, *44*, 2604.
90. Swartsz, J. E.; Bunnett, J. F. *J. Org. Chem.* **1979**, *44*, 4673.
91. Santhanam, K. S. V.; Bard, A. J. *J. Am. Chem. Soc.* **1968**, *90*, 1118.
92. Fry, A. J.; Krieger, R. L. *J. Org. Chem.* **1976**, *41*, 54.
93. Kim, J. K.; Bunnett, J. F. *J. Am. Chem. Soc.* **1970**, *92*, 7463.
94. Peover, M. E.; White, B. S. *Electrochem. Acta* **1966**, *11*, 1061.
95. Scamehorn, R. G.; Bunnett, J. F. *J. Org. Chem.* **1977**, *42*, 1449.

96. Kyger, R. G.; Lorano, J. P.; Stevens, N. P.; Herron, N. R. *J. Am. Chem. Soc.* **1977**, *99*, 7589.
97. Kornblum, N. *Ang. Chem. Int. Ed.* **1975**, *14*, 734.
98. Schwerzel, R. E.; Caldwell, R. A. *J. Am. Chem. Soc.* **1973**, *95*, 1382.
99. Hay, J. V.; Wolfe, J. F. *J. Am. Chem. Soc.* **1975**, *97*, 3702.
100. Strauss, M. J. *Chem. Rev.* **1970**, *70*, 667.
101. Galli, C.; Bunnett, J. F. *J. Am. Chem. Soc.*, **1981**, *103*, 7140.
102. Tremelling, M. J.; Bunnett, J. F. *J. Am. Chem. Soc.* **1980**, *102*, 7375.
103. Bunnett, J. F.; Gloor, B. F., quoted in ref. 102.
104. Komin, A. P.; Wolfe, J. F. *J. Org. Chem.* **1977**, *42*, 2481.
105. Rossi, R. A.; Bunnett, J. F. *J. Org. Chem.* **1973**, *38*, 3020.
106. Semmelhack, M. F.; Bargar, T. J. *Am. Chem. Soc.* **1980**, *102*, 7765.
107. Rossi, R. A.; Pierini, A. B. *J. Org. Chem.* **1980**, *45*, 2914.
108. Pierini, A. B.; Pěneňory, A. B.; Rossi, R. A., unpublished results.
109. Scamehorn, R. G.; Hardacre, J., quoted in ref. 101.
110. Grimshaw, J.; Trocha-Grimshaw, J. *J. Electroanal. Chem.* **1974**, *56*, 443.
111. Nadjo, L.; Saveant, J. M. *J. Electroanal. Chem.* **1971**, *30*, 41.
112. Margel, S.; Levy, M. *J. Electroanal. Chem.* **1974**, *56*, 259.
113. Andrieux, C. P.; Dumas-Bouchiat, J. M.; Saveant, J. M. *J. Electroanal. Chem.* **1978**, *87*, 55.
114. Andrieux, C. P.; Blocman, C.; Dumas-Bouchiat, J. M.; Saveant, J. M. *J. Am. Chem. Soc.* **1979**, *101*, 3431.
115. Andrieux, C. P.; Blocman, C.; Saveant, J. M. *J. Electroanal. Chem.* **1979**, *105*, 413.
116. Dorfman, L. M. *Acc. Chem. Res.* **1970**, *3*, 224.
117. Shimozato, Y.; Shimada, K.; Szwarc, M. *J. Am. Chem. Soc.* **1975**, *97*, 5831.
118. Bartak, D. E.; Houser, K. J.; Rudy, B. C.; Hawley, M. D. *J. Am. Chem. Soc.* **1972**, *94*, 7526.
119. Alwair, K.; Grimshaw, J. *J. Chem. Soc., Perkin 2* **1973**, 1150.
120. "Handbook of Chemistry and Physics," 61st ed.; CRC Press: Boca Raton, FL, 1981; p. 4243.
121. Oostveen, E. A.; van der Plas, H. C. *Recl. Trav. Chim. Pays-Bas* **1979**, *98*, 441.
122. Bunnett, J. F.; Traber, R. P. *J. Org. Chem.* **1978**, *43*, 1867.
123. Wolfe, J. F.; Green, J. C.; Hudlicky, T. *J. Org. Chem.* **1972**, *37*, 3199.
124. Hay, J. V.; Wolfe, J. F. *J. Am. Chem. Soc.* **1975**, *97*, 3702.
125. Komin, A. P.; Wolfe, J. F. *J. Org. Chem.* **1977**, *42*, 2481.
126. Wolfe, J. F.; Sleevi, M. C.; Goehring, R. R. *J. Am. Chem. Soc.* **1980**, *102*, 3646.
127. "Handbook of Chemistry and Physics," 61st ed.; CRC Press: Boca Raton, FL, 1981; p. F-233.

The Termination Step

A reaction that produces an unreactive species from a reactive intermediate of the chain propagation steps is a termination step. Some termination steps depend on the method used to provoke the $S_{RN}1$ reaction, while others depend on the reaction conditions and intermediates involved. There are others that are intrinsic termination steps characteristic of the mechanism.

The nature of the termination steps, as well as that of the initiation steps of the $S_{RN}1$ reaction, remains obscure in many cases because termination steps produce very small concentrations of products, and they are difficult to detect by the usual analytic techniques. In this chapter we will discuss, in general terms, some possible termination steps.

Termination Steps That Depend on the $S_{RN}1$ Initiation Reaction

If initiation of the $S_{RN}1$ reaction is by solvated electrons in liquid ammonia, the same solvated electrons can act as termination agents, reacting with the intermediates of the chain (Reactions 1, 2).

$$Ar^{\cdot} + e^{-} \longrightarrow Ar^{-} \xrightarrow{NH_3} ArH \qquad (1)$$

$$(ArNu)^{\overline{\cdot}} + e^{-} \longrightarrow (ArNu)^{2-} \xrightarrow{NH_3} ArNuH_2 \qquad (2)$$

Both products of these termination steps are found in reactions stimulated by solvated electrons in liquid ammonia. ArH and $ArNuH_2$ are the reduction products of the substrate ArX, and the substitution product ArNu, respectively. There is evidence that $ArNuH_2$ actually comes from the reduction of the radical anion intermediate $(ArNu)^{\overline{\cdot}}$ rather than from ArNu.

Reduction can occur at the Nu moiety, for example, 1-phenyl-2-propanol is obtained in the reaction of halobenzenes with acetone enolate ion (1), or at the aromatic moiety as when dihydronaphthylacetones are obtained in the reaction of 1-halonaphthalene with acetone enolate ion (2).

0065-7719/83/0178-0239$06.00/1

Competition between the reduction of the aryl radical to arene ArH (Reaction 1) and the coupling with the nucleophile will determine the relative amount of reduction versus substitution products. This ratio has been shown to be strongly dependent on the leaving group of the substrate.

Competition between Reactions 2 and 3 determines the relative amount of ArNu and $ArNuH_2$.

$$(ArNu)^{\cdot -} + ArX \longrightarrow ArNu + (ArX)^{\cdot -} \qquad (3)$$

If the reduction potential of ArNu is more positive than the reduction potential of ArX, the electron transfer Reaction 3 will be slow compared with Reaction 2. The importance of Reaction 2 will increase and the overall yield of substitution product will decrease.

In electrochemically induced reactions, an important termination step is the heterogeneous reduction of aryl radicals formed near the surface of the electrode (Reaction 4).

$$Ar^{\cdot} + e^{-}_{electrode} \longrightarrow Ar^{-} \xrightarrow{SH} ArH \qquad (4)$$

If decomposition of the radical anion $(ArX)^{\cdot -}$ is fast, the aryl radical will be produced near the electrode and will be reduced to aryl anion.

In photostimulated or spontaneous $S_{RN}1$ reactions, the initiation step occurs by an electron transfer reaction from the nucleophile to the substrate. A possible termination step is radical–radical coupling, provided that decomposition of the initial radical anion is fast enough that the aryl radical and the radical derived from the nucleophile cannot diffuse away. That is $k_{react} > k_{diff}$ (Reaction 5).

$$ArX + Nu^{-} \longrightarrow \left[(ArX)^{\cdot -} Nu^{\cdot} \right] \xrightarrow{k_{react}} (Ar^{\cdot} X^{-} Nu^{\cdot}) \qquad (5)$$
$$\text{caged ion pair}$$
$$\downarrow k_{diff} \qquad\qquad \downarrow$$
$$(ArX)^{\cdot -} + Nu^{\cdot} \qquad ArNu + X^{-}$$

Although the substitution product is formed by Reaction 5, the process is not a chain mechanism.

A similar substitution pathway is proposed for the reaction of diphenyliodonium ions with carbanionic nucleophiles. Alternatively, coupling of free aryl radicals with free radicals derived from the nucleophile (Nu$^{\cdot}$) amounts to a termination step (Reaction 6), even though it may not be a major reaction pathway.

$$Ar^{\cdot} + Nu^{\cdot} \longrightarrow ArNu \qquad (6)$$

Because this termination step and the chain reaction produce the same substitution product, there is no way to detect its occurrence.

Termination Steps Independent of the Nature of the Initiation Step

General Discussion. Reactions that consume the radical and radical anion intermediates of $S_{RN}1$ propagation steps become termination steps. Although these termination steps may depend on the reaction conditions (solvent, temperature, concentration, etc.) they do not usually depend on the nature of the initiation step. Reactions 7–14 summarize these possibilities.

$$Ar^{\cdot} + (ArX)^{\overline{\cdot}} \longrightarrow Ar^{-} + ArX \qquad (7)$$

$$Ar^{\cdot} + (ArNu)^{\overline{\cdot}} \longrightarrow Ar^{-} + ArNu \qquad (8)$$

$$Ar^{\cdot} + Nu^{-} \longrightarrow Ar^{-} + Nu^{\cdot} \qquad (9)$$

$$(ArNu)^{\overline{\cdot}} \xrightarrow{SH} ArNuH^{\cdot} \qquad (10)$$

$$2\,(ArNu)^{\overline{\cdot}} \longrightarrow ArNu + ArNu^{=} \qquad (11)$$

$$Ar^{\cdot} + (ArNu)^{\overline{\cdot}} \longrightarrow (ArNuAr)^{-} \qquad (12)$$

$$Ar^{\cdot} + Ar^{\cdot} \longrightarrow Ar-Ar \qquad (13)$$

$$Ar^{\cdot} + Ar'-Z \longrightarrow Ar'{\overset{\displaystyle Z}{\underset{\displaystyle Ar}{<}}} \qquad (14)$$

In addition to Reactions 7–14, fragmentation of radical anion intermediates is also a termination step when the species formed are unreactive in propagating the chain, as already discussed. Moreover, hydrogen-atom abstraction from the solvent or from other species present in the reaction are also termination steps, and in some systems are so important that they may inhibit the reaction. The latter possibility will be discussed in the next section.

In a quantitative study of the photostimulated reaction of potassium diethyl phosphite with iodobenzene in DMSO, it was determined that termination steps that are second order in propagating radicals (Reaction 13) are not very important, but may play some role (3). Biaryls have never been found in $S_{RN}1$ reactions in amounts large enough to be detected by ordinary analytical techniques (GLC, NMR, etc.). However, the formation of a dimer, which probably results from Reaction 14, was observed in the photostimulated and metal-stimulated reactions of acetone enolate ion with 2-chloroquinoline (4, 5).

A dimerization rate of 3×10^8 M^{-1} s^{-1} has been assigned to aryl radicals at ambient temperature (6), whereas phenyl radical addition to chlorobenzene was found to be on the order of 10^6 M^{-1} s^{-1} (7). Considering the difference in concentration of radicals and ArX, Reaction 14 appears more likely than Reaction 13 as a termination step. The rate of Reaction 14 seems to depend little on the aromatic substituent (8).

Reactions 7 and 8 cannot be distinguished from each other, but they have been studied independently in the electrochemical reduction of haloarenes (9).

Electron transfer from the nucleophile to the aryl radical (Reaction 9) is thought to prevent p-nitrophenyl radicals from coupling with certain carbanions. The high electron affinity of the p-nitrophenyl radical and the low oxidation potential of the anion means that electron transfer is favored over coupling (10). Electron transfer is also favored over coupling whenever the interaction between the orbitals involved is poor during the early stages of approach of the nucleophile to the radical (11).

The fact that reduction products derived from protonation of radical anions have not been found in spontaneous or photostimulated reactions may be due to the inefficiency of Reaction 10 as a termination step. Rate constants for protonation of radical anions are known to depend on the pK_a of the solvent and on the type of radical anion. In methanol, for instance, the rate of Reaction 15 is 1.7×10^5 M^{-1} s^{-1}, while in 2-propanol the rate is 0.72×10^5 M^{-1} s^{-1} (12).

$$(Ph–Ph)^{\cdot -} + ROH \longrightarrow Ph–PhH^{\cdot} \qquad (15)$$

Products derived from the addition of aryl radicals to radical anions (Reaction 12) have never been found, even though phenyl radicals are known to add to naphthalene radical anions (13) and that aliphatic radicals also react with radical anions (14).

Hydrogen Atom Abstraction. Photostimulated reactions of aryl halides with ketone enolate ions afford substitution products in high yields. The reduction product, ArH, is usually formed in very small amounts (1–5% yield). However, when the substituents in positions 2 and 6 of the aryl ring are large, or when the nucleophile is bulky, substitution decreases and reduction increases (Reaction 16).

Reactivity of the aryl radical is apparently reduced by steric hindrance. Higher hydrogen atom donor ability of the substrates and nucleophiles, also enhances reduction over substitution.

It has been reported that with cyclic ketones the ratio of substitution to reduction products depends on the number of carbon atoms in the ketone. Remarkably, cycloalkanones with even numbers (4, 6, 8) give

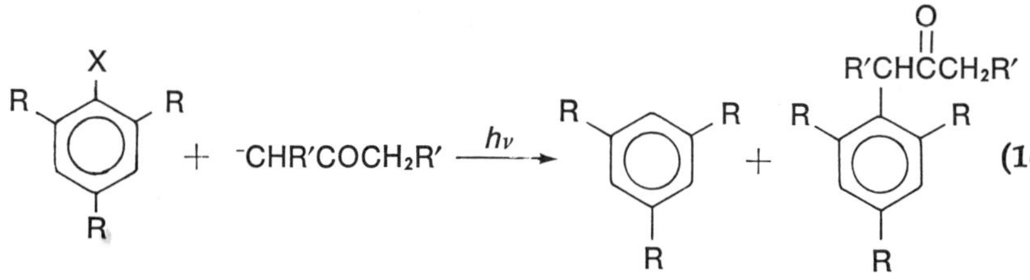

$$R\text{-}\underset{R}{\overset{X}{\bigcirc}}\text{-}R + \ ^-CHR'COCH_2R' \xrightarrow{h\nu} R\text{-}\underset{R}{\overset{R}{\bigcirc}} + R\text{-}\underset{R}{\overset{R}{\bigcirc}}\text{-}R \quad (16)$$

X	R	R'	hν(min)	%	%	Ref.
I	Me	H	32	10	82	15
Br	Et	H	120	22	70	15
Br	isoPr	H	150	7	2	15
I	isoPr	H	180	37	16	15
Br	H	Me	70	trace	80	16
Br	Me	Me	130	20	14	16
I	Me	Me	130	25	24	16

good yields of α-phenylation products (72–85% yield) with only 3–6% of benzene formation. However, cycloalkanones with *odd* numbers form α-phenylketones in lower yields (58–64%) and benzene increases to 18–28% (16).

In a study of intramolecular $S_{RN}1$ reactions, good yields of cyclization products were obtained with substrates such as **1** (17). However with substrates like **2**, bearing a β-hydrogen atom, it was found that the principal product is the β–γ unsaturated ketone, presumably from isomerization of the α–β isomer. With substrate **3** only the reduction product was found.

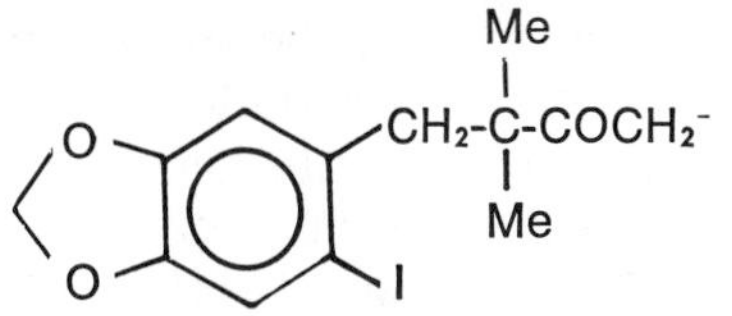

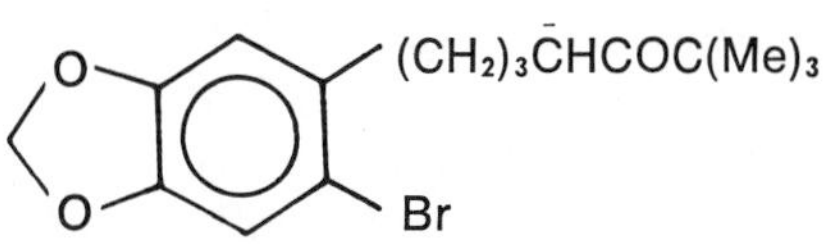

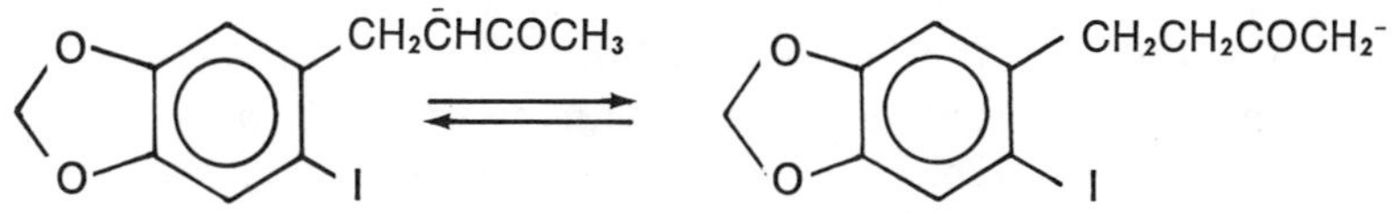

It was suggested that formation of unsaturated ketones with concomitant replacement of the leaving group by hydrogen is due to intramolecular hydrogen atom transfer from the β-position of the enolate to
the aryl radical (Reaction 17).

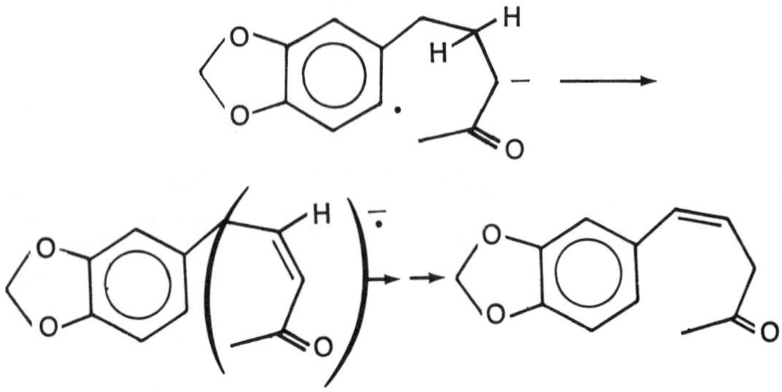

(17)

The dideuterated analogue substrate **4** produces the unsaturated
ketone **5** (Reaction 18) (*17*).

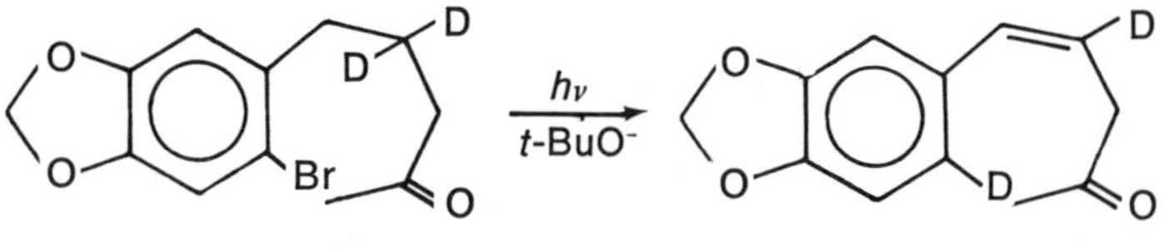

(18)

4   **5**

β-Hydrogen atom abstraction from ketones by aryl radicals was
demonstrated in intermolecular reactions. For instance, the enolate ion
of *t*-butyl-2-methylpropanoate reacted with *p*-bromoanisole upon irradiation to give the substitution product in 5% yield; the major product
was anisole (35–50%). *t*-Butylacetate enolate ion gave the substitution
product (67%) and disubstitution product (29%) under irradiation (*17*).

Under irradiation, 2,4-dimethyl-3-pentanone enolate ion **6** reacts
sluggishly with iodobenzene to form the substitution product **7** (32%),
benzene (20%), and a dimeric ketone **8** (20%) (Reaction 19) (*18*).

It was suggested that phenyl radical, besides coupling with the
nucleophile (Reaction 20), abstracts a β-hydrogen atom, as shown in
Reaction 21, thereby forming benzene and the radical anion **9**.

Although the radical anion **9** could follow the $S_{RN}1$ chain propagation, transfering its odd electron to iodobenzene, the sluggishness of the
reaction indicates that **9**, being the radical anion of an α–β unsaturated

$$PhI + {}^-CMe_2COCHMe_2 \xrightarrow{h\nu} PhCMe_2COCHMe_2 + PhH + \qquad (19)$$
$$\mathbf{6} \hspace{7cm} \mathbf{7}$$

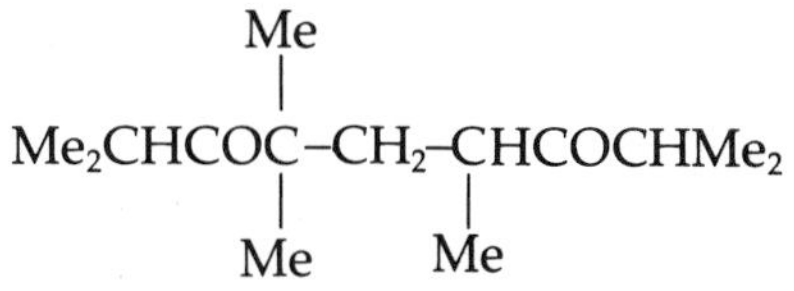

$$\begin{array}{c} Me \\ | \\ Me_2CHCOC-CH_2-CHCOCHMe_2 \\ | \qquad | \\ Me \qquad Me \\ \mathbf{8} \end{array}$$

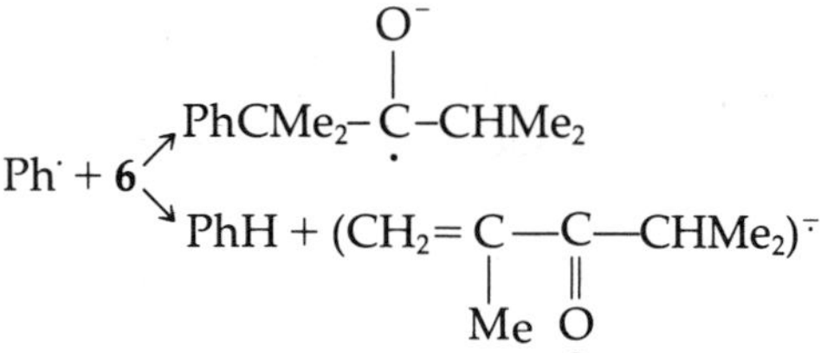

$$Ph^{\cdot} + \mathbf{6} \begin{cases} \nearrow \begin{array}{c} O^- \\ | \\ PhCMe_2-\overset{\cdot}{C}-CHMe_2 \end{array} & (20) \\ \\ \searrow PhH + (CH_2= \overset{|}{\underset{Me}{C}} - \overset{\parallel}{\underset{O}{C}} - CHMe_2)^{\cdot -} & (21) \\ \hspace{5cm} \mathbf{9} \end{cases}$$

ketone, is probably too stable to transfer the odd electron to the substrate. Disproportionation of **9** (Reaction 22) is the suggested reaction pathway, and this reaction is a termination step. The α–β unsaturated ketone **10** reacts with the nucleophile **6** to give **8** (Reaction 23). The dianion **11** is protonated by the solvent ammonia to give **6** (Reaction 24) (*18*).

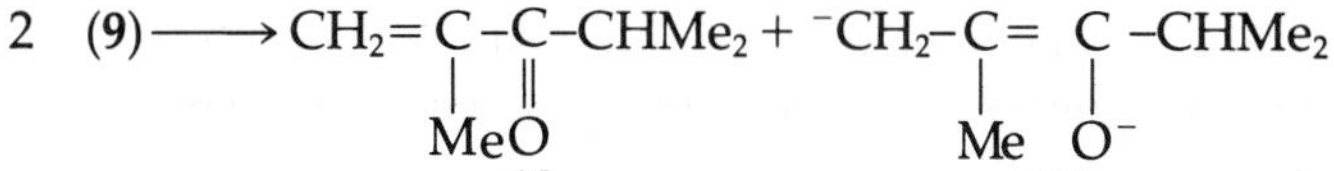

$$2 \ (\mathbf{9}) \longrightarrow CH_2= \overset{|}{\underset{MeO}{C}}-\overset{\parallel}{\underset{}{C}}-CHMe_2 + {}^-CH_2-\overset{}{\underset{Me}{C}}= \overset{|}{\underset{O^-}{C}} -CHMe_2 \qquad (22)$$
$$\hspace{3.5cm} \mathbf{10} \hspace{4cm} \mathbf{11}$$

$$\mathbf{10} + \mathbf{6} \longrightarrow \mathbf{8} \qquad (23)$$

$$\mathbf{11} \xrightarrow{NH_3} \mathbf{6} \qquad (24)$$

In contrast with these results, it was found that the nucleophile **6** reacts with 2-bromopyridine (*19*) and 2-chloroquinoline (*4*) to give the expected substitution products unaccompanied by significant amounts of dimer **8**.

The fact that dimer **8** is not formed in the latter reaction may be due to the higher electrophilicity of the aromatic radical involved, or to the increased electron acceptor ability of the substrates. The latter would increase the rate of Reaction 25 and thus Reaction 21 would become a propagation step, rather than a termination step.

$$ArX + \mathbf{9} \longrightarrow (ArX)^{\cdot -} + \mathbf{10} \qquad (25)$$

Ar = 2-bromopyridine or 2-chloroquinoline

All these examples indicate that when the enolate anion has a β-hydrogen atom, abstraction competes seriously with the coupling reaction with phenyl derivatives as substrates.

In the photostimulated reaction of 1-halonaphthalenes with butanethiol in ammonia, a substantial amount of naphthalene (31%), but only 10% of the substitution product was formed and the rest was starting material. Under the same reaction conditions, but with the sodium salt ($C_4H_9S^- Na^+$), naphthalene (18%) and the substitution product (81%) were obtained (2). The low yield of substitution product in the former case may be explained by incomplete ionization of the butanethiol to butanethiolate in ammonia (Reaction 26). The residual butanethiol then acts as a hydrogen atom donor (Reaction 27).

$$n\text{-Bu-SH} + NH_3 \rightleftharpoons n\text{-Bu-S}^- + NH_4^+ \tag{26}$$

$$(27)$$

The importance of the solvent as a hydrogen atom donor is obvious. Ammonia is the most commonly used solvent, and it is also the poorest hydrogen atom donor. In a study of the reaction of 1-naphthyl radicals with benzenethiolate ions in aprotic solvents, it was determined that the hydrogen donor ability follows the order DMSO ~ acetonitrile < DMF (20).

Literature Cited

1. Rossi, R. A.; Bunnett, J. F. *J. Am. Chem. Soc.* **1972**, *94*, 683.
2. Rossi, R. A.; de Rossi, R. H.; López, A. F. *J. Am. Chem. Soc.* **1976**, *98*, 1252.
3. Hoz, S.; Bunnett, J. F. *J. Am. Chem. Soc.* **1977**, *99*, 4690.
4. Hay, J. V.; Wolfe, J. F. *J. Am. Chem. Soc.* **1975**, *97*, 3702.
5. Hay, J. V.; Hudlicky, T.; Wolfe, J. F. *J. Am. Chem. Soc.* **1975**, *97*, 374.
6. MacLachlan, A.; McCarthy, R. L. *J. Am. Chem. Soc.* **1962**, *84*, 2519.
7. Kryger, R. G.; Lorand, J. P.; Stevens, N. R.; Herron, N. R. *J. Am. Chem. Soc.* **1977**, *99*, 7589.
8. Madhavan, U.; Schuler, R. H.; Fessenden, R. M. *J. Am. Chem. Soc.* **1978**, *100*, 888.
9. M'Halla, F.; Pinson, J.; Saveant, J. M. *J. Am. Chem. Soc.* **1980**, *102*, 4120.
10. Russell, G. A.; Metcalfe, A. R. *J. Am. Chem. Soc.* **1979**, *101*, 2359.
11. Eberson, L.; Blum, Z.; Helgee, B.; Wyberg, K. *Tetrahedron* **1978**, *34*, 731.
12. Dorfman, L. M. *Acc. Chem. Res.* **1970**, *3*, 224.
13. Sargent, G. D. *Tetrahedron Lett.* **1971**, 3279.
14. Garst, J. F. *Acc. Chem. Res.* **1971**, *4*, 400.
15. Bunnett, J. F.; Sundberg, J. E. *Chem. Pharm. Bull.* **1975**, *23*, 2620.
16. Bunnett, J. F.; Sundberg, J. E. *J. Org. Chem.* **1976**, *41*, 1702.
17. Semmelhack, M. F.; Bargar, T. M. *J. Org. Chem.* **1977**, *42*, 1481.

18. Wolfe, J. F.; Moon, M. P.; Sleevi, M. C.; Bunnett, J. F.; Bard, R. R. *J. Org. Chem.* **1978,** *43,* 1019.
19. Komin, A. P.; Wolfe, J. F. *J. Org. Chem.* **1977,** *42,* 2481.
20. Helgee, B.; Parker, V. D. *Acta Chem. Scand. Ser. B* **1980,** *34,* 129.

Other Related Mechanisms

Other reactions have mechanisms that resemble the $S_{RN}1$ mechanism in some aspects. Thus it is important to know the most important characteristics of these reactions as well as their similarities and differences. Therefore, this chapter introduces data which may be of interest. The reader is referred to the original literature for more extensive discussion.

Photonucleophilic Aromatic Substitution

The subject of photonucleophilic aromatic substitution has been frequently and extensively reviewed (1–4), but a clear picture about the multiplicity of mechanisms of these reactions was not established until 1979 (5).

Salient features of photonucleophilic aromatic substitution reactions include the finding that orientation in the excited state differs from that in the ground state (6) and that hydrogen, which is not a leaving group in ground state nucleophilic substitution, can be substituted if there is an oxidizing agent present (oxygen or other) (1–4). This is so for aromatic and heteroaromatic compounds (7).

Studies from several laboratories have identified a range of leaving groups and substituting nucleophiles. Leaving groups include phosphate, sulfate, RO^-, NO_2^-, I^-, Br^-, Cl^-, F^-, SO_3^{2-}, N_2, and H^-. Reacting nucleophiles include water, alcohols, amines (primary, secondary, and tertiary), pyridines, HO^-, RO^-, CN^-, H^-, CH_3^-, RCO_2^-, and NO_2^-.

Systematic variation of the aromatic nucleus, the leaving group, the substituent and attacking reagent in numerous aromatic photosubstitution reactions has led to four rules describing orientation (2).

Orientation Rules. NITRO GROUP ACTIVATION OF THE META POSITION (8). The preference for m-activation is shown in Reactions 1–4 (1). The examples show clearly that the nitro group activates the *ortho* and *para* positions in the thermal reaction, but the *meta* position in the photostimulated reactions.

The same rule applies to naphthalene derivatives. Thus, in a study of isomeric fluoronitronaphthalenes and methoxynitronaphthalenes

0065-7719/83/0178-0249$09.50/1

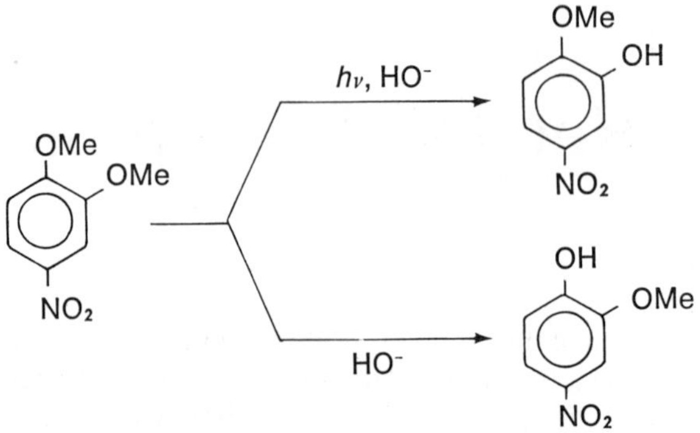

with hydroxide or methoxide ions, photosubstitution was found to proceed cleanly and smoothly when there was a *meta*-relationship between the groups (9).

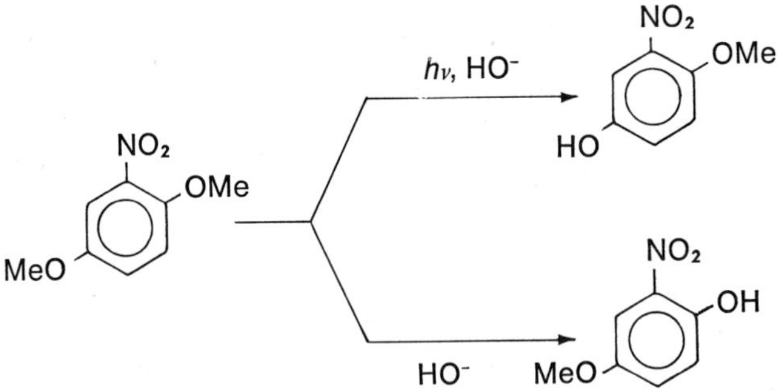

ALKOXY SUBSTITUENT ACTIVATION OF THE *ORTHO* AND *PARA* POSITIONS. The halogen substituent in 2-halo-4-nitroanisoles is readily replaced upon irradiation in ammonia (10–11). Likewise, 1-amino-2-methoxyanthraquinone is formed from irradiation of 2-methoxyanthraquinone in ammonia (12).

That methoxy is an activating group and orients substitution follows from the fact that illumination of 1-methoxynaphthalene in a medium containing cyanide ion gives a good yield of 1-cyano-4-methoxynaphthalene (and only very little 1-cyanonaphthalene). Photocyanation of naphthalene itself proceeds in only low yields. As further evidence, the three dimethoxybenzenes (1–3) undergo smooth photocyanation, yielding in all cases the products to be expected from the *ortho* and *para* activation. In addition, the reaction of 5-nitro-1,3-dimethoxybenzene with cyanide ion gives products which indicate that *ortho* and *para* activation by two methoxy groups may compete with *meta* activation by a nitro group (Reaction 5).

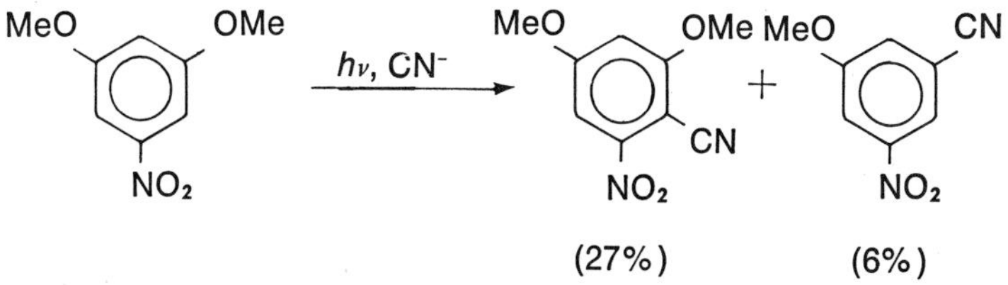

$$(5)$$

BICYCLIC AND TRICYCLIC AROMATIC COMPOUNDS PREFER SUBSTITUTION IN THE α POSITION (*13–15*). The preference for substitution in the α position has not been satisfactorily rationalized. In azulene, considerations of the charge densities for the various electronic states could explain the results. However, this is not the case for other polycyclic compounds.

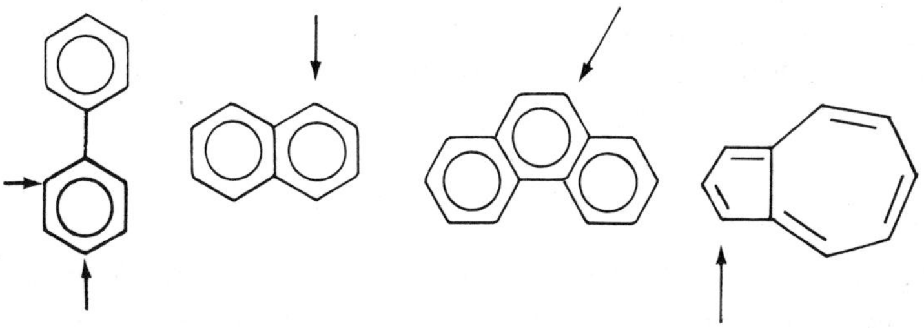

MERGING STABILIZATION DURING PRODUCT FORMATION. This rule is related to the different selectivity of nucleophiles, and has been called the rule of merging stabilization during product formation. An example of this rule is the reaction of 1-methoxy-4-nitronaphthalene with methylamine and cyanide ion. The former gives *N*-methyl-1-amino-4-nitronaphthalene (Reaction 6), whereas the second forms 1-cyano-4-methoxynaphthalene (Reaction 7) (*16*).

The reaction of cyanide ion is not unexpected because a good leaving group, $-NO_2$, is activated by $-OMe$, but the amine behaves anomalously in going to the other position. In the product, and from the point of view of resonance stabilization, this situation is more attractive than that which would result from substitution of the nitro group by amine (or methoxy by cyano). Thus it was suggested that the electron donating or withdrawing effect that the incoming reagent will have when it becomes fully attached to the aromatic ring influences the product forming step (*2*).

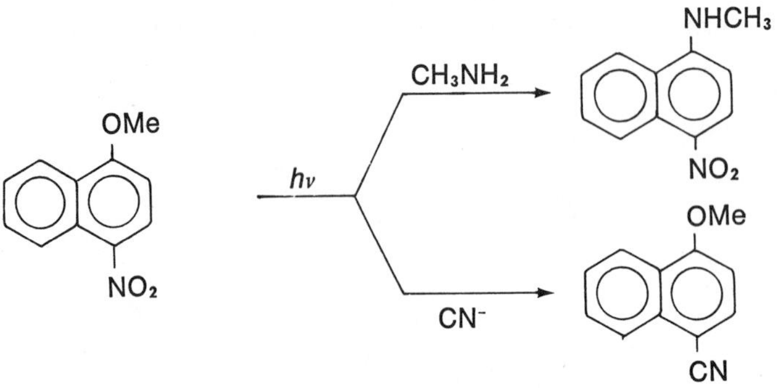

Reaction Mechanism. Four totally different mechanisms have been recognized in photostimulated reactions of aromatic substrates with nucleophiles (5).

The $S_{RN}1$ mechanism belongs in this group of reactions, and it was suggested that it should be renamed $S_{R^-N}1$ to indicate that there are radical anion intermediates involved (Reaction 8).

Aromatic substrates with electron donating groups form cation radicals as primary species and react by a mechanism named $S_{R^+N}1$ (Ar^*) (Reaction 9) to create a consistent connection with the $S_{R^-N}1$ mechanism (Reaction 8).

Some reactions involve a primary photodissociation into an aromatic cation and a leaving group brought about by illumination. This mechanism was denoted S_N1 (Ar^*) (Reaction 10). Photosubstitution of aromatic diazonium ion compounds probably proceeds along this reaction pathway, which, on kinetic grounds, was tentatively adopted for the photocyanation of 2-nitrofurans (17).

The fourth mechanism is denoted S_N2 (Ar^*) (Reaction 11) and involves the formation of σ complexes of excited aromatic substrates.

On the basis of these classifications it is now possible to choose reaction partners and conditions so that photosubstitution will follow a well defined pathway. Thus, in water medium, $S_{R^+N}1$ (Ar^*) reactions (and possibly S_N1 (Ar^*) processes) can be expected to occur provided that the aromatic compound is chosen appropriately. On the other hand, S_N2 (Ar^*) processes are expected predominantly in less polar media (5).

THE $S_{R^+N}1$ (AR*) MECHANISM. Photocyanation of isomeric haloanisoles and other methoxy-substituted compounds obeys second order kinetics, but it has been argued that direct interaction of cyanide with the excited

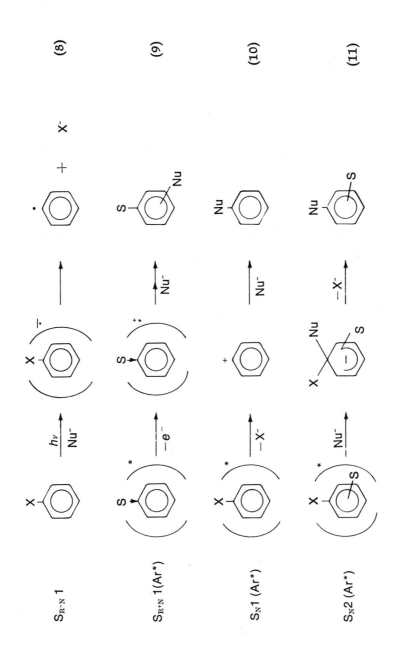

state of the anisole derivative is unlikely. Calculated charge densities in excited singlet and triplet states of anisole show that the methoxy group in the excited molecule is still an electron donating substituent, and even more so than in the ground state (18).

It was proposed that photoionization is the primary step (19), a result which is in agreement with anodic cyanation of dimethoxybenzenes (20) and trimethoxybenzenes (21) that form the same products as in the photochemical reaction.

Calculations of a number of aromatic radical cations indicate that the distribution of positive charge is completely consistent with the distribution of products found, thus confirming the suggestion that ionization of the substrate is the primary step that follows excitation. Evidence for the formation of radical cations and solvated electrons from triplet excited anisoles was obtained by laser spectroscopic experiments (5).

The proposed mechanism is found in Reactions 12 and 13. The radical cation derived from the product may abstract one electron from the substrate, thus setting up a chain reaction (Reaction 14).

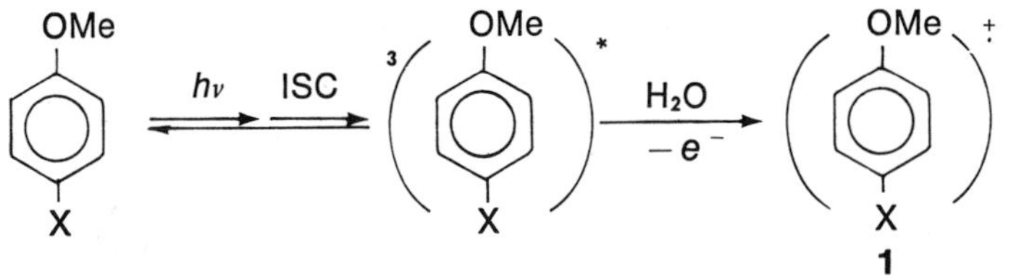

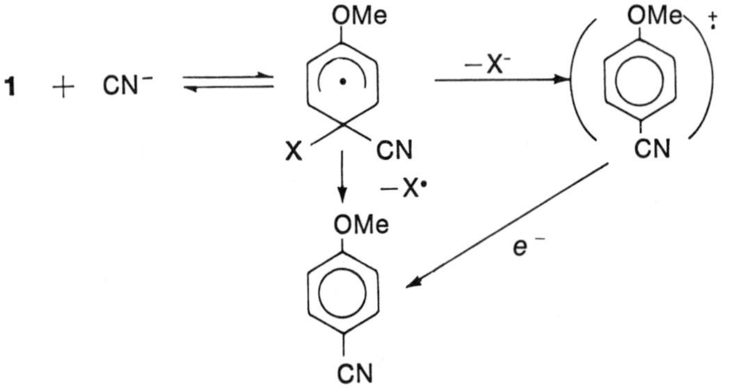

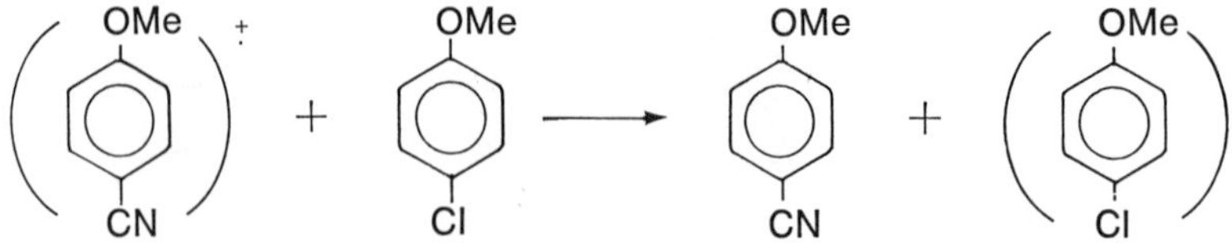

The likelihood of this process will not only depend on the concentration of the starting material but also on the relative ionization potentials of products and starting material. The quantum yield of photocyanation of 4-chloroanisole depends on the concentration of starting materials and, at high concentration, it exceeds unity.

THE S_N2 (AR*) REACTION. Description of the pathway(s) in the S_N2 (Ar*) reaction, which is promoted by electron-attracting substituents and in particular nitro groups, is still partly tentative. The prominent role of the nitro group, which has both a strong activating and *meta*-directing influence, still remains a challenge for future research (5). The reaction proceeds in most cases via a triplet state (22, 23), although in a few cases a singlet mechanism seems to occur (23, 24).

Quantitative studies of 1-methoxy-3-nitronaphthalene or 1-fluoro-3-nitronaphthalene with hydroxide and methylamine, and the effect of sensitizers and quenchers led to the suggested kinetic scheme of Figure 1 for these reactions (25).

In Figure 1, S_1 represents the first singlet excited state of the aromatic substrate, which results from light absorption. From this state the molecule can be deactivated to the ground state, or, through intersystem crossing (ISC), is converted to T_1, (first triplet state) with a quantum yield for the ISC (Φ_{ISC}). The triplet state can be deactivated in a radiative or nonradiative process represented by k_d, quenched by an added substance Q with a rate k_q, quenched by the nucleophile N with a rate k_s, or reacted with the nucleophile to form the products with a rate k_p.

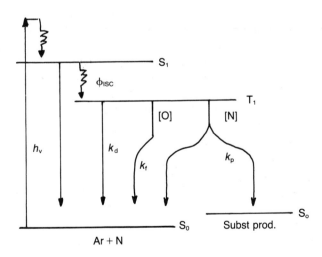

Figure 1. Simplified kinetic scheme for bimolecular nucleophilic aromatic photosubstitution via the triplet state (S_N2 $^3Ar*$). (Reproduced, with permission, from Ref. 25. Copyright 1977, Weizmann Science Press.)

Table I
Rate Constants of Some Nucleophilic Photosubstitution Naphthalene Derivatives

Naphthalene Derivatives	Nu^-	Φ_{ISC}	k_p	k_s	$\tau = k_d^{-1}$
1-F-3-NO$_2$	OH$^-$	0.8	9.6 10^7	8.8 10^7	2.3 10^{-7}
1-F-3-NO$_2$	MeNH$_2$	0.8	1.2 10^8	2.3 10^8	2.3 10^{-7}
1-F-6-NO$_2$	OH$^-$	0.8	1.0 10^8	1.5 10^8	2.5 10^{-7}
1-MeO-3-NO$_2$	·OH$^-$	0.2	7.1 10^8	7.1 10^8	3.8 10^{-9}
1-MeO-3-NO$_2$	MeNH$_2$	0.2	1.2 10^9	1.2 10^9	2.5 10^{-9}

The quantum yield for the nucleophilic substitution is given by Equation 15. The data reported are summarized in Table I. 1-Methoxy-3-nitronaphthalene is about 10 times more reactive than the 1-fluoro derivative, thus agreeing with orientation rule 2.

$$\Phi^{-1} = \Phi_{ISC}^{-1} \frac{k_s + k_p}{k_p} + \frac{k_d}{k_p\,(N)} + \frac{k_f\,(Q)}{k_p\,(N)} \tag{15}$$

In order to minimize all complications and simplify the interpretation of the results, symmetrical exchange of methoxide and cyanide ions was studied (5).

In the photochemical methoxide exchange of the three isomeric nitroanisoles in methanol (Reactions 16–18) and 1,2-dimethoxy-4-nitrobenzene (Reactions 19–20) results for Φ were obtained. (Conditions: λ, 313 nm; air saturated solution; and room temperature.)

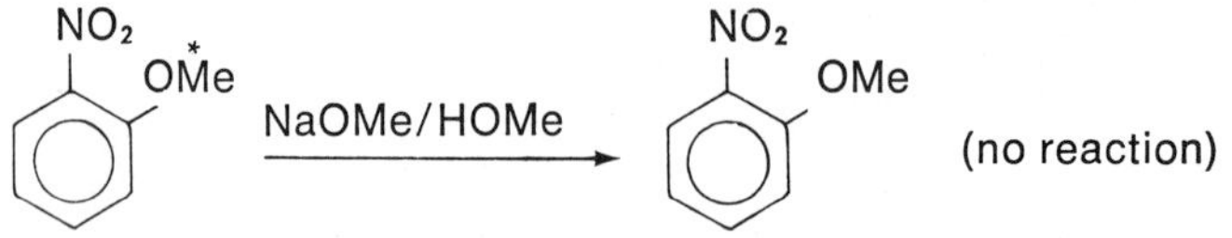

(16)

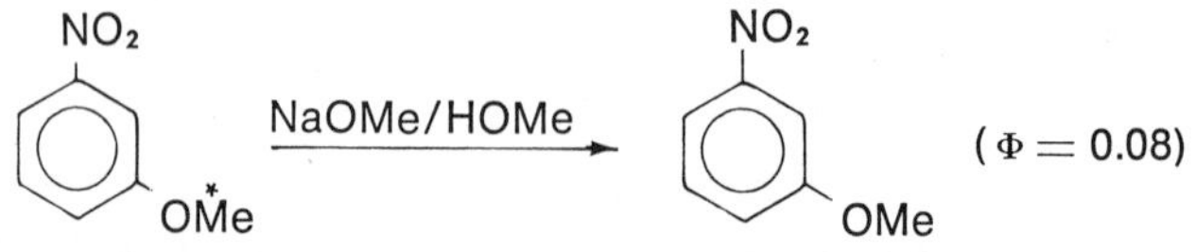

(17)

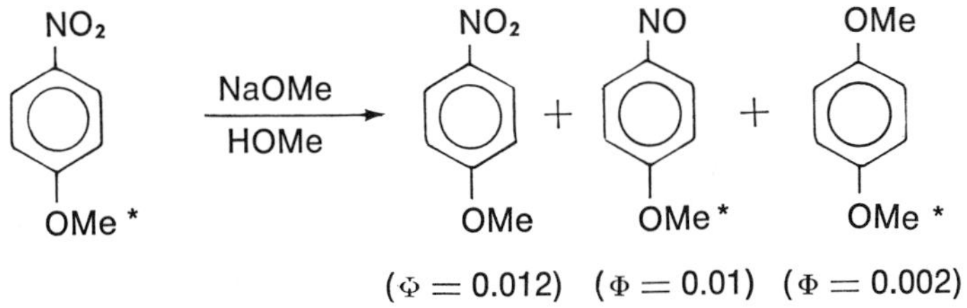

$$(\Phi = 0.012) \quad (\Phi = 0.01) \quad (\Phi = 0.002)$$

$$(18)$$

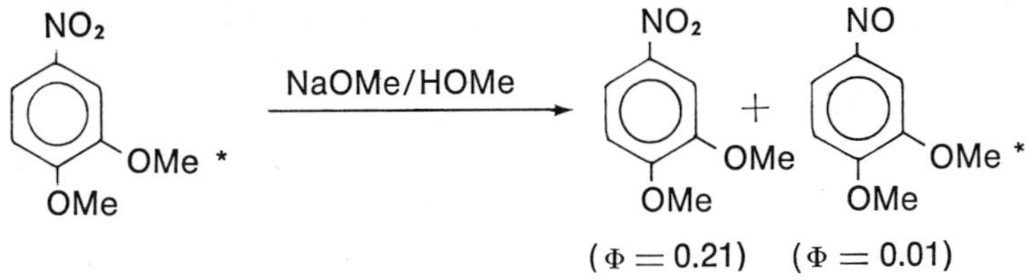

$$(\Phi = 0.21) \quad (\Phi = 0.01)$$

$$(19)$$

$$(\Phi = 0.018) \quad (\Phi = 0.01)$$

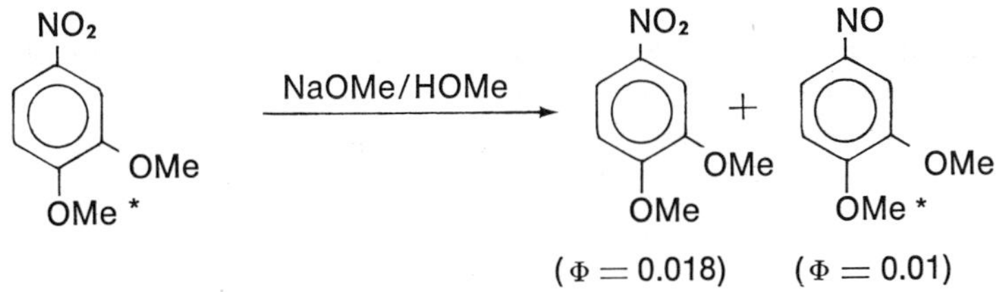

$$(20)$$

In 1,2-dimethoxy-4-benzene different lifetimes of the reacting species as a possible complicating factor is excluded. The results show clearly the *meta* activation by the nitro group.

The fact that electrolysis (anodic oxidation) of 1,2-dimethoxy-4-benzene in NaOMe/HOMe or NaOH/H$_2$O did not lead to reaction, that electrolysis leading to the anion (cathodic reduction) of 1,2-dimethoxy-4-benzene in a NaOMe/HOMe gave the nitroso product but no exchange and that anodic oxidation of *m*-nitroanisole in NaOH/H$_2$O gave no hydroxylation corroborated that these reactions are not of the radical cation (or radical anion) type. In addition, there is no correlation between the substitution pattern of 1,2-dimethoxy-4-nitrobenzene and the INDO and CNDO/2 calculated charge densities of its radical cation, and laser flash photolysis experiments confirm the formation of *m*-nitroanisole triplet, while *o*-nitroanisole and *p*-nitroanisole give photoionization of the triplet in competition with its radiationless decay (26).

Indications for differences in mechanistic behavior of *o*-nitroanisole and *p*-nitroanisole on the one hand and *m*-nitroanisole on the other, are also found in the photocyanation of these compounds in aqueous solutions. Photocyanation and anodic cyanation of *o*-nitroanisole and

p-nitroanisole gave the same products in the same ratio, but this is not the case with the *meta* isomer.

The *meta* orientation of the nitro group was rationalized using the energy difference between the excited state and the ground state surface as a dynamic reactivity index for the case of formation of a σ complex from the excited triplet state and the nucleophile.

The course of a photochemical reaction is determined by both the shape of the potential energy surface of the excited state and the shape of the potential energy surface of the ground state. Both surfaces are functions of the geometry of the states. The probability of internal conversion is related to the width of the energy gap between the surfaces. When the nucleophile interacts with the aromatic triplet with possible formation of an exciplet, it can either decompose into its components, or be converted to the σ complex ground state surface. Internal conversion is easier when the σ complex ground state surface is of high energy content and the excited state surface is of low energy content.

Calculations by the CNDO/2 method carried out for nitroanisoles and 2,4-dimethoxynitrobenzene indicated that the *meta* position is the one where the energy gap between ground state and excited state surfaces is smaller, in agreement with the experimental results (5).

$S_{RN}1$ Reaction in Aliphatic Systems

The p-Nitrobenzyl System. This mechanism was recognized simultaneously and independently (27, 28) and has been reviewed several times (29, 30). Historically, p-nitrobenzylic compounds were the first examples found to react by a radical-chain mechanism, particularly in substitution reactions with anions of nitroalkanes. Nitroalkane anions are ambident nucleophiles that can be alkylated at either oxygen or carbon sites (Reactions 21, 22).

$$RR'\overset{-}{C}NO_2 \longleftrightarrow RR'C=NO_2^- \qquad (21)$$

$$\downarrow \begin{array}{c} R''CH_2X \\ and/or \end{array}$$

$$\begin{array}{cc} RR'C-NO_2 & RR'C=\overset{+}{N}\diagdown\begin{array}{l}O^- \\ O-CH_2R''\end{array} \quad (22) \\ | \\ CH_2 \\ | \\ R'' \end{array}$$

C–alkylation   O–alkylation

The C-alkylation product is now known to come from the $S_{RN}1$ mechanism, whereas the O-alkylation product represents the S_N2 mechanism. There are some exceptions where alkylation by the $S_{RN}1$ mechanism in highly strained systems occurs preferentially at the oxygen (31–32).

The mechanism formulated is found in Reactions 23–27.

The competition $S_{RN}1$ vs. S_N2 or C-alkylation vs. O-alkylation depends greatly on the leaving group (Table II) (29). S_N2 reactions seem to be favored with good leaving groups, whereas $S_{RN}1$ reactions predominate with the poorest leaving groups.

Initially, the most striking mechanistic evidence was the strong inhibition of C-alkylation and depression of the overall rate when electron acceptors such as m-dinitrobenzene or p-dinitrobenzene were added (33, 34). Later, much more evidence about the mechanism was obtained,

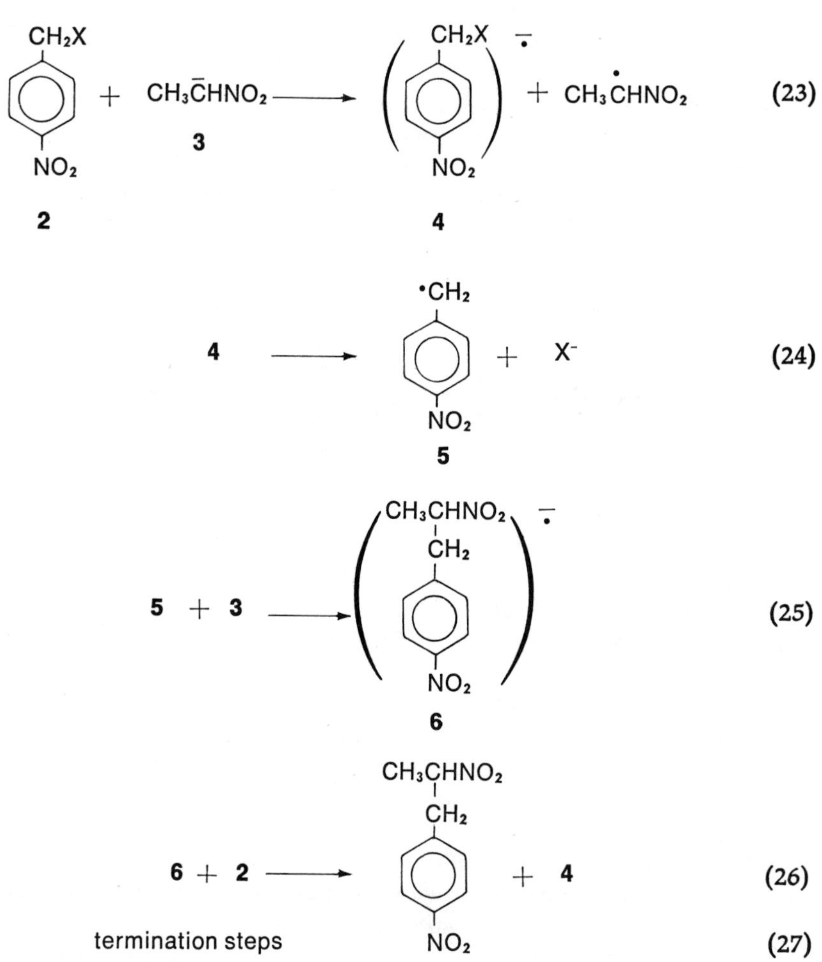

Table II

Nature of the Reaction of $p\text{-}O_2NC_6H_4CH_2X$ with the Lithium Salt of 2-Nitropropane in DMF as a Function of X

X	C-alkylate (%)	O-alkylate (%)
$^+NMe_3$	93	0
$C_6Cl_5CO_2$	93	0
Cl	92	6
Ts	40	32
Br	17	65
I	7	81

Source: Reproduced, with permission, from Ref. 29. Copyright 1975.

including detection of intermediates by physical methods. An example is the reaction of benzenethiolate or methanethiolate ions with p-nitrobenzyl chloride in ethanol, yielding the corresponding p-nitrobenzyl sulfides in 97% and 68% yield respectively (35). When the reactions were performed in an ESR cell, the signal of radical anions 7 and 8 were detected. However, the reaction of benzenethiolate ion appears to be mainly S_N2 while methanethiolate reacts by both mechanisms.

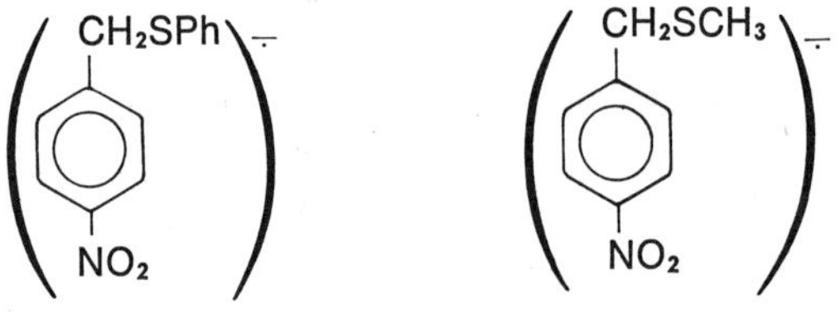

7   8

Tertiary carbons are unreactive in S_N2 reactions, yet p-nitrocumyl chloride was found to react readily with several nucleophiles. Examples are shown in Scheme I (36, 37), and the reaction mechanism is $S_{RN}1$, not S_N2.

The reactions of Scheme I were carried out in DMSO or DMF with illumination from an ordinary 20 W fluorescent light. Also, $S_{RN}1$ substitution in α, p-dinitrocumene by nitronate ions was carried out successfully under phase-transfer conditions (38).

It is remarkable that nucleophiles such as CN^-, $PhSO_2^-$, and NO_2^-, which are unreactive in reactions with aryl radicals, are so good in reacting at aliphatic sites. As previously discussed, slow electron transfer from the radical anion of the substitution product to an aromatic substrate may account for the difference in reactivity of $S_{RN}1$ aromatic versus

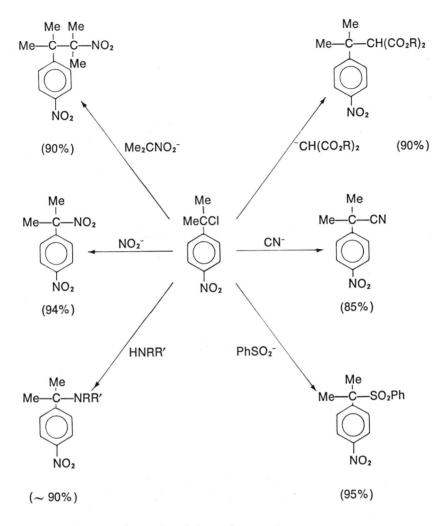

Scheme I

$S_{RN}1$ aliphatic systems. Compare for instance, Reactions 28–29 with 30–31.

The nucleophilic Reaction 28 forms a radical anion intermediate **9** with a reduction potential for the couple $PhNO_2/(PhNO_2)^-$ of -1.06 V (in DMF vs. SCE) (39), whereas the reduction potential of the couple $PhCl/(PhCl)^-$ is -2.78 V (in DMF vs. SCE) (40). Thus Reaction 29 is thermodynamically unfavorable by 1.72 V or about 40 kcal/mol, and the rate of electron transfer is probably very slow.

On the other hand nucleophilic Reaction 30 forms a radical anion

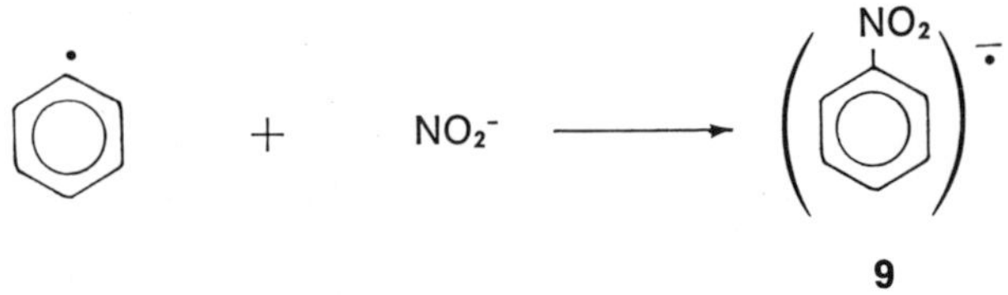

$$(28)$$

$$\mathbf{9}$$

$$(29)$$

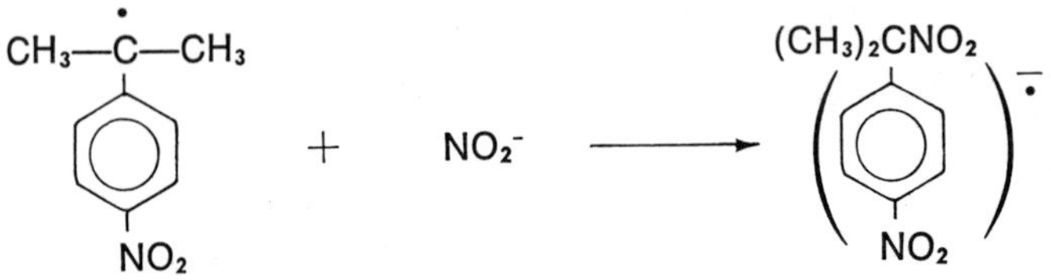

$$(30)$$

$$\mathbf{10}$$

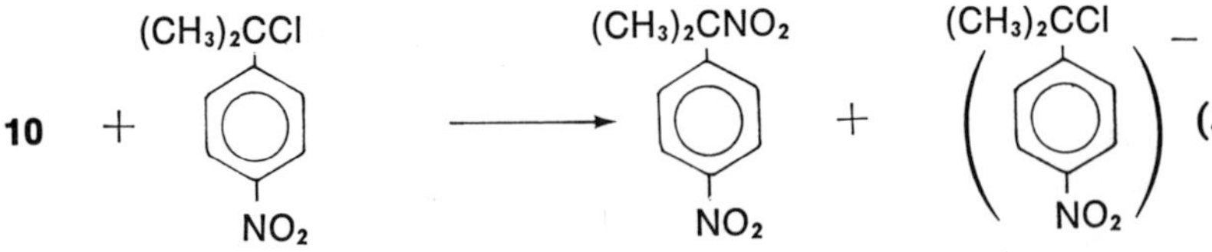

$$(31)$$

intermediate **10,** whose parent compound probably has about the same reduction potential as the substrate, p-nitrocumyl chloride, because the nitro or chloride groups are separated by an sp^3 carbon atom from the aromatic ring. Therefore, the reduction potential of both compounds is expected to be about the same. Reaction 31 is probably very fast and consequently the chain is sustained.

$S_{RN}1$ reactions of p-nitrocumyl derivatives seem to be reversible because the nucleophiles employed are also good leaving groups under appropriate conditions. Some examples are shown in Reaction 32 (*29*), where the leaving group, X, is found to react as a nucleophile also.

The reversibility of the coupling of anions with cumyl radicals is further demonstrated in the Reaction 33 (*41*).

Products **11, 12,** and **13** are likely to come from Reaction 34, which represents the reverse of the coupling reaction of 2-nitropropane anion with *p*-nitrocumyl radical. This fragmentation is favored by electron withdrawing groups in the aromatic ring (*42*).

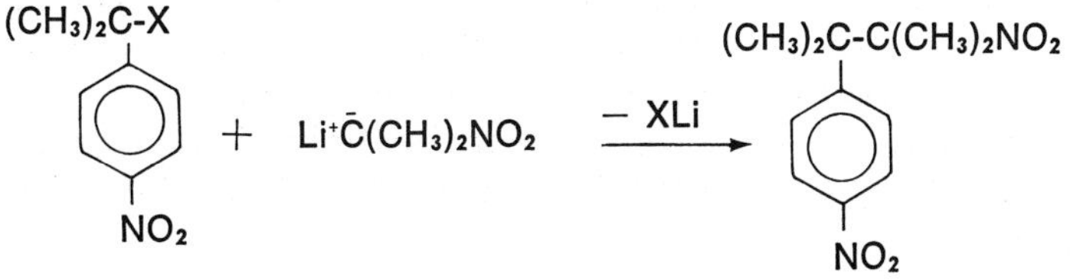

$$(32)$$

X = NO$_2$, PhSO$_2$, N$_3$,
1-methyl-2-naphthoxide.

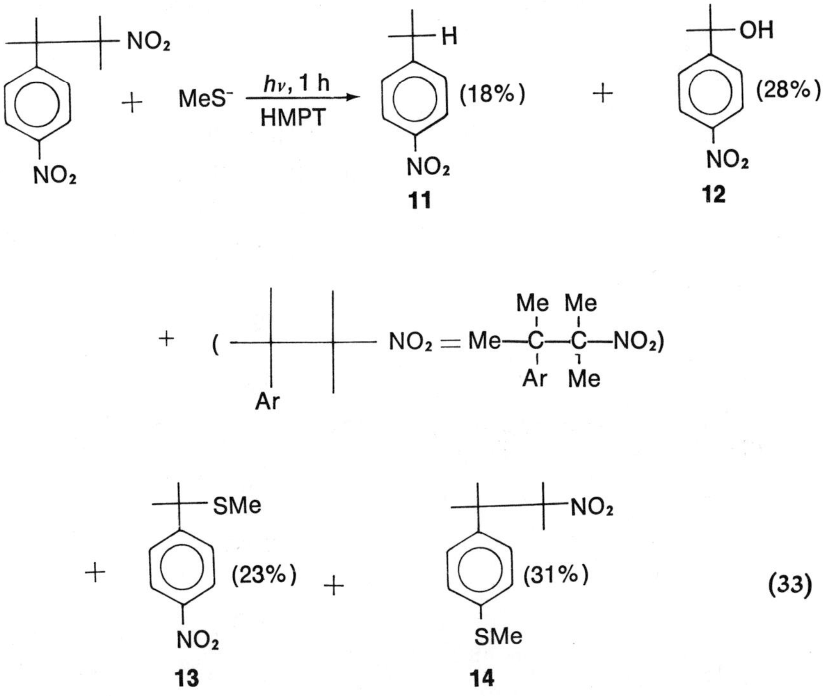

11 (18%)   12 (28%)

Me Me
NO$_2$ = Me—C—C—NO$_2$)
Ar Me

13 (23%)   14 (31%)

$$(33)$$

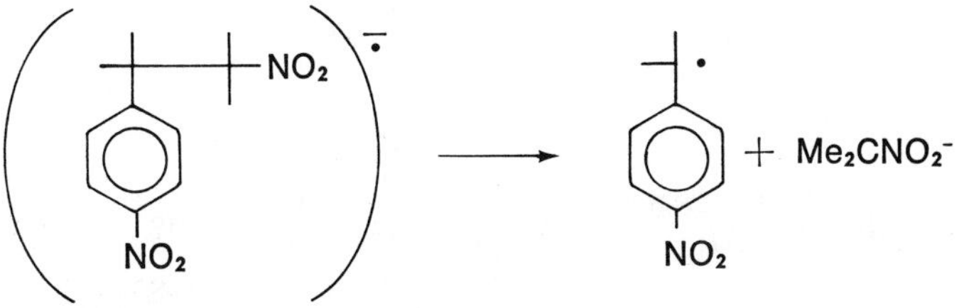

$$(34)$$

Other Arylalkane Systems. When the $S_{RN}1$ reaction was discovered, it appeared that the nitro group in the aromatic ring was absolutely necessary for the reaction to occur. However, in the last few years it has become evident that the mechanism can take place with various substrates provided that the proper conditions are chosen.

The substrate 4-(1-methyl-1-nitroethyl)pyridine **15** reacts with several nucleophiles to give excellent yields of products, as shown in Reactions 35–38 (*43*).

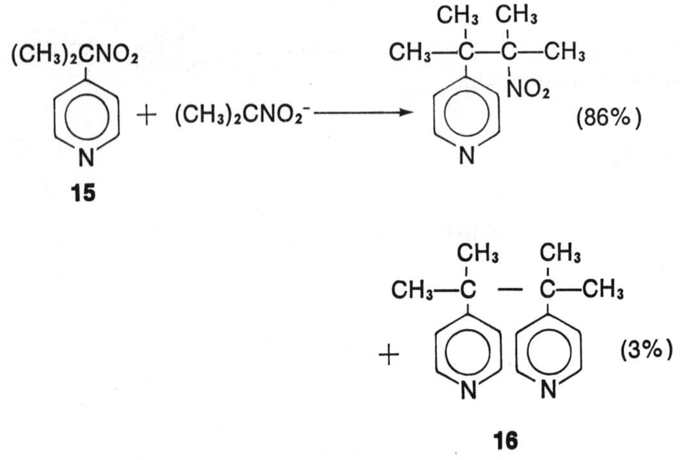

$(CH_3)_2CNO_2$

15 + $(CH_3)_2CNO_2^-$ ⟶ (86%)

+ (3%) (35)

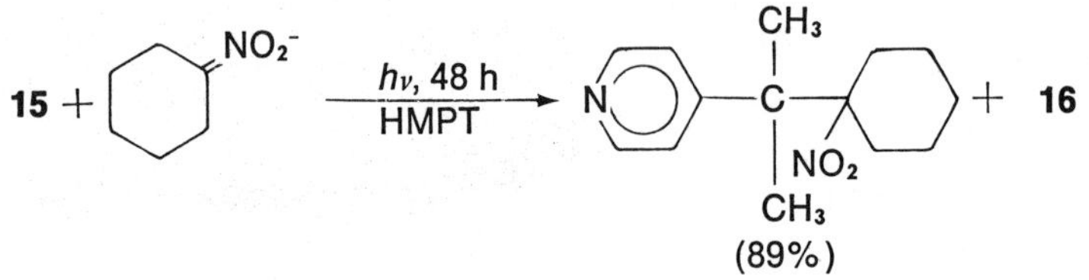

16

15 + [cyclohexyl–NO_2^-] $\xrightarrow[\text{HMPT}]{h\nu,\ 48\ h}$ (89%) + **16** (36)

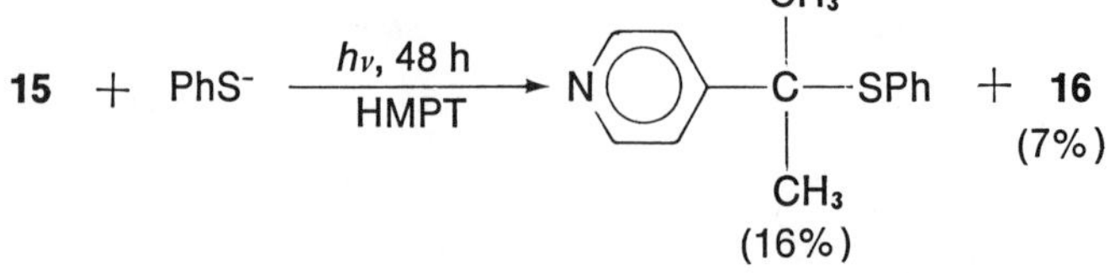

15 + PhS⁻ $\xrightarrow[\text{HMPT}]{h\nu,\ 48\ h}$ —C—SPh (16%) + **16** (7%) (37)

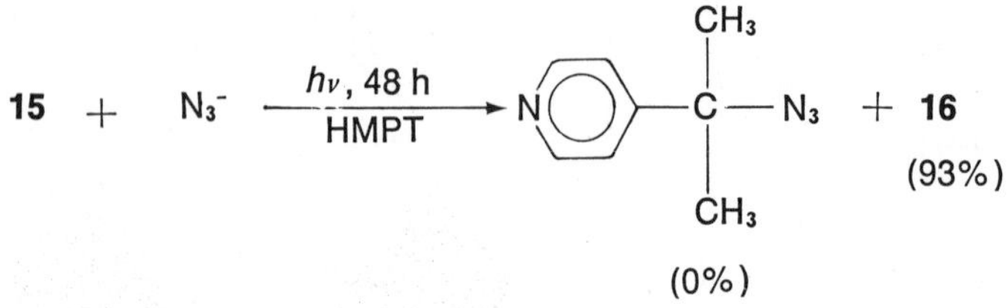

15 + N₃⁻ $\xrightarrow[\text{HMPT}]{h\nu,\ 48\ h}$ —C—N₃ (0%) + **16** (93%) (38)

Formation of the substitution product arises from $S_{RN}1$ reactions whereas the dimer is formed in a termination step (Reaction 39).

Based on the yields in Reactions 35–38, it appears that the nucleophilic reactivity is carbanion $> PhS^- > N_3^-$, but, kinetic data for these reactions has not been reported.

Even nitrocumene, which originally appeared to be unreactive in DMSO or DMF, was later found to react in HMPT (Reaction 40) (44).

Other substrates that react with the lithium salts of nitroalkanes are the derivatives of IV, where S represents p-CN, p-PhSO$_2$, p-Ph–CO, or 3,5-(CF$_3$)$_2$.

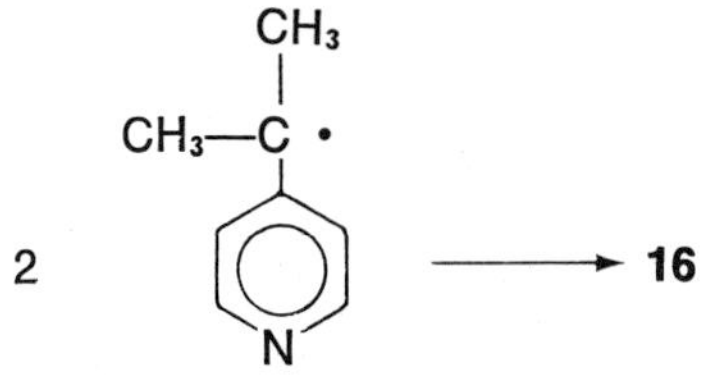

$$\text{(39)}$$

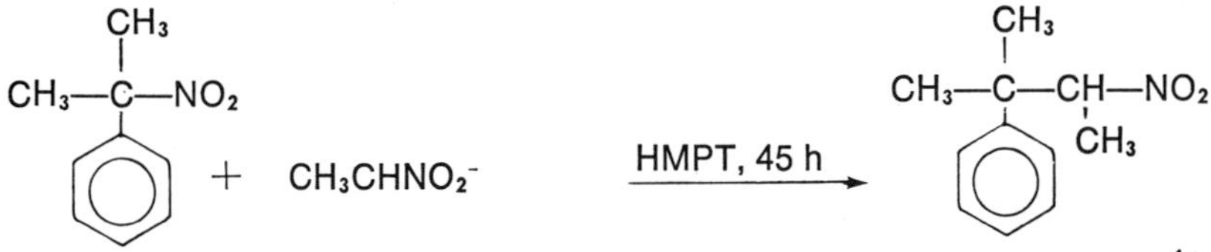

$$\text{(40)}$$

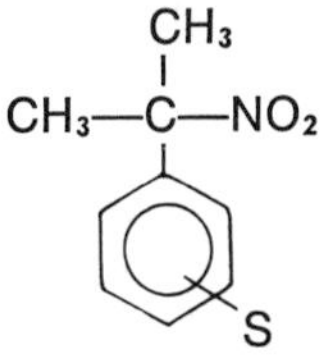

IV

Several substrates bearing cyano groups also react with various nucleophiles (45).

Interestingly, in many of these reactions Reduction Product **19** is found, which probably indicates that the radical anion **17** suffers competitive reactions, such as electron transfer (Reaction 41) and fragmentation (Reaction 42). The radical **18** formed by the fragmentation of **17** abstracts a hydrogen atom from a suitable donor and is reduced to **19** (Reaction 43).

Fragmentation of the radical anion **17** following Reaction 42 is demonstrated further in the reaction of substrates of similar type, such as **20**, with methanethiolate ion (Reactions 44–45).

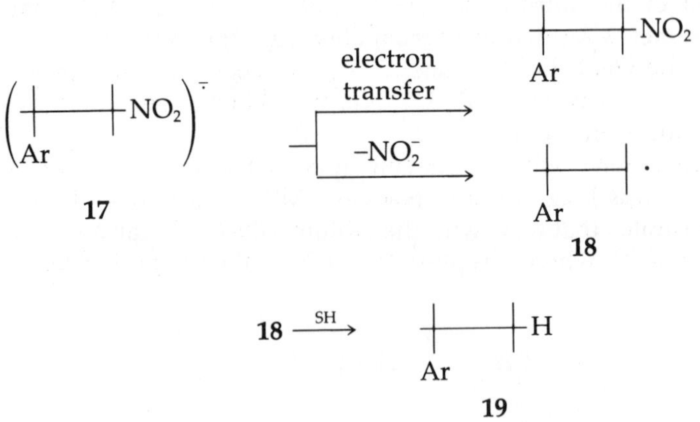

$$\left(\underset{Ar}{\boxed{}} NO_2\right)^{\bar{\cdot}} \quad \xrightarrow[\text{transfer}]{\text{electron}}$$

17

$$\underset{Ar}{\boxed{}} NO_2 \qquad (41)$$

$$\xrightarrow{-NO_2^-}$$

$$\underset{Ar}{\boxed{}} \cdot \qquad (42)$$

18

$$18 \xrightarrow{\text{SH}} \underset{Ar}{\boxed{}} H \qquad (43)$$

19

Reaction in either DMF or HMPT involves the same radical anion intermediate, which fragments to give radical **23**. Abstraction of a hydrogen
atom from methanethiolate forms the reduction product **21** and the radical anion **24** (Reaction 46), whereas addition to methanethiolate gives
the substition product **22** (Reaction 47) (*40, 42*).

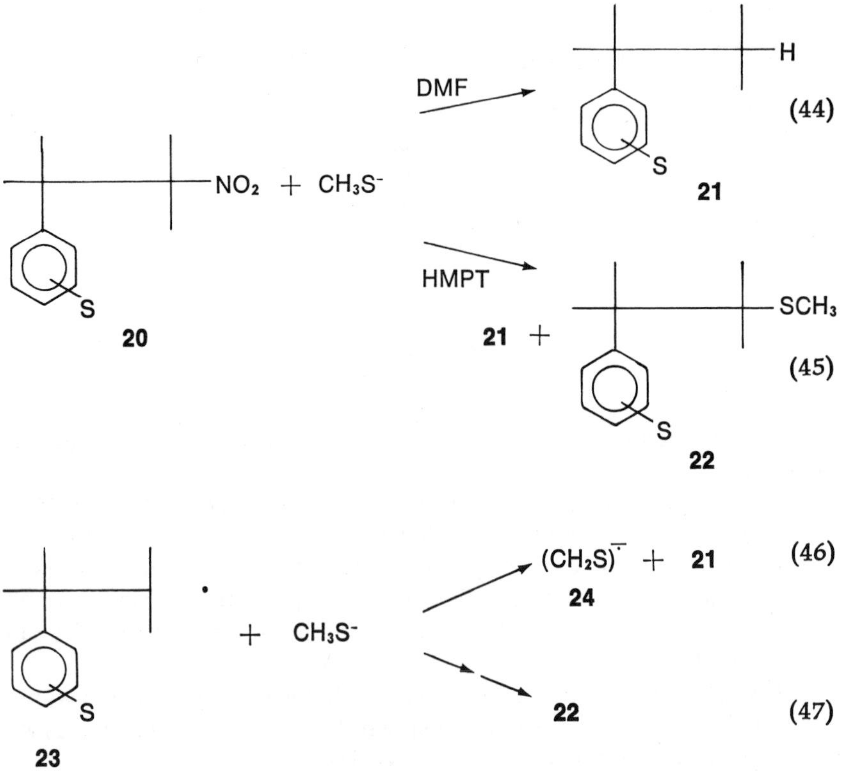

$$20 \qquad NO_2 + CH_3S^- \qquad \xrightarrow{\text{DMF}} \qquad 21 \qquad H \qquad (44)$$

$$\xrightarrow{\text{HMPT}} \qquad 21 + \qquad 22 \qquad SCH_3 \qquad (45)$$

$$23 \qquad + \quad CH_3S^- \qquad \nearrow \quad (CH_2S)^{\bar{\cdot}} + 21 \qquad (46)$$
$$24$$
$$\searrow \quad 22 \qquad (47)$$

When the substrate has two leaving groups, such as *p*-dichloromethylnitrobenzene **25**, reaction with 2-nitropropane ion **26** as nucleophile led to three products, **27**, **28**, and **29** (Reaction 48), by way of two consecutive reactions, Reactions 49 and 50 (*46*).

Reaction 49 is a $S_{RN}1$ reaction. The elimination reaction in Reaction 50 is also a radical chain reaction. The mechanism suggested for Reaction 50 based on the experimental evidence is given in Reactions 51–56 (*46*).

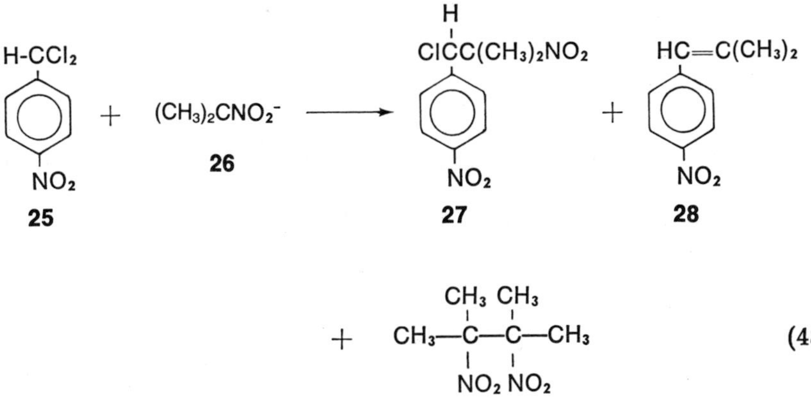

25 **26** **27** **28** **29** (48)

$$\mathbf{25} + \mathbf{26} \longrightarrow \mathbf{27} + Cl^- \qquad (49)$$

$$\mathbf{27} + 2\ \mathbf{26} \longrightarrow \mathbf{28} + \mathbf{29} + Cl^- + NO_2^- \qquad (50)$$

30 **31** (51)

32 (52)

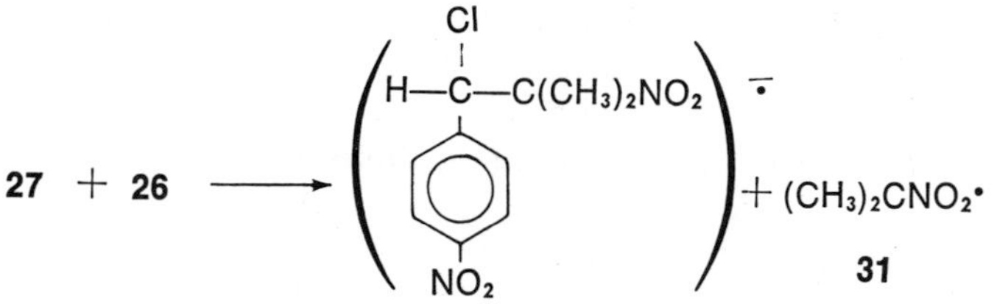

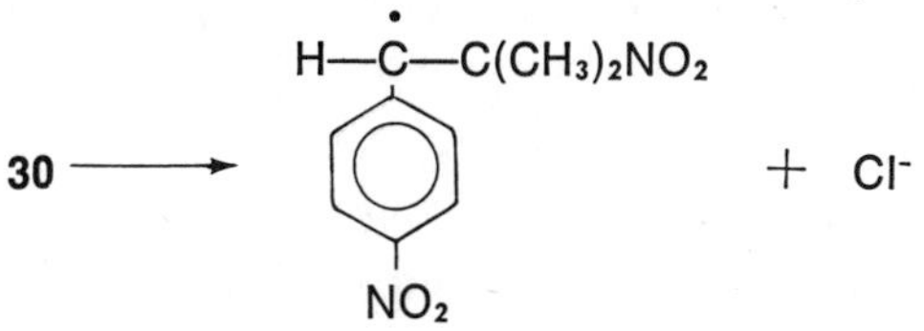

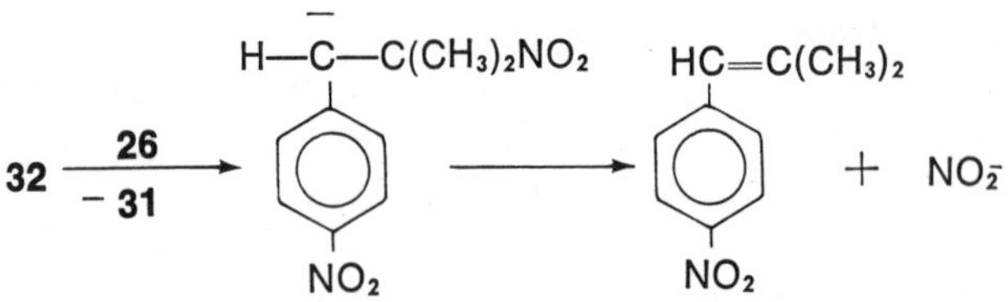

$$32 \xrightarrow[-31]{26} \underset{NO_2}{\overset{H-\overset{-}{C}-C(CH_3)_2NO_2}{\bigcirc}} \longrightarrow \underset{NO_2}{\overset{HC=C(CH_3)_2}{\bigcirc}} + NO_2^- \quad (53)$$

$$\underset{33}{}$$

$$32 \longrightarrow 33 + NO_2 \cdot \xrightarrow[-31]{26} 33 + NO_2^- \quad (54)$$

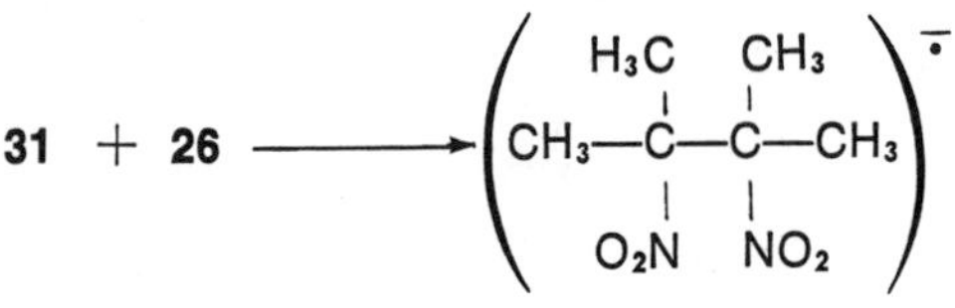

$$31 + 26 \longrightarrow \left(\begin{array}{c} H_3C \quad CH_3 \\ CH_3-\overset{|}{\underset{|}{C}}-\overset{|}{\underset{|}{C}}-CH_3 \\ O_2N \quad NO_2 \end{array} \right)^{\overline{\cdot}} \quad (55)$$

$$\underset{34}{}$$

$$34 + 27 \longrightarrow 30 + 29 \quad (56)$$

Slow elimination of chloride ion from **30** is supported by the fact that under certain conditions **27** can be isolated. The overall mechanism was denoted E$_{RC}$1, standing for *Elimination, Radical Chain, Unimolecular (46)*.

Entrainment. Entrainment is particularly important in these reactions, and, in many cases, is spectacular. For instance, α, p-dinitrocumene does not react with sodium azide during 48 h in the dark. In contrast, the lithium salt of 2-nitropropane reacts readily with the same substrate, and the reaction is complete in 3 h. When the substrate (1 mol) is treated with sodium azide (2 mol) in the presence of the lithium salt of 2-nitropropane (0.1 mol), all the α, p-dinitrocumene is consumed in 3 h, and 97% yield of pure p-nitrocumylazide is formed (*47*).

The reaction of α, p-dinitrocumene with sodium benzenesulfinate provides another clear example of entrainment. In the dark only 8% reaction occurs in 96 h. However, when 5 mol % of the lithium salt of 2-nitropropane is added, the reaction goes to completion in 4 h and gives 95% yield of the p-nitrocumylphenylsulfone (*47*).

Aliphatic Systems. Anions of nitrocompounds can react with α-nitrocarbonyl compounds (esters and ketones), α-nitronitriles, and α, α-dinitro compounds, displacing the nitro group, as in Reactions 57–59 (*48, 49*).

$$Me_2C-CO_2R + Me_2CNO_2^- \longrightarrow Me_2-\overset{\overset{\displaystyle NO_2}{\displaystyle |}}{C}-\overset{\overset{}{}}{\underset{\underset{\displaystyle CO_2R}{\displaystyle |}}{C}Me_2} + NO_2^- \quad (57)$$

$$\underset{NO_2}{\overset{|}{Me_2C}}$$

R = Et, Ph

R = Et (95% yield)
R = Ph (82% yield)

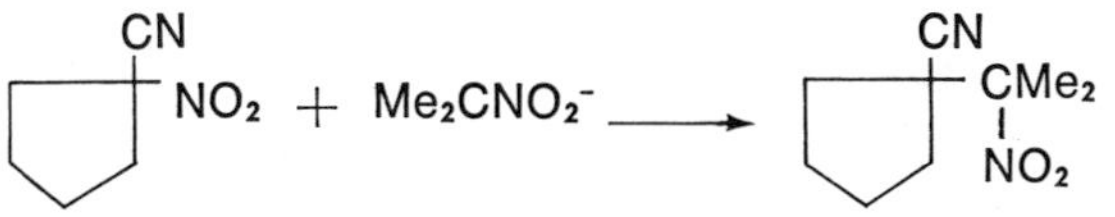

$$\text{(58)}$$

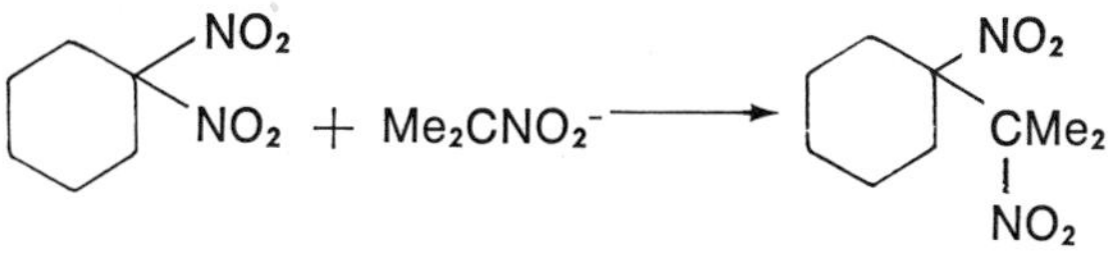

$$\text{(59)}$$

   The anions of nitroalkanes are not unique in displacing the nitro group from α,α-dinitro compounds. Thus 2,2-dinitropropane reacts, albeit slowly, with sodium benzenesulfinate to give the corresponding sulfone in 81% yield (Reaction 60) (50). Other examples are shown in Reactions 61–66 (30, 50–52).

   Reactions of this kind constitute the basis of the synthesis of α,β-unsaturated nitriles, ketones, and esters, where the first stage is the interaction of the sodium salt of RCHY-CO$_2$Et (Y = CN, COMe, CO$_2$Et, or SO$_2$Ar), with α-halo or α-nitro derivatives, such as R$_1$R$_2$C(NO$_2$)$_2$ in HMPT or DMF. Subsequent heating of the product formed, R$_1$R$_2$C(NO$_2$)–CR(Y)CO$_2$Et, in HMPT at 120°C leads to the formation of R$_1$R$_2$C=CRY (53, 54).

   The reduction of R$_1$R$_2$C(NO$_2$)–CR(SO$_2$Ar)CO$_2$Et with sodium sulfide in DMF at room temperature yields R$_1$R$_2$C=CRCO$_2$Et (55).

   The reactions shown in Reactions 61–66 indicate that NO$_2$ is a poor leaving group compared with the others represented, but it can be displaced as the examples in Reactions 67–68 show (56).

   The lithium salts of several acetylenides react with 2-halo-2-nitropropanes in THF or THF/hexane to give significant amounts (25–50%) of coupling products (Reaction 69) (57).

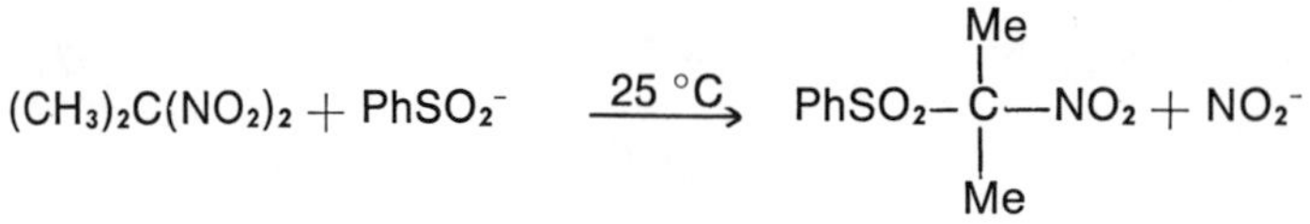

$$(CH_3)_2C(NO_2)_2 + PhSO_2^- \xrightarrow{25\ °C} \underset{\underset{Me}{|}}{\overset{\overset{Me}{|}}{PhSO_2-C-NO_2}} + NO_2^- \qquad (60)$$

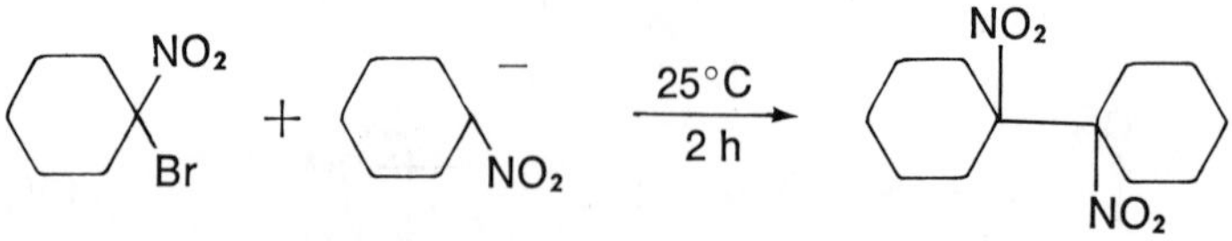

(89%) (30)

(61)

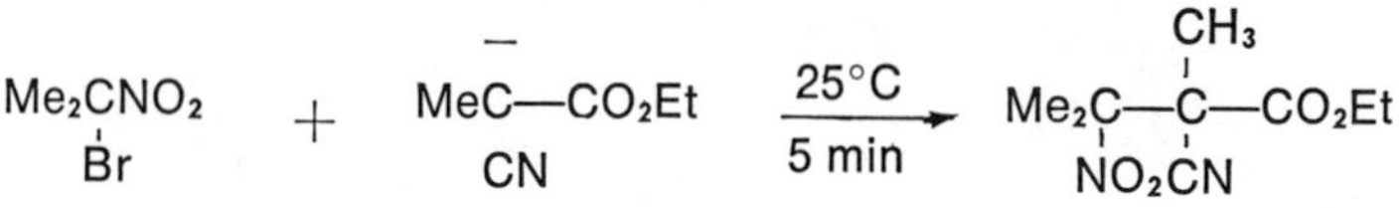

$$\underset{\underset{Br}{|}}{Me_2CNO_2} + \underset{CN}{MeC-CO_2Et} \xrightarrow[5\ min]{25°C} \underset{\underset{NO_2CN}{|}}{\overset{\overset{CH_3}{|}}{Me_2C-C-CO_2Et}} \qquad (62)$$

(78%) (30)

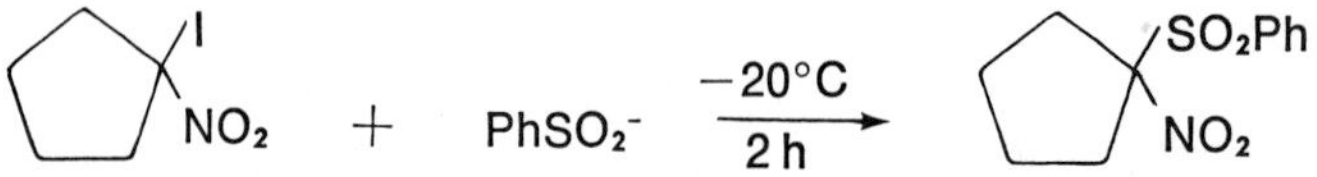

(63)
(93%) (50)

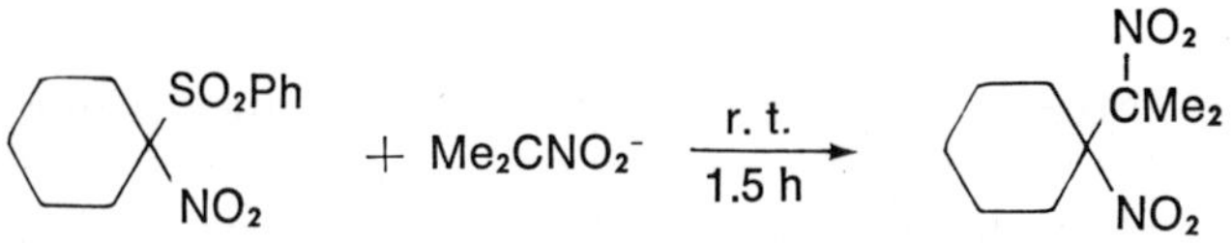

(85%) (51)
(64)

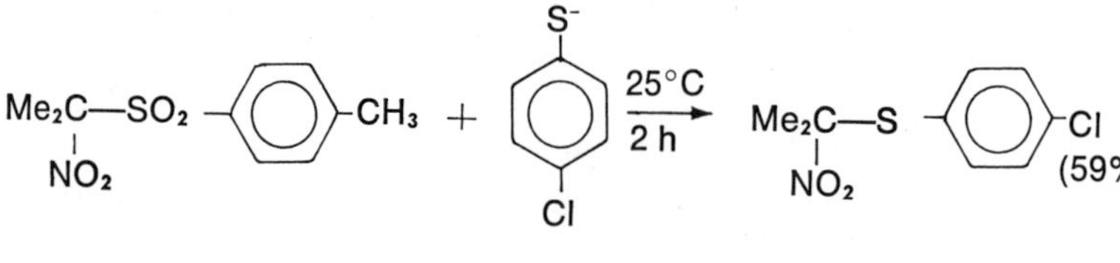

(51)
(59%)

(65)

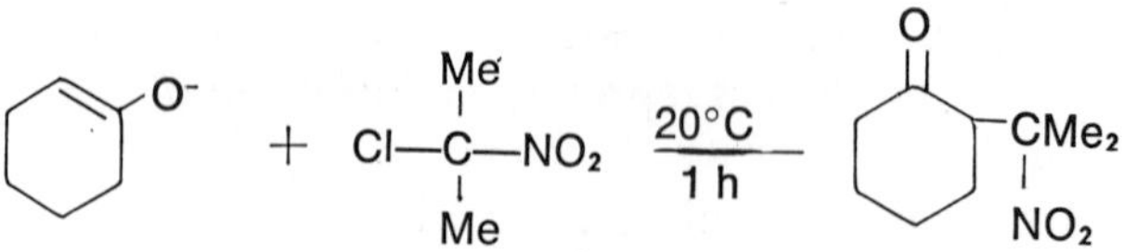

(46%)   (51)

(66)

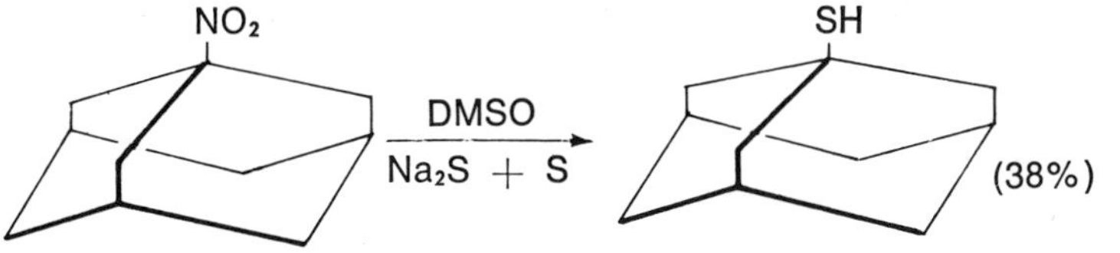

$$R-C \equiv C\ Li + Me_2C{\overset{X}{\underset{NO_2}{\diagdown}}} \longrightarrow R-C \equiv C-\underset{\underset{NO_2}{|}}{C}Me_2 + XLi \qquad (69)$$

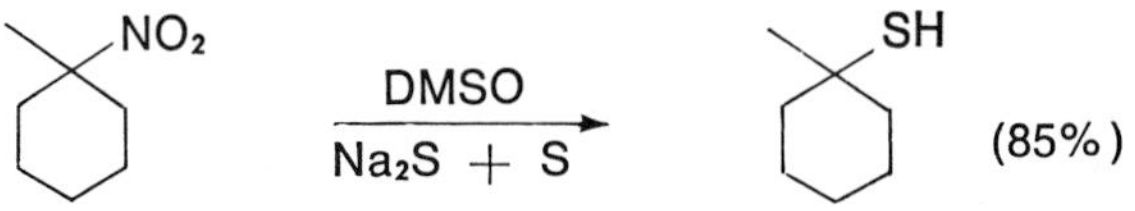

However, the mechanism is probably not a free radical chain. The mechanism proposed involves C–C bond formation in a solvent cage (Reaction 70 (57)).

$$R-C \equiv C^- + R_2C{\overset{X}{\underset{NO_2}{\diagdown}}} \longrightarrow \left[R-C \equiv C \cdot + R_2C{\overset{X}{\underset{NO_2^-}{\diagdown}}} \right] \qquad (70)$$

$$\downarrow$$

$$\left[R-C \equiv C \cdot + R_2\dot{C}NO_2 + X^- \right]$$

$$\downarrow$$

$$R-C \equiv C - \underset{\underset{NO_2}{|}}{C}R_2 + X^-$$

This mechanism, denoted $S_{ET}2$, requires that the radical anion $R_2C(X)NO_2^-$ dissociate to $R_2(NO_2)C \cdot$ and X^- before the initially formed radical and radical anion have diffused apart (57).

Several alkylmercury halides react with the anion of 2-nitropropane or nitrocyclohexane in a light-catalyzed process at 25°C to give the coupling products, shown in Reactions 71 and 72 (58).

The reaction rates decrease from R = benzyl to R = alkyl. For this reason, it was suggested that resonance stabilization of the radical R formed in Reaction 73 plays an important role.

$$R—Hg—X + Me_2CNO_2^- \xrightarrow{h\nu} \underset{\underset{NO_2}{|}}{R—CMe_2} + X^- + Hg° \qquad (71)$$

$$R = \text{(2-methylcyclohexanone structure)}$$

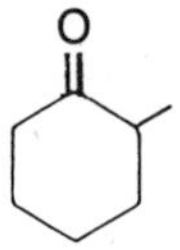

68% (1 h)

$$= PhCH_2 \qquad\qquad\qquad 99\% \text{ (2 h)}$$

$$= \text{cyclohexyl} \qquad\qquad 76\% \text{ (34 h)}$$

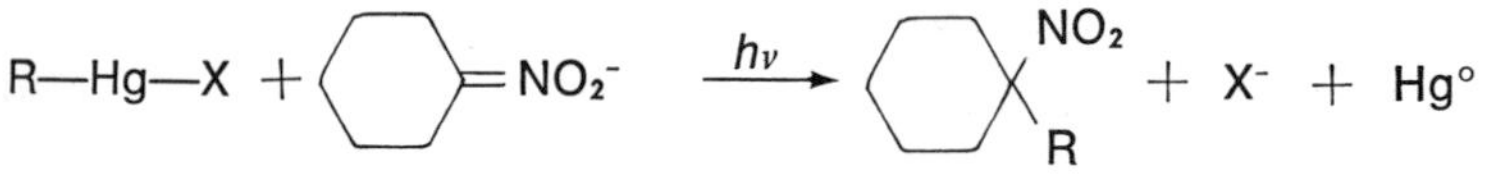

$$R—Hg—X + \text{(cyclohexylidene)}=NO_2^- \xrightarrow{h\nu} \text{(cyclohexane ring with } NO_2 \text{ and } R) + X^- + Hg° \qquad (72)$$

$$R = PhCH_2 \qquad\qquad\qquad 87\% \text{ (2 h)}$$

$$R = \text{cyclohexyl} \qquad\qquad 84\% \text{ (60 h)}$$

$$(R–Hg–X)^{\overline{\cdot}} \longrightarrow R^{\cdot} + X^- + Hg° \qquad (73)$$

Attempts to couple phenyl or vinylmercury halides with nitronate anions were unsuccessful. This result is consistent with the failure of these anions to undergo aromatic $S_{RN}1$ reactions (58).

The $S_{RN}1'$ Mechanism. The reaction of p-nitrophenyl allyl chloride **35** with the lithium salt of 2-nitropropane gives 70% of the rearranged substition product **36** (Reaction 74) (59).

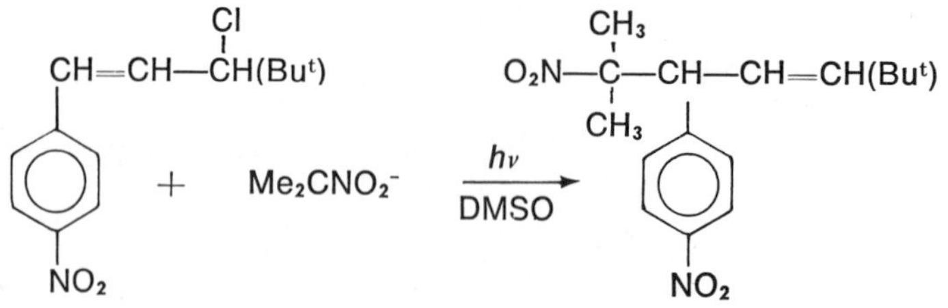

$$\underset{\underset{NO_2}{|}}{\underset{|}{\overset{\overset{Cl}{|}}{CH=CH—CH(Bu^t)}}} + Me_2CNO_2^- \xrightarrow[DMSO]{h\nu} \underset{\underset{NO_2}{|}}{\overset{\overset{CH_3}{\underset{|}{O_2N—C—CH—CH=CH(Bu^t)}}}{}} \qquad (74)$$

35   **36** (70% yield)

Other anions such as diethyl methylmalonate and p-toluenesulfinate were also shown to undergo the same type of reaction. The mechanism proposed is sketched in Scheme II, and was termed the $S_{RN}1'$ mechanism for its resemblance with S_N1' reactions (59).

The predominance of the $S_{RN}1'$ mechanism in these systems was attributed to steric hindrance by the t-butyl group which hinders the simple S_N2 process.

$S_{RN}1$ Reaction in Vinylic Systems

Vinyl halides are unreactive toward nucleophiles (60), resembling aryl halides in many respects. The possibility that they might react with nucleophiles by the $S_{RN}1$ mechanism was tested (61).

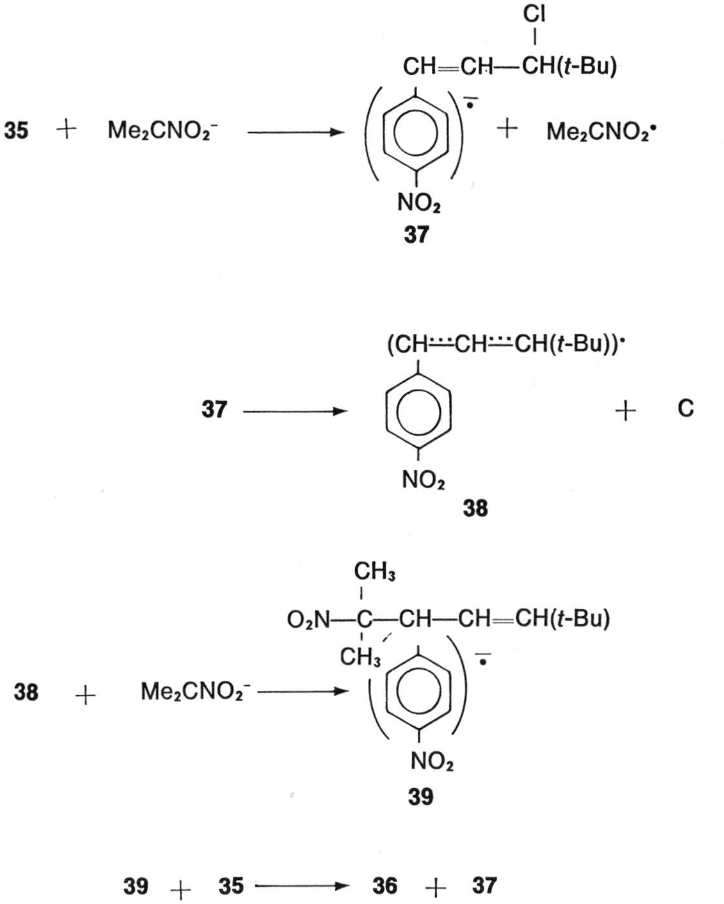

$$\mathbf{35} \;+\; Me_2CNO_2^- \longrightarrow \underset{\substack{\mathbf{37}}}{\left(\underset{NO_2}{\overset{\overset{\displaystyle Cl}{\overset{|}{CH=CH-CH(t\text{-}Bu)}}}{\bigcirc}}\right)^{\!\!\!\cdot\,-}} \;+\; Me_2CNO_2^{\bullet} \qquad (75)$$

$$\mathbf{37} \longrightarrow \underset{\substack{\mathbf{38}}}{\underset{NO_2}{\overset{\overset{\displaystyle (CH\cdots CH\cdots CH(t\text{-}Bu))^{\bullet}}{|}}{\bigcirc}}} \;+\; C \qquad (76)$$

$$\mathbf{38} \;+\; Me_2CNO_2^- \longrightarrow \underset{\substack{\mathbf{39}}}{\left(\underset{NO_2}{\overset{\overset{\displaystyle O_2N-\underset{\underset{CH_3}{|}}{\overset{\overset{CH_3}{|}}{C}}-CH-CH=CH(t\text{-}Bu)}{|}}{\bigcirc}}\right)^{\!\!\!\cdot\,-}} \qquad (77)$$

$$\mathbf{39} \;+\; \mathbf{35} \longrightarrow \mathbf{36} \;+\; \mathbf{37} \qquad (78)$$

Scheme II

β-Bromostyrene **40** reacted with potassium acetone enolate ion in liquid ammonia in a pyrex flask exposed to 350-nm radiation during 5 h forming the substitution product **41** in 48% yield and a tautomer **42** in 34% yield (Reaction 79).

$$PhHC = CHBr + {}^-CH_2COCH_3 \xrightarrow{h\nu} PhHC = CH{-}CH_2{-}CO{-}CH_3 + \quad (79)$$

$$\textbf{40} \qquad\qquad\qquad\qquad\qquad\qquad \textbf{41}$$

$$PhCH_2{-}CH = CH{-}CO{-}CH_3$$

$$\textbf{42}$$

When the same reactants were treated with potassium metal, the products had the same carbon skeleton as in the photostimulated reaction, but represented lower stages of oxidation (Reaction 80).

$$\textbf{40} + {}^-CH_2COCH_3 \xrightarrow{K,\ NH_3} \qquad\qquad\qquad\qquad OH$$

$$PhCH_2CH_2CH_2COCH_3 + PhCH_2CH_2CH_2 \overset{|}{C}HCH_3 \qquad (80)$$

$$(63\%\ \text{yield}) \qquad\qquad\qquad (15\%\ \text{yield})$$

3,3-Dimethyl-1-iodoindene **43** reacted with acetone enolate ion in liquid ammonia under irradiation during 75 min to form mainly **44**, and a tautomer thereof **45** in a total yield of 66% (Reaction 81). In the dark there was no reaction after 1 h. A solution of **43** irradiated with excess benzenethiol for 90 min gives about 20% yield of the sulfide **46** (Reaction 82).

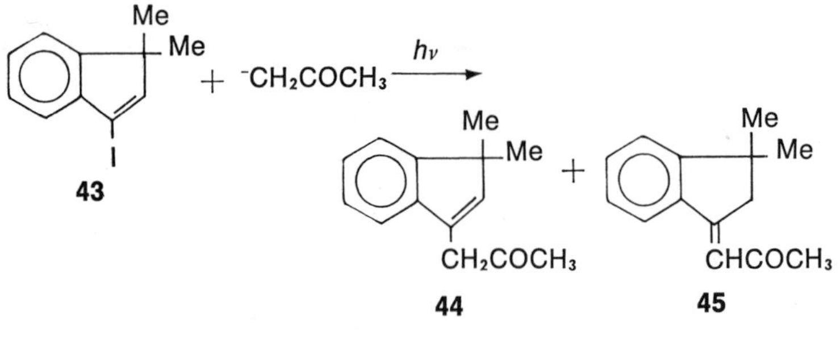

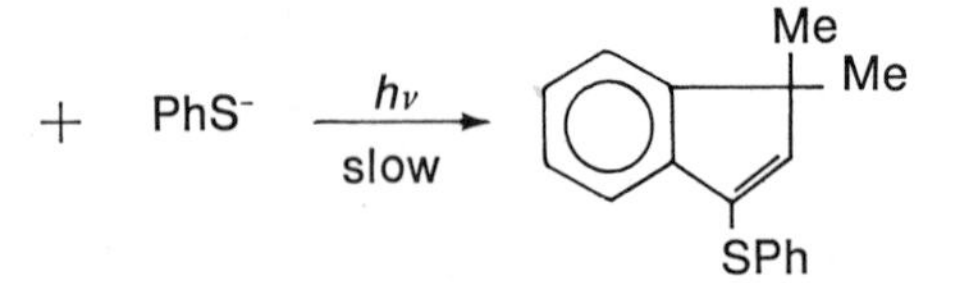

$$(81)$$

$$(82)$$

2-Iodo-2-norbornane and 1-iodocyclopentene irradiated with acetone enolate ion formed the normal substitution product or the isomer-
eric α, β-unsaturated ketone.

The previous examples, and others reported elsewhere (61), suggest
that the reactions occur by the $S_{RN}1$ mechanism, but the reactivity is, on
the whole, lower than with aryl halides.

Although the mechanism was not investigated, it is noteworthy that
the reactions of Ph_2Z^- (Z being As or P) nucleophiles with vinyl halides
reported by Aguiar and colleagues may well be further examples of
vinylic $S_{RN}1$ reactions. They are summarized as follows: either cis or trans
β-bromostyrene reacts with Ph_2AsLi in THF to give the corresponding
substitution product with retention of configuration (Reaction 83) (62),
or cis or trans-1,2-dichloroethylene reacts with diphenylarasenide ion
giving the stereospecific products (Reactions 84–88) (63).

Lithium diphenylphosphide also reacts with the haloalkene to give
substitution products of retained configuration (64, 65).

$$PhHC = CHBr + Ph_2AsLi \xrightarrow{THF} PhHC = CHAsPh_2 + BrLi \qquad (83)$$

cis $\longrightarrow$ cis

trans $\longrightarrow$ trans

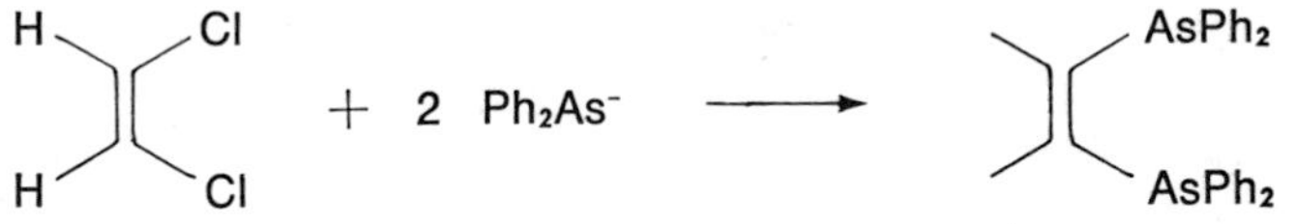

(84)

(61%)

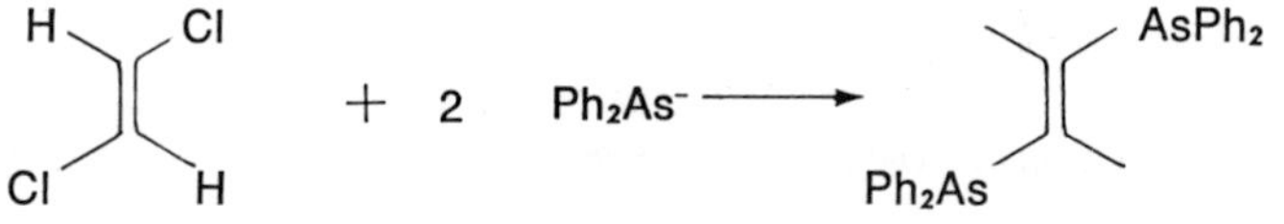

(85)

(10%)

Miscellaneous Reactions Involving Electron Transfer

Although $S_{RN}1$ reactions represent a new mechanism in aromatic
chemistry, the fact that many organic reactions involve electron transfer
with concomitant formation of radicals and radical anions has been
known for a long time (66). However, in many instances it has been
difficult to establish whether a particular radical or radical anion is the

intermediate in the major reaction pathway leading to product. Another general point is that electron transfer is not restricted to nucleophilic aromatic substitution but has been invoked in electrophilic reactions as well (67).

Electron Transfer in Reactions of Nitroaromatic Compounds. A mechanism was proposed for aromatic nucleophilic substitution in which σ complex formation proceeds through prior radical anion formation (68) as shown for the case of p-nitroanisole and sodium methoxide (Reaction 86).

$$\tag{86}$$

Although the formation of radical anions by the interaction of nitroaromatic compounds with bases is a well-known process (69–72), it is also true that the yield of radical anions is usually very small (~ 1%), and ESR detection of radical anions does not necessarily mean that they are intermediates in the major reaction pathway (73).

It has been claimed (74) that the radical anion formed by the interaction of hydroxide ion with o-dinitrobenzene and p-dinitrobenzene in DMSO to form o-nitrophenol and p-nitrophenol, respectively, may in fact be intermediates in the substitution reaction.

Reactions 87–91 represent the mechanism suggested to account for spectroscopic observations (ESR and UV-visible) (74, 75).

Steps 89 or 90 and 91 were suggested as alternative routes to the substitution product. Kinetic analysis of the system is in agreement with any one of the proposed mechanisms. The $S_{RN}1$ mechanism was discarded because the decomposition of the radical anion is first order in hydroxide ion, a result which is inconsistent with the $S_{RN}1$ mechanism, provided that the decomposition of the radical anion **48** is the rate determining step in the whole process (Reaction 92) (74, 75).

As far as we know, there are no other examples in the literature indicating that radical anions react with nucleophiles.

The Superoxide Ion as Nucleophile. The reaction of superoxide ion with aromatic substrates seems to be preceded by an electron transfer reaction.

By bubbling oxygen into a solution of p-nitrochlorobenzene radical anion in ammonia, 33% of p-nitrophenol is formed and traces of 4,4'-dinitrobiphenyl (76). Also, potassium superoxide reacts with p-nitrochlorobenzene in HMPT lithium 76% of p-nitrophenol.

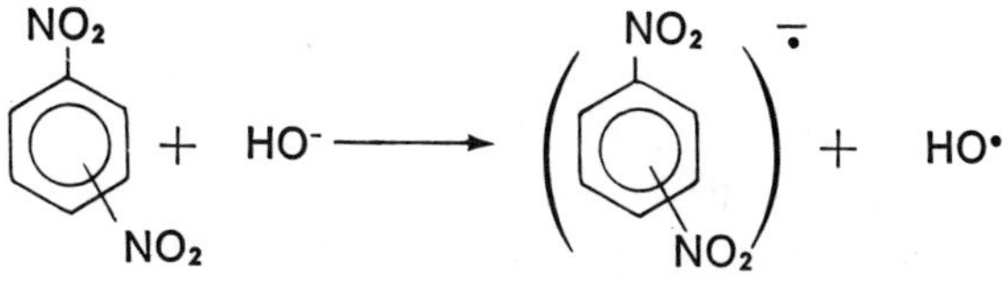

$$47 + HO^- \longrightarrow 48 \qquad (87)$$

$$48 + HO^- \longrightarrow 49 \qquad (88)$$

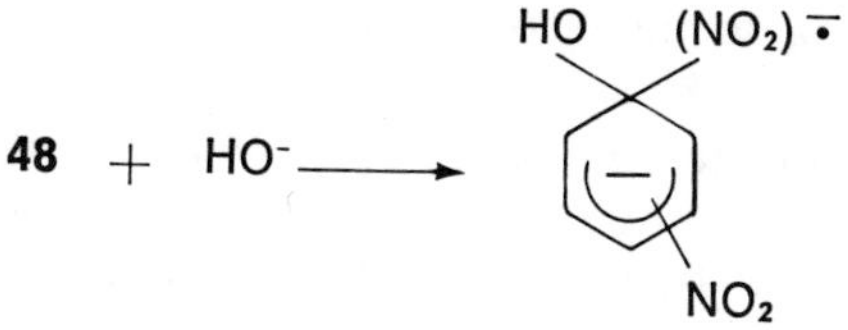

$$49 \longrightarrow 50 + \cdot NO_2^{2-} \qquad (89)$$

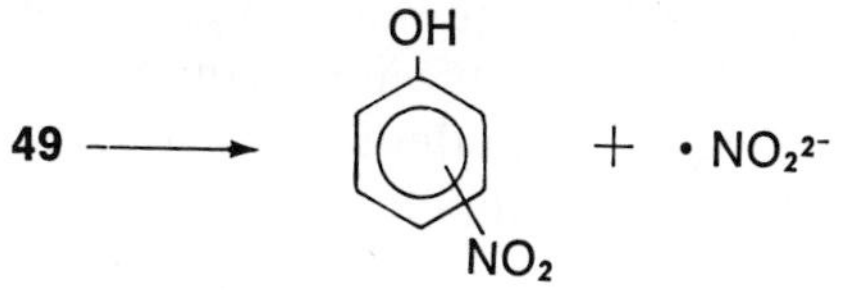

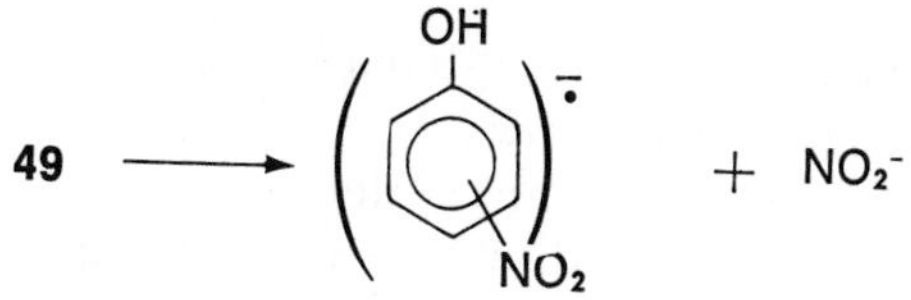

$$49 \longrightarrow 51 + NO_2^- \qquad (90)$$

$$51 + 47 \longrightarrow 48 + 50 \qquad (91)$$

$$48 \longrightarrow + NO_2^- \qquad (92)$$

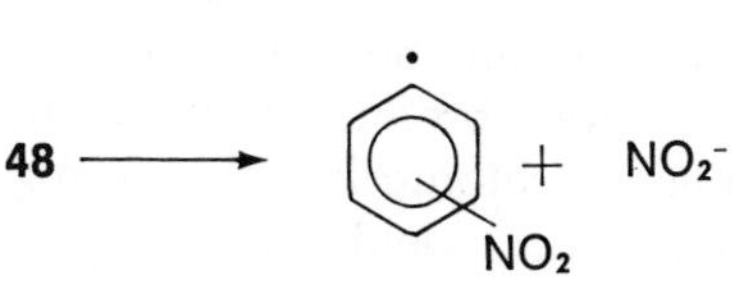

Similarly, in the gas phase, the negative chemical ionization spectra of 4-bromobenzophenone, 4-nitrochlorobenzene and 2,4-dinitrochlorobenzene, were each replaced by the spectra of the corresponding phenoxide ions when superoxide ion was generated in the system (76).

The reaction of superoxide ion with nitro-substituted aromatic halides occurs via an electron transfer from the superoxide to the substituted benzenes to yield radical anions **52**. They are subsequently scavenged by molecular oxygen ultimately to give nitrophenols (Reactions 93–94) (77).

The reactions of 3-bromoquinoline, 2-chloroquinoline, and 1-bromoisoquinoline with potassium superoxide were studied in the presence of 18-crown-6-ether hoping to increase the solubility of the nucleophile (78).

The reaction of 3-bromoquinoline was unsuccessful in benzene, while in DMSO 3-hydroxyquinoline was obtained in reasonable yield.

2-Chloroquinoline and 1-bromoisoquinoline are more reactive than the 3-bromo derivative. The mechanism suggested is similar to Reactions 93–94. A mechanism involving the formation of 3-quinolyl radical was discarded because no 3,3'-biquinoline was isolated (78). The mechanism for 2-chloroquinoline and 1-bromoisoquinoline is less clear since both compounds are known to undergo a $S_N Ar$ substitution easily, and this process can certainly be competitive with the electron transfer.

Photochemistry of Aryllithium Compounds. Irradiation of solutions of phenyllithium in ethyl ether gave over 80% yield of biphenyl and metallic lithium (79, 80). The coupling is very specific; for instance, 2-naphthyllithium gave exclusively 2,2'-binaphthyl. The mechanism proposed is shown in Reactions 95–96 (79, 80).

A dimer of phenyllithium undergoes homolysis of the phenyl–lithium bond, and the aryl radical thus formed attacks the other C–Li bond (Reaction 95) forming biphenyl radical anion and lithium cation. Since this radical anion is unstable, lithium metal and biphenyl result (Reaction 96).

It was also pointed out that direct formation of biphenyl on synchronous generation and coupling of two aryl radicals from the dimer would account for these results.

Photoreduction with Sodium Borohydride. In the irradiation of solutions ($\lambda = 254$ nm) of chloro, bromo, and iodobenzene with sodium borohydride in 6% aqueous acetonitrile, the sole organic product was benzene, formed in quantitative yield, with quantum yields of 0.5, 5.7 and 7.5 respectively (81). In addition, a gas evolved was shown to be mainly hydrogen with traces of diborane. The solution contained halide ion, and a white solid which gave a positive test for borate.

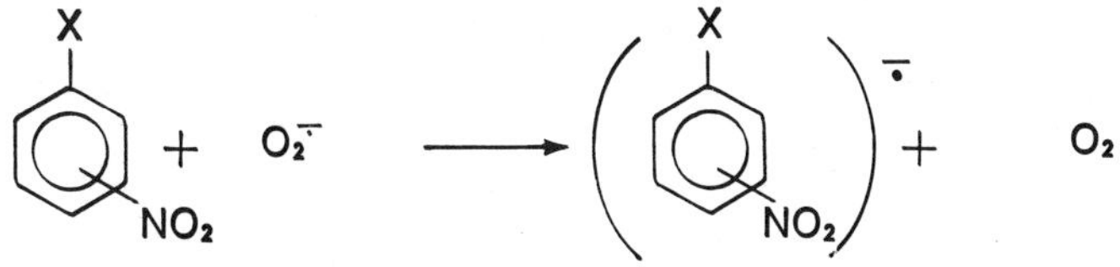

$$\text{52}$$

(93)

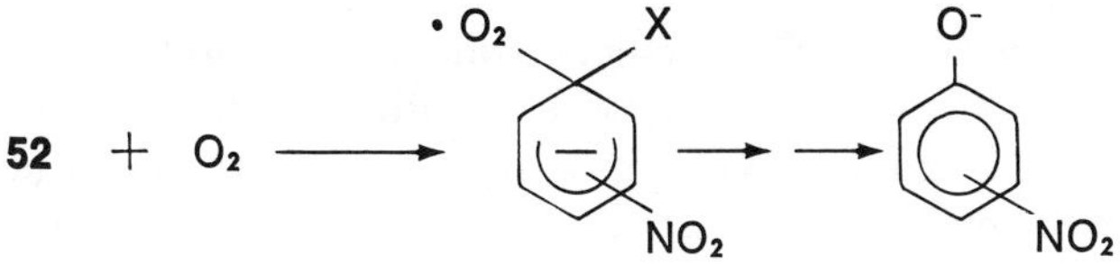

$$\text{52} + O_2 \longrightarrow \longrightarrow \longrightarrow$$

(94)

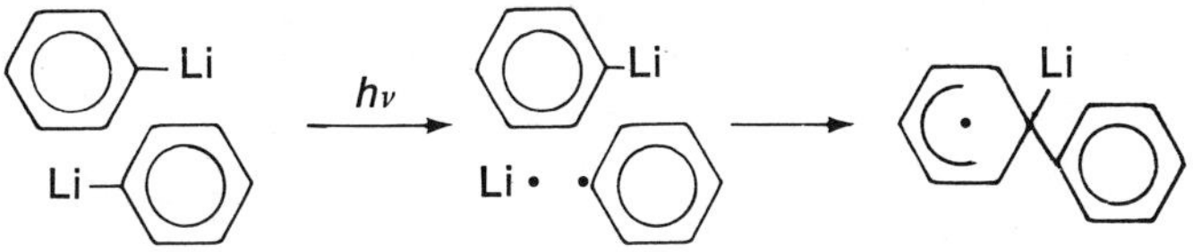

$$\xrightarrow{h\nu} \longrightarrow$$

53

(95)

$$\text{53} \longrightarrow \longrightarrow + Li^\circ$$

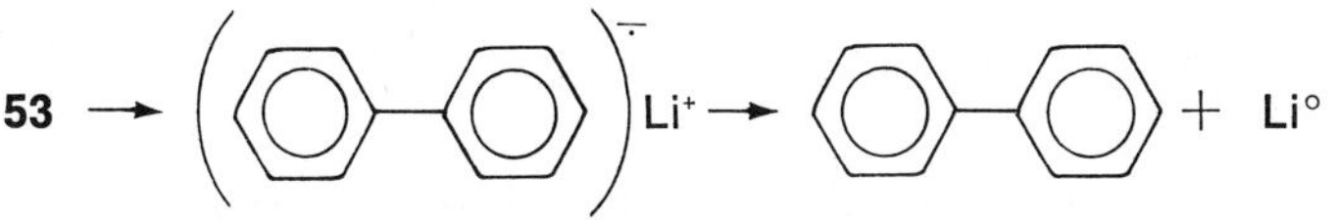

(96)

To accommodate these results, the mechanism of Reactions 97–102 was proposed (*81*)

$$PhX \xrightarrow{h\nu} (PhX)^* \longrightarrow Ph^\cdot + X^\cdot \qquad (97)$$

$$X^\cdot + BH_4^- \longrightarrow BH_3^{\cdot -} + X^- + H^+ \qquad (98)$$

$$Ph^\cdot + BH_4^- \longrightarrow BH_3^{\cdot -} + PhH \qquad (99)$$

$$PhX + BH_3^{\cdot -} \longrightarrow Ph^\cdot + BH_3 + X^- \qquad (100)$$

$$2\,BH_3 \longrightarrow B_2H_6 \xrightarrow{H_2O} borate + H_2 \qquad (101)$$

$$Ph^\cdot + X^\cdot \longrightarrow PhX \qquad (102)$$

where Reaction 97 is the initiation step, Reactions 99–100 are the propagation steps of a chain mechanism, and Reaction 102 is a termination step.

The radical chain nature of the mechanism was corroborated by the reaction of iodobenzene with sodium borohydride in liquid ammonia and acetonitrile to which sodium metal was added to initiate the reaction. The yield observed, based on the sodium metal consumed, was 675%! Similar photoreductions of halobenzenes, 1- and 2-bromonaphthalenes and 9-bromophenanthrene are normally quantitative but are inhibited by acrylonitrile.

The photoreduction of 3- and 4-chlorobiphenyl with sodium borohydride in acetonitrile–water (10:1), proceeds with a quantum yield of 0.013 and 0.0024 respectively, giving biphenyl as the sole product (82). Experiments carried out in the presence of deuterated reagents led the authors to suggest that 4-chlorobiphenyl reacts by way of Reactions 97–102, but that the mechanism of reduction of 3-chlorobiphenyl could be represented by Reactions 103–105 (82).

The nature of the intermediates involved was not defined (82).

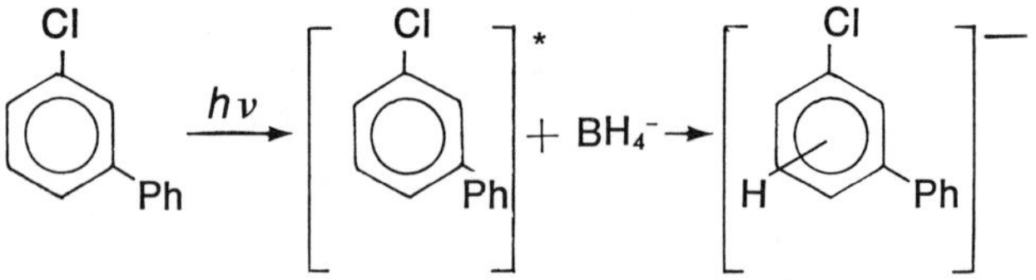

$$\text{(103)}$$

$$54 + H^+ \longrightarrow \text{(3-chloro-dihydrobiphenyl)} \quad \text{(104)}$$
$$\textbf{55}$$

$$55 \xrightarrow{-HCl} \text{biphenyl} \quad \text{(105)}$$

Literature Cited

1. Cornelisse, J.; de Gunst, G. P.; Havinga, E. *Adv. Phys. Org. Chem.* **1975,** *11,* 225.
2. Cornelisse, J. *Pure Appl. Chem.* **1975,** *41,* 433.
3. Havinga, E.; Cornelisse, J. *Pure Appl. Chem.* **1976,** *47,* 1.
4. Havinga, E.; Cornelisse, J. *Chem. Rev.* **1975,** *75,* 353.
5. Cornelisse, J.; Lodder, G.; Havinga, E. *Rev. Chem. Intermed.* **1979,** *2,* 231.
6. Havinga, E.; de Jongh, R. O.; Dorst, W. *Recl. Trav. Chim. Pays-Bas* **1956,** *75,* 378.

7. Odenhof, C.; Cornelisse, J. *Recl. Trav. Chim. Pays-Bas* **1978**, *97*, 35.
8. Lok, C. M.; Havinga, E. *Proc. K. Ned. Akad. Wet. B: Phys. Sci.* **1974**, *77*, 15.
9. Lammers, J. G.; Cornelisse, J. *Isr. J. Chem.* **1977**, *16*, 299.
10. Nijhoff, D. F.; Havinga, E. *Tetrahedron Lett.* **1965**, 4199.
11. Brasem, P.; Lammers, J. G.; Lugtenbury, J.; Cornelisse, J.; Havinga, E. *Tetrahedron Lett.* **1972**, 685.
12. Griffith, J.; Hawkins, C. *J. Chem. Soc., Chem. Commun.* **1973**, 111.
13. Vink, J. A. J.; Lok, C. M.; Cornelisse, J.; Havinga, E. *J. Chem. Soc., Chem. Commun.* **1972**, 710.
14. Lok, C. M.; den Boer, M. E.; Cornelisse, J.; Havinga, E. *Tetrahedron* **1973**, *29*, 867.
15. Vink, J. A. J.; Verheydt, P. L.; Cornelisse, J.; Havinga, H. *Tetrahedron* **1972**, *28*, 5081.
16. Letsinger, R. L.; Hautala, R. R. *Tetrahedron Lett.* **1969**, 4205.
17. Groen, M. B.; Havinga, E. *Mol. Photochem.* **1974**, *6*, 9.
18. Lodder, G.; Havinga, E. *Tetrahedron* **1972**, *28*, 5583.
19. Nilson, S. *Acta Chem. Scand.* **1973**, *27*, 329.
20. Andreades, S.; Zhnow, E. W. *J. Am. Chem. Soc.* **1969**, *91*, 4181.
21. Eberson, L.; Helgee, B. *Acta Chem. Scand. Ser. B* **1975**, *29*, 451.
22. Letsinger, R. L.; Steller, K. E. *Tetrahedron Lett.* **1969**, 1401.
23. Beijersbergen van Henegouwan, G. M. J.; Havinga, E. *Recl. Trav. Chim. Pays-Bas* **1970**, *89*, 907.
24. van Vliet, A.; Kronenberg, M. E.; Cornelisse, J.; Havinga, E. *Tetrahedron* **1970**, *26*, 1061.
25. Lammers, J. G.; Tamminga, J. J.; Cornelisse, J.; Havinga, E. *Isr. J. Chem.* **1977**, *16*, 304.
26. Varma, C. A. G. O.; Plantenga, F. L.; van den Ende, C. A. M.; van Zeyl, P. H. M.; Tamminga, J. J.; Cornelisse, J. *Chem. Phys.* **1977**, *22*, 475.
27. Kornblum, N.; Michel, R. E.; Kerber, R. C. *J. Am. Chem. Soc.* **1966**, *88*, 5662.
28. Russell, G. A.; Danen, W. C. *J. Am. Chem. Soc.* **1966**, *88*, 5663.
29. Kornblum, N. *Angew. Chem. Int. Ed. Eng.* **1975**, *14*, 734.
30. Beletskaya, I. P.; Drozd, V. N. *Russ. Chem. Rev. Eng. Transl.* **1979**, *48*, 431.
31. Norris, R. K.; Randles, D. *Aust. J. Chem.* **1976**, *29*, 2621.
32. *Ibid.*, **1979**, *32*, 1487.
33. Kerber, R. C.; Urry, G. W.; Kornblum, N. *J. Am. Chem. Soc.* **1964**, *86*, 3904.
34. *Ibid.*, **1965**, *87*, 4520.
35. Russell, G. A.; Pecoraro, J. P. *J. Am. Chem. Soc.* **1979**, *101*, 3331.
36. Kornblum, N.; Davies, T. M.; Earl, G. W.; Holy, N. T.; Kerber, R. C.; Musser, M. T.; Snow, D. H. *J. Am. Chem. Soc.* **1967**, *89*, 725.
37. Kornblum, N.; Stuchel, F. W. *J. Am. Chem. Soc.* **1970**, *92*, 1804.
38. Burt, B. L.; Freeman, D. J.; Gray, P. G.; Norris, R. K.; Randles, D. *Tetrahedron Lett.* **1977**, 3063.
39. Danen, W. C.; Kensler, T. T.; Lawless, J. G.; Marcus, M. F.; Hawley, M. D. *J. Phys. Chem.* **1969**, *73*, 4389.
40. Andrieux, C. P.; Blocman, C.; Dumas-Bouchiat, J. M.; Saveant, J. M. *J. Am. Chem. Soc.* **1979**, *101*, 3431.
41. Kornblum, N.; Carlson, S. C.; Smith, R. G. *J. Am. Chem. Soc.* **1979**, *101*, 647.
42. Kornblum, N.; Widmer, J.; Carlson, S. C. *J. Am. Chem. Soc.* **1979**, *101*, 658.
43. Feuer, H.; Doty, J. K.; Kornblum, N. *J. Heterocycl. Chem.* **1978**, *15*, 1419.
44. Kornblum, N.; Carlson, S. C.; Widmer, J.; Fifolt, M. J.; Newton, B. N.; Smith, R. G. *J. Org. Chem.* **1978**, *43*, 1394.
45. Kornblum, N.; Fifolt, M. J. *J. Org. Chem.* **1980**, *45*, 360.
46. Freeman, D. J.; Norris, R. K. *Aust. J. Chem.* **1976**, *29*, 2631.
47. Kornblum, N.; Swiger, R. T.; Earl, G. W.; Pinnick, H. W.; Stuchal, F. W. *J. Am. Chem. Soc.* **1970**, *92*, 5513.
48. Kornblum, N.; Boyd, S. D.; Stuchal, F. W. *J. Am. Chem. Soc.* **1970**, *92*, 5783.
49. Kornblum, N.; Boyd, S. D. *J. Am. Chem. Soc.* **1970**, *92*, 5784.
50. Kornblum, N.; Kestner, M. M.; Boyd, S. D.; Cattran, L. C.; *J. Am. Chem. Soc.* **1973**, *95*, 3356.

51. Kornblum, N.; Boyd, S. D.; Ono, N. *J. Am. Chem. Soc.* **1974**, *96*, 2580.
52. Russell, G. A.; Jawdosiuk, M.; Ros, F. *J. Am. Chem. Soc.* **1979**, *101*, 3378.
53. Ono, N.; Eto, H.; Tamura, R.; Hayami, J.; Kaji, A. *Chem. Lett.* **1976**, 757.
54. Ono, N.; Tamura, R.; Hayami, J.; Kaji, A. *Chem. Lett.* **1977**, 189.
55. Kornblum, N.; Carlson, S. C.; Smith, R. G.; *J. Am. Chem. Soc.* **1978**, *100*, 289.
56. Kornblum, N.; Widmer, J. *J. Am. Chem. Soc.* **1978**, *100*, 7086.
57. Russell, G. A.; Jawdosiuk, M.; Makosza, M. *J. Am. Chem. Soc.* **1979**, *101*, 2355.
58. Russell, G. A.; Hershberger, J.; Owens, K. *J. Am. Chem. Soc.* **1979**, *101*, 1312.
59. Barker, S. D.; Norris, R. K. *Tetrahedron Lett.* **1979**, 973.
60. March, J. "Advanced Organic Chemistry," 2nd ed.; McGraw-Hill Kogakusa: Tokyo, 1977; p. 317.
61. Bunnett, J. F.; Creary, X.; Sundberg, J. E. *J. Org. Chem.* **1976**, *41*, 1707.
62. Aguiar, A. M.; Archibald, T. G. *J. Org. Chem.* **1967**, *32*, 2627.
63. Aguiar, A. M.; Mague, J. T.; Aguiar, H. J.; Archibald, T. G.; Prejean, G. *J. Org. Chem.* **1968**, *33*, 1681.
64. Aguiar, A. M.; Daigle, D. J. *J. Am. Chem. Soc.* **1964**, *86*, 2299.
65. Aguiar, A. M.; Daigle, D. J. *J. Org. Chem.* **1965**, *30*, 2826, 3527.
66. Bilevich, K. A.; Okhlobystin, O. Y. *Russ. Chem. Rev. Eng. Trans.* **1968**, *37*, 954.
67. Perrin, C. L. *J. Am. Chem. Soc.* **1977**, *99*, 5516.
68. Shein, S. M.; Bryukhovetskaya, L. V.; Starichenko, V. F.; Iranova, T. M. *Org. React. Tartu* **1969**, *6*, 1087.
69. Russell, G. A.; Janzen, E. G. *J. Am. Chem. Soc.* **1962**, *84*, 4153.
70. Russell, G. A.; Janzen, E. G.; Strom, E. T. *J. Am. Chem. Soc.* **1964**, *86*, 1807.
71. Fendler, E. J.; Fendler, J. H.; Arthur, N. L.; Griffin, C. E. *J. Org. Chem.* **1972**, *37*, 812.
72. Relles, H. M.; Johnson, D. S.; Manello, J. S. *J. Am. Chem. Soc.* **1977**, *99*, 6677.
73. Bernasconi, C. F. *Chimia* **1980**, *34*, 1.
74. Abe, T.; Ikegami, Y. *Bull. Chem. Soc. Jpn.* **1976**, *49*, 3227.
75. *Ibid.*, **1978**, *51*, 196.
76. Levonowich, P. F.; Tannenbaum, H. P.; Dougherty, R. C. *J. Chem. Soc., Chem. Commun.* **1975**, 597.
77. Frimer, A.; Rosenthal, I. *Tetrahedron Lett.* **1976**, 2809.
78. Yamaguchi, T.; van der Plas, H. C. *Recl. Trav. Chim. Pays-Bas* **1977**, *96*, 89.
79. van Tamelen, E. E.; Brauman, J. I.; Ellis, L. E. *J. Am. Chem. Soc.* **1965**, *87*, 4964.
80. van Tamelen, E. E.; Brauman, J. I.; Ellis, L. E. *J. Am. Chem. Soc.* **1971**, *93*, 6141.
81. Barltrop, J. A.; Bradbury, D. *J. Am. Chem. Soc.* **1973**, *95*, 5085.
82. Tsujimoto, K.; Tasaka, S.; Ohashi, M. *J. Chem. Soc., Chem. Commun.* **1975**, 758.

Appendix: Reactions in Liquid Ammonia

Most aromatic $S_{RN}1$ reactions have been conducted in liquid ammonia as solvent. This is a very convenient medium for the reactions because it is cheap, easy to purify, and easy to eliminate once the reaction is over. However, it has a low boiling point ($-33°C$), therefore some special techniques have to be used.

The inconvenience of working at this low temperature may not be as serious as it first appears. We offer here some experimental details that may be helpful to those interested in this type of work.

The basic equipment is illustrated in Figure 1. Two one-liter three-necked round bottom flasks are used (A and B), each equipped with a Dewar-type condenser (C). L_1–L_5 are stopcocks and all the connections are made of latex tubing.

The apparatus is dried prior to the introduction of the solvent into the distillation flask.

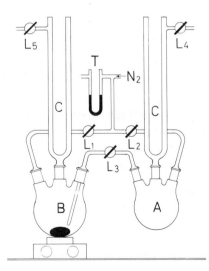

Figure 1. Schematic of the equipment used for reactions in liquid ammonia.

0065-7719/83/0178-0283$06.00/1
© 1983 American Chemical Society

Drying

The system is connected through L_5 to a vacuum pump and evacuated (L_1, L_2, and L_4 closed while L_3 and L_5 remain open). At the same time the glass is flamed with a Bunsen type of burner. L_5 is then closed and nitrogen is let in by carefully opening L_1 (or L_2). Once the system is full of nitrogen, as indicated by the bubbling of the Hg contained in tube T, L_1 and L_2 are closed and L_5 opened to repeat the process two or three times.

Distillation of the Ammonia

Once the drying is complete and the whole system is full of nitrogen, L_1, L_3, and L_5 are closed and L_2 and L_4 are opened. The left hand arm of Flask A is disconnected and liquid ammonia from an inverted commercial tank is introduced into A to half its total volume. Small pieces of sodium metal are added to the ammonia until the blue color persists, and A and B are interconnected again. L_4 is closed, L_3 and L_5 are opened and L_5 is connected to a mercury trap of the type of T. Flask B is submerged in a cooling system, and the ammonia is allowed to evaporate from A and condense in B. The distillation may be helped with a slow stream of warm air from a blower gun directed at A.

Once the required amount of solvent has distilled into B, L_1 and L_4 are opened, and L_2 and L_3 closed. The reagents are introduced from the right hand arm of Flask B, and the distillation tube is replaced by a glass stopper. The nitrogen is kept flowing all the time.

Preparation of the Nucleophile

In some cases the nucleophiles are added directly, but in most cases they are formed in situ either from the conjugate acid by an acid–base reaction with a strong base, or by reaction with an alkali metal, usually sodium or potassium.

The most frequently used bases are potassium or sodium amide (lithium amide is quite insoluble in liquid ammonia), or potassium or sodium t-butoxide.

Amide ion is formed by the reaction of the appropriate metal dissolved in ammonia, catalyzed by ferric nitrate or platinum black. When a weaker base is required, t-butoxide is used, and is added directly to the ammonia. t-Butoxide ion can also be prepared from the reaction of t-butyl alcohol and the metal, catalyzed by ferric nitrate (caution: this is a violent reaction).

Occasionally the nucleophile is formed from the reaction of an appropriate precursor and an alkali metal. For instance potassium diethyl phosphite can be formed by the reaction of diethyl phosphonate and potassium metal in ammonia (Reaction 1).

$$(EtO)_2P(O)H + K \xrightarrow{NH_3} (EtO)_2PO^- + \frac{1}{2} H_2 + K^+ \qquad (1)$$

In this case the required amount of metal is dissolved in ammonia and diethyl phosphonate is added dropwise until the blue color disappears. Thus the metal serves as an indicator.

The reactions of several substrates with solvated electrons from dissolution of alkali metals may also be used, as in Reactions 2–4.

$$ArZ-ZAr + 2\ e^- \longrightarrow 2\ ArZ^- \qquad (2)$$

$$ArZAr + 2\ e^- + NH_3 \longrightarrow ArZ^- + NH_2^- + ArH \qquad (3)$$

$$Ar_3Y + 2\ e^- + NH_3 \longrightarrow Ar_2Y^- + NH_2^- + ArH \qquad (4)$$

Z = S, Se, Te
Y = P, As, Sb, Bi

The preparation of nucleophiles following Reactions 2–4 is more convenient than their formation from the corresponding conjugate acids because these acids are unstable, whereas the substrates used in Reactions 2–4 are stable and easy to handle.

The precursors are dissolved in ammonia and small pieces of the alkali metal are added until the blue color of the solution persists. In some cases the nucleophile may react further with any metal present in excess. For example, when Z is Te or Y is Bi care must be taken to add the stoichiometric amounts of metal to avoid destruction of the nucleophile. If the metal has been added in excess, a small amount of the precursor is added to eliminate the excess of alkali metal.

In the case of Reaction 3 or 4, an equivalent of an acid (usually t-butyl alcohol) is added to neutralize the amide ion formed.

Photostimulated Reactions

There are several commercial photochemical reactors, but the one we used was built in our laboratories. It is made of aluminum, and has four high pressure water-refrigerated UV-lamps, that can be lighted independently. When the reactions are carried out in boiling liquid ammonia, the external walls of the reaction flask need to be periodically washed with acetone or alcohol to prevent frosting.

Cooling System

The condenser wells are filled up with a cooling mixture of dry ice and 2-propanol, the temperature being maintained by adding pieces of dry ice. The same procedure is used to refrigerate (externally) the reaction flask while distilling the ammonia. Liquid nitrogen can also be used, but in this case, care must be taken to avoid a sudden decrease in temperature, which can provoke a decrease in pressure inside the system and draw air back into it.

Isolation of Products

To take samples before the reaction is over, a J-shaped tube, shown in Figure 2, is introduced into the reaction flask. A sample of the solution is collected and is carefully discharged into water. The water is then extracted with an organic solvent, usually diethyl ether, and the solution is analyzed by GC. Unfortunately this procedure is not very convenient for quantitative analysis because some material is usually lost during the handling of the sample, but it is a useful way to follow the reaction progress.

After the reaction is finished, about 10 mL of water, or 1 g of an ammonium salt is added to quench the reaction and the ammonia is allowed to evaporate through the T tube connected to L_5, keeping the stream of nitrogen flowing if the products are air sensitive, or directly opening the flask to the air if they are not.

When the products are volatile (say benzene, for example), it is convenient to add diethyl ether (about 100 mL) to the ammonia before

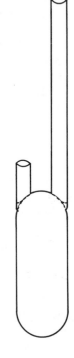

Figure 2. J-shaped tube used for taking samples before completion of the reaction.

starting the evaporation of the solvent, otherwise some material is lost. When the products are to be analyzed quantitatively by GC, it is convenient to add an internal standard together with the diethyl ether. It is not advisable to accelerate the evaporation by means of external heating because more material will be lost.

INDEX

INDEX

A

Acetamide anion 63
Acetone enolate anion 13
  coupling with naphthyl radical . 153f
  coupling with phenyl radical ... 153f
  formation 19
  reaction with 3-bromothiophene 16
  reaction with 1-chloronaph-
    thalene 15
Acetone enolate ion
  entrainment 226
  reaction with phenyl diethyl
    phosphate ester 105
  selectivity toward 2-chloro-
    quinoline 233
  and substituted benzenes 103t
Acetonitrile, reaction rate with
  solvated electrons 201
Acetophenone
  fragmentation rate 208t
  reduction potential 208t
Acetophenone enolate 34
Acetophenone enolate ion 20
3-Acetyl-2-methyl indene
  synthesis 113
Aldehyde enolate ions 37
Aliphatic system reactions 258–73
  entrainment 268
Alkali metal catalyst, with heteroaromatic compounds 121
Alkali metal iodide, potential-
  sweep voltammogram 203
Alkali metal stimulation 13–17
  ketone enolate ions 18t
Alkali metals, dissolution in
  liquid ammonia 5
Alkali phenoxides, solubility in
  ammonia 80
Alkanethiolate ions 88–92, 218–22
  coupling with aryl radical 159f
  photostimulated reactions with
    haloaromatic compounds .. 91t
Alkoxides 79

Alkyl aryl sulfides, cleavage 104
α-Cyano carbanions
  photostimulated reactions 47–50
  stimulation with solvated
    electrons 43
α-Diketones 87
α,α-Diarylketone 17
Amide ions
  formation 284
  reaction with 3-bromothiophene 117
  reaction with 2-bromothiophene 119
  reaction with iodotrimethyl-
    benzenes 161
  reactions 59–62
Amines as nucleophiles 62
3-Aminothiophene, formation 61
5-Amino-1,2,4-trimethylbenzenes,
  formation 4
6-Amino-1,2,4 trimethylbenzenes,
  formation 4
Ammonia
  distillation 284
  liquid 173
    using in reactions 283–87
  reaction of benzene derivatives 101–17
  substrate reactivity 123
  suitability 127
Ammonium benzenethiolate 119
Aniline, formation 225
Anionic photoionization 172
m-Anisidine, preparation 59
Anisole 102
p-Anisyl phenyl telluride, carbon–
  tellurium bonds 212
p-Anisylpropenide, coupling 150
(p-Anisyl)propenide, phenylation . 12
Antibonding molecular orbitals,
  low lying 154
Antimony nucleophiles 76–77
  reaction with dichloroalkanes .. 76
Arenes, reduction potential 215f
Aromatic nucleophilic substitutions
  mechanisms 1–6
  mechanistic steps 6–11

Aromatic substrates
  with electron-withdrawing
    groups 2
  nitrogen as leaving group 1
Arsenic nucleophiles 72–76
Aryl alkyl sulfide radical anion,
  stability 221
Aryl bromides, halogen–metal
  exchange mechanism 56
Aryl carbanions, formation 1
Aryl cations, formation 1
Aryl diethyl phosphate esters,
  reaction with sodium
  naphthalene 105
Aryl diethyl phosphates, reaction
  with amide ion 60
Aryl halide fragmentation 134
Aryl halide reactions,
  ketone: alcohol ratios 166t
Aryl halides
  decomposition rates 180
  photostimulated reaction with
    N,N-disubstituted amide
    enolate ions 41t
  reaction with amide ion 60t
  reduction reaction 180
  unactivated 2
Aryl iodides
  fragmentation 169
  reaction with dialkyl phosphite
    ions 68
  reaction with potassium pina-
    colone enolate in DMSO .. 134t
  reactions with potassium pina-
    colone enolate, rate 133t
Aryl phenyl sulfides 80
Aryl phenyl telluride radical anion,
  energy changes 157f
Aryl radical reduction, leaving
  group dependence 240
Aryl radicals 143
  character 193
  coupling with alkanethiolate ... 159f
  coupling with ketone enolate
    anions 151–54
  coupling with a nucleophile ...192–94
  dimerization rate 242
  coupling with a nucleophile,
    energy changes 146f
  interaction with a nonconju-
    gated nucleophile 144, 145
  reaction with nucleophiles 234
  reaction with oxygen nucleo-
    philes 80
  reduction potential 177
  singly occupied molecular orbital 143
Aryl rings, scrambling 214f

Aryl substrate, photostimulated
  reaction with ketone enolate
  ions 23t–25t
Aryl sulfide ions 80–88
Aryl trapping agents 229
Arylalkane systems 264–68
Aryldiazonium salts, reduction ... 65
Aryldiazonium tetrafluoroborate,
  reaction with potassium iodide 162
Arylketones, with enolates 20
Aryllithium compounds, photo-
  chemistry 278
Aryne mechanism 59
4-Azaindole synthesis 111t
Azaoxindoles, synthesis 38
Azido group, reaction with aryl
  radicals 63

B

Benzene, formation 164
Benzene derivatives
  reaction in ammonia 101–17
  substrates with two leaving
    groups 113–17
Benzenediazonium ion, reaction
  with sodium dithionite 128
Benzenesulfinate ion reaction 127
Benzenethiolate ion
  reaction with 3-bromoiso-
    quinoline 129
  reaction with haloarenes 83t
  reaction with m-halobenzenes .. 81
  reaction with nitrobenzyl
    chloride · 260
Benzo(b)furan synthesis 112
Benzonitrile
  electrochemical studies 106
  formation 181
  fragmentation rate 208t
  reduction potential 208t
Benzophenone
  fragmentation rate 208t
  reduction potential 208t
Benzyne intermediates 1
Benzyne mechanism 1
β-hydrogen atom, abstraction
  competition 246
Bond energies, low carbon–
  nucleophile 212–16
4-Bromoacetophenone, electrochemical reduction 86, 87t
p-Bromoanisole
  photostimulated reaction 37
  reaction with diphenylstibide ion 76
Bromobenzene
  dark reaction in ammonia 132

Bromobenzene—*Continued*
photostimulated reaction with
   sodium azide 63
reaction with acetone enolate .. 231*f*
reaction with carbanions 14*t*
reaction with pinacolone enolate
   ion 131
reaction with potassium
   diphenylphosphinite 71
reaction with potassium pinacolone enolate in DMSO .. 134*t*
reactivity in DMSO 132
reactivity with pinacolone
   enolate in DMSO 133
Bromobenzene formation, phosphanion nucleophiles 72*t*
Bromobenzene and iodobenzene,
   competition experiments 34
4-Bromobenzophenone
cyclic voltammetry 42
electrochemical reduction 84
electrolysis 42
electrolysis in DMSO 223
Bromobiphenyl photochemistry ... 171
m-Bromoiodobenzene 68
reaction with diethyl phosphite . 200*f*
p-Bromoiodobenzene 93
3-Bromoisoquinoline, reaction with
   benzenethiolate ion 129
Bromomesitylene, reaction with
   tributylstannyllithium 55
1-Bromonaphthalene, reaction with
   sodium alkanethiolate 91
p-Bromonitrobenzene, cyclic
   voltammogram 137
Bromothiophene
photostimulated reaction 49
reaction with acetone enolate ion 16
reaction with amide ions 61, 117
reaction with cyanomethyl anion 119
reaction with ketone enolate
   anions 164
reaction with potassium amide . 62*t*
p-Bromotoluene, entrainment
   with *p*-iodotoluene 227
Bromotoluene, reaction with potassium diphenyl phosphide 67
t-Butoxide ion 79
t-Butyl acetate, photostimulated
   reactions 36
n-Butyl phenylphosphonite 70
t-Butyl-2-methylpropanoate, reaction with *p*-bromoanisole 244

C

Carbanionic nucleophiles 101

Carbanions, hydrocarbon-derived,
   coupling with phenyl
   radicals 149–50
Carbanions, reaction with bromobenzene 14*t*
Carbanions, reaction with halobenzenes 161
Carbanions, spontaneous electron
   transfer 163
Carbon–cyanide bond
   fragmentation 48
Carbon nucleophiles 11–49
conjugated hydrocarbons 11–13
ketone enolate anions 13–36
ring closure reactions 41–47
Carbon–nucleophile bond energies,
   low 212–16
Catalysts 183
Cephalotaxinone, synthesis 29
Chain propagation steps 7*f*, 187–238
competing reactions 205–25
Charge-transfer complex 172
Chemically induced dynamic
   nuclear polarization 67
Chloride ion, coupling with a
   phenyl radical 154*f*
Chlorobenzene
cyclic voltammetry 170
dark reaction 163
reaction with pyrrolyl anion ... 63
reduction 181, 183
Chlorobenzene radical,
   decomposition 154
4-Chlorobenzonitrile, electrochemical reduction 181, 184*f*
4-Chlorobenzophenone 102
arsine products 74
reaction with diphenyl arsenide
   ion 163
substitution products 213
N-(*p*-Chlorobenzyl)-*N*-methyl-
   acetamide enolate ion 40
m-Chloroiodobenzene 68
reaction with benzenethiolate .. 81
1-Chloronaphthalene
photostimulated reaction 48
reaction with acetamide anion .. 63
reaction with sodium alkanethiolate 91
2-Chloropyrimidine, reaction with
   acetone and pinacolone
   enolate ions 121
2-2-Chloroquinoline
cyclic voltammetry 197, 199*f*
reaction with acetone enolate
   ions 30
reaction with potassium diphenyl
   phosphide 67

2-2Chloroquinoline—*Continued*
   substitution products 213
2-Chlorothiophene 155
p-Chlorotoluenes, reaction with
   tributylstannyllithium 55
CIDNP—*See* Chemically induced
   dynamic nuclear polarization
Cine substitution 51
Comparative reactivity 233–36
Competing reactions 205–25
Conjugated hydrocarbon
   nucleophiles 11–13
Cooling system 285
Coulombic repulsion 143
Coupling position 149
   prediction 158
Cyanide ion 42
Cyanomethyl anion 119, 216
   photostimulated arylation 49*t*
   photostimulated reactions 47
   reactivity 47
Cyclic unsaturated ketones,
   reduction 50
Cyclic voltammetry
   2-chloroquinoline 30
   quinoline, reduction 33
Cyclization products, relative
   yields 30
Cycloalkanones, dependence on
   carbon atoms in ketone 243

D

Dark reactions of substrates with
   one leaving group 101–13
Decomposition of a radical
   anion 188–92
Dialkyl phosphite ions 68
Dialkylarsenide ions 72
Diarsines 72
Diaryl sulfides
   cleavage 104
   preparation 80
Diarylarsenide ions 72
Diazonium salts 1
   decomposition 63
o-Dibromobenzene, photostimulated reaction 27, 28
m-Dibromobenzene, reaction with
   potassium diethyl phosphite . 129
2,6-Dibromopyridine, reaction with
   pinacolone enolate ion 121
α-Dicarbonyl compounds, enolate
   anions 21

Dichloroalkanes, reaction with
   stibides 76
2,6-Dichloropyridine, reaction with
   pinacolone enolate ion 121
Diethyl arylphosphonate radical
   anion, electron transfer rates . 210
Diethyl phosphite ion
   as nucleophile 198
   electrochemical reaction 184*f*
   entrainment 225
   reaction with *m*-bromoiodo-
     benzene 200*f*
Dihaloarenes 93
   reaction with phenyl telluride
     ion 96*t*
   reactions with phenyl selenide . 94*t*
Dihalobenzenes
   chlorine leaving group 210
   reactions with arsenide ions ... 72
p-Dihalobenzenes, reaction with
   ketone enolate ions 28
m-Diiodobenzene 68
Dimer formation 241
1,2-Dimethoxy-4-benzene,
   anodic oxidation 257
3,3-Dimethyl-1-iodoindene, reaction with acetone enolate ion . 274
Dimethylstibide ion 76
2,4-Dimethyl-3-pentanone, reaction with iodobenzene 244
Diorganophosphides, reaction with
   alkyl halides 67
Dioxygen 229
   reaction with phenyl radicals .. 232
Diphenyl arsenide ion
   aryl radical coupling 213
   photostimulated reactions with
     haloarenes 75*t*
   reaction with 4-chlorobenzo-
     phenone 102
   reaction with *p*-chlorotoluene .. 102
Diphenyl ether, decomposition ... 104
Diphenyl phosphide ion 67–68
   bond dissociation energy 214
   reaction with *p*-bromotoluene .. 102
   reaction with iodotoluenes 102
   substitution reactions 216
Diphenyl selenide
   decomposition 104
   formation 95
Diphenyl stibide ion 76, 102
   scrambling of aryl rings 213
Diphenyl sulfide
   decomposition 104
   formation 88

Diphenyl telluride, cleavage 104
Diphenylated products 11
1,2-Diphenylethane, formation ..44, 176
Diphenyliodonium ion 101
Dipolar aprotic solvents 192
Direct electron transfer 162
   from amide ions to iodobenzenes 163
Disodium selenide, formation 95
Disodium telluride
   formation 99
   reaction with iodobenzene 99
Disproportionation 245
Disproportionation zone 181
Distillation of ammonia 284
N,N-Disubstituted amide 79
N,N-Disubstituted amide enolate
   ions 37–39
Disubstitution products, from
   dihalobenzenes 207
Dithienylation products 16

E

ECE zone 180
Electrochemical reduction 86
Electrochemically generated
   reactions 30–34
Electrochemically induced reactions, hypothetical zones 180f
Electrode stimulated
   reactions 135–40, 140t, 176–85
   competition reactions 178
Electron ejection to solvent 198–205
Electron stimulated reactions,
   solvated 102–6
Electron transfer
   direct 162
      from amide ions to iodo-
        benzenes 163
   miscellaneous reactions 276–80
   photoassisted 172
   spontaneous 163
   to excited states 172
   to substrate 194–98
Electron transfer rates, solvent
   effect 212
Electron transfer reactions
   in chain propagation 194–205
   mechanism 194
   rate constant 195
Electron transfer steps 8
Electrons, photoejection 171
Electron-withdrawing groups,
   mechanism of nucleophilic
   substitution 1

Elimination, radical chain,
   unimolecular 268
Enolate anion, with a β-hydrogen
   atom 246
Enolates from arylketones 20
Entrainment 225–27
   requirements 227
Entrainment effect 226
Ester enolate ions 36, 37, 79
Ester enolates, β-hydrogens 37
Ethanethiolate ion 88
Excited states, electron transfer ... 172

F

First-order perturbation 144
Fluorenone
   fragmentation rate 208t
   reduction potential 208t
Fluorobenzene radical anion,
   decomposition 155
Fluorohalo compounds 208
m-Fluoroiodobenzene 68
   reaction with acetone enolate
      ions 29
   reaction with benzenethiolate .. 81
p-Fluorotoluenes, reaction with
   tributylstannyllithium 55
Fragmentation 158
   aryl halides 134
   of radical anions 216–24
   solvent effect 212
Frontier molecular orbitals 143

G

Germanium nucleophiles 51
Group IVA nucleophiles 11–58
Group VA nucleophiles 59–79
Group VIA nucleophiles 79–99

H

Haloacetophenones, electrochemical reduction 189
Haloanisoles, isomeric,
   photocyanation 252
o-Haloanilines 108
   reaction with amide ions 59
p-Haloanisoles 74
Haloarene radical anions
   decomposition rate 207
   reduction potential 207
Haloarenes, electrochemical
   reduction 87t

Haloarenes
    fragmentation rate 208t
    reaction with phenyl telluride
      ion 96t
    reaction with benzenethiolate
      ion 83t
    reactions with phenyl selenide . 94t
    reduction potential 208t
Haloaromatic compounds, photo-
    stimulated reaction with
      thiolate ions 219t
Haloaryl radical 27
Halobenzenes 106
    photostimulated reaction 65
    photostimulated reactions with
      cyanomethyl 195t
    reaction with arsenide ions 72
    reaction with phenoxide ion 80
    reduction reaction 181
    solvated electron effects 104
Halobenzophenone stability,
    solvent and temperature
      effects 191t
Halogen mobility
    pyridine family 126t
    substrates with two leaving
      groups 114
Haloiodobenzenes 116, 117
2-Halothiophene anions, reaction
    with ketone enolate ions 164
2-Halothiophenes, reactivity 165
p-Halotoluenes, reaction with
    tributylstannyllithium 56t
5-Halouracils 155
Hetaryl substrates, photostimulated
    reaction with ketone enolate
      ions 23t-25t
Heterocyclic compounds, reactions
    in ammonia 117-23
Homolytic bond dissociation 170-71
Hydrocarbon-derived carbanions,
    coupling with phenyl
      radicals 149-50
Hydrocarbons, polycyclic, reac-
    tions in ammonia 117
Hydrogen atom abstraction 79, 126
    formation of reactive inter-
      mediates 224-25
    rates 179
    as termination step 242

I

Indole synthesis 108, 110t
    from aldehyde enolate ions 111
Inhibition 227-33

Initiation step 7f, 161-85
    photostimulated reactions 170-76
    spontaneous reactions 161-63
Intramolecular cyclization 108
2-Iodo-1,3-dimethylbenzene 60
Iodine 123
Iodoacetophenones, electro-
    chemical reduction 189
p-Iodoaniline, photostimulated
    reaction 37
p-Iodoanisole 74
    photostimulated reaction 97
Iodoarenes, photostimulated reac-
    tions with potassium dialkyl
      phosphites 70t
Iodobenzene
    absorption 174
    charge-transfer complex 174
    decomposition 155
    decomposition 155
    photostimulated reaction with
      benzenethiolate 88
    reaction with acetone enolate
      ions 34
    reaction with pinacolone
      enolate ion 233
    reaction with pinacolone
      enolate, nitrobenzene effects 235f
    reaction with potassium diethyl
      phosphite 130t
    reaction with potassium O,O-
      diethyl thiophosphite 71
    reactivity with pinacolone
      enolate in DMSO 133
    spectra 175f
Iodobenzene and bromobenzene,
    competition experiments 34
Iodobenzene formation, phos-
    phanion nucleophiles 72t
2-Iodomesitylene 34
1-Iodonaphthalene, photostimulated
    reaction with phenylmethane-
      thiolate 175
p-Iodonitrobenzene 136
    cyclic voltammetric behavior ... 135
    cyclic voltammogram 42, 138f
p-Iodophenyltrimethylammonium
    ion, electron transfer
      fragmentation 211
2-Iodothiophene, reactions with
    ammonium benzenethiolate .. 119
Iodotoluenes, reaction with
    diphenyl phosphide 102
Iodotoluenes, reaction with lithium
    diphenylphosphide 67
p-Iodotoluene, reaction with
    potassium diphenyl phosphide 67

Iodotrimethylbenzenes, reaction
with amide ion 161
Isolation of products 286
Isomeric haloanisoles, photocyanation 252

K

Ketone 79
Ketone enolate anion
reaction with iodobenzene 101
coupling with aryl radicals 151–54
Ketone enolate anion
nucleophiles 13–36
alkali metal stimulation 13–17
photostimulated reactions 17–30
reaction with disubstituted
substrates 26–29
ring closure reactions 29, 30
Ketone enolate ions
alkali metal stimulation 18t
reaction with aryl halides,
features 107
reaction with halobenzenes 106
Ketone enolates 13
Ketones, with beta-hydrogens 20
Kim–Bunnett mechanism 9

L

Leaving group expulsion 207
Leaving groups 123
Liquid ammonia 173
reactions 283–87
apparatus 283f
Lithium diphenylphosphide, reaction with iodotoluenes 67
Low energy antibonding molecular
orbital 49
Low lying antibonding molecular
orbitals 154

M

Marcus' theory 194
Mechanistic characteristics 101
Metal iodide, alkali, potential-
sweep voltammogram 203
Methanethiolate, hydrogen atom
abstraction 266
Methanethiolate, reaction with
nitrobenzyl chloride 260
Methoxide ions as catalysts 81
Methoxy 250
Methoxyacetone enolate ion,
reaction with bromobenzene .. 222

Methyl chloride radical anion,
dissociation 156
2-Methylpiperidine 62
2-Methyl-2-nitrosopropene 229
MO—See Molecular orbitals
Molecular collisions, forces 143
Molecular orbital
considerations 16, 143–59
Molecular orbitals
antibonding, low lying 154
energy levels 146f
pi 145
sigma 145

N

Naphthalene 183
Naphthalene derivatives, rate
constants 256t
1-Naphthyl phenyl telluride anions,
dissociation 97
Naphthyl radical, coupling with
acetone enolate anion 153f
1-Naphthyl radical reactions,
rate constants 135t
1-Naphthyl radicals, coupling rates 194
1-Naphthyl radicals vs. phenyl
radicals 221
Nitrile α-hydrogens 43
Nitrite ion
as nucleophile 65, 66
photostimulated reaction with
halobenzenes 65
Nitroalkane anions 258, 269
Nitroanisoles, mechanistic behavior 257
Nitroaromatic compounds, electron
transfer reactions 276
Nitrobenzene
fragmentation rate 208t
reduction potential 208t
Nitrobenzene formation 62
in tetraethylammonium iodide .. 136
p-Nitrobenzonitrile, cyclic
voltammograms 138f
p-Nitrobenzonitrile radical anion . 137
4-Nitrobenzyl compounds, decom-
position rate 217, 217t
Nitrobenzyl halide radical, decom-
position rates 218t
Nitrobenzyl halides, fragmentation
rate 217
p-Nitrobenzyl system 258–63
Nitrocumene 265
p-Nitrocumylazide, formation 268
Nitrogen, as leaving group 1
Nitrogen nucleophiles 59–67
Nitrohalobenzenes, reactivity 188

Nucleophiles
  coupled with aryl radicals 192–94
Nucleophiles
  from Group IVA 11–58
  from Group VA elements 59–79
  from Group VIA elements 79–99
  ketone enolate ions 125
  ketone enolates 13–36
  with low carbon–nucleophile
    bond energies 212–16
  superoxide ion 276, 278
Nucleophile preparation 284

O

Oxindole isolation 38
Oxindoles, synthesis 38
Oxygen nucleophiles 79–80, 158

P

Pentadienide anion 149
Pentadienide ion
  coupling with phenyl radical ... 150
  photostimulation 12
Perturbational molecular orbital
    method 143–49
Phenanthrene 183
Phenoxide ion 79–80
Phenyl-azo-triphenylmethane,
    decomposition 65
Phenyl aryl sulfide radical anion,
    electron-transfer rates 210
Phenyl diethyl phosphate ester,
    reaction iwth acetone enolate
    ion 105
Phenyl halides, treatment with
    potassium metal 165
Phenyl radical(s)
  coupling with acetone enolate
    anion 153f
  coupling with a chloride ion ... 154f
  coupling with hydrocarbon-
    derived carbanions 149–50
  coupling with methyl anion 145
  coupling with a methyl anion,
    energy changes 148f
  coupling with a pentadienide
    ion 150, 151f
  preparation 65
  reaction with benzenethiolate
    ions 219
Phenyl selenide ion 92
  reaction with bromoiodo-
    benzenes 117
  reaction with haloarenes 94t

Phenyl selenide ion—*Continued*
  substrate reactivity 124
Phenyl telluride ion
  reaction with bromobenzene ... 95
  reaction with haloarenes 96t
  scrambling of aryl rings 157
Phenylacetone 102
  formation 222
  pi molecular orbitals 152
Phenylacetonitrile anion, com-
    petitive reactions 216
Phneylacetonitrile radical anion
  adducts with substrates 201
  products 200
Phenylated products 11
Phenyllithium, photochemistry ... 278
Phenylmercury halides, coupling . 272
Phenylmethanethiolate ion 175, 220
Phneylpentadienes, isomer
    formation 12
Phenylpentenes 12
Phenylsodium, formation 53
4-Phenylthiobenzophenone,
    formation 137
4-Phenylthioisoquinoline,
    preparation 81
Phenyltrimethylsilane, reduction
    in ammonia 102
1-Phenyl-2-propanol 102
Phosphanion nucleophiles
  bromobenzene formation 72t
  iodobenzene formation 72t
Phosphanions, photostimulated
    reaction with phenyl halide .. 71t
Phosphonites 70
Phosphorus nucleophiles 67–72
Photoassisted electron transfer ... 172
Photochemical reactor 285
Photocyanation of isomeric
    haloanisoles 252
Photodissociation, mechanism 171
Photoejection of electrons 171
Photonucleophilic aromatic
    substitution 249–58
  alkoxy substituent activation ... 250
  features 249
  leaving groups 249
  nitro group activation 249
  nucleophiles 249
  orientation rules 249–52
  photoionization 254
  reaction mechanism 252–58
  stabilization during product
    formation 251
  triplet state kinetics 255f
Photostimulated reactions .17–30, 170–76

Photostimulated reactions—*Continued*
ketone enolate ions 23*t*-25*t*
substrates with one leaving
   group 106–8
unsymmetrical dialkyl ketones .23–26
Pi molecular orbitals 145
1-Picolyl-anion 39–41
Pinacolone enolate ion 26, 34
PMO—*See* Perturbational molecular orbital method
Polycyclic hydrocarbons
photostimulated reactions in
   ammonia 120*t*
reactions in ammonia 117
Potassium anilide 61
Potassium diethyl phosphite 119
Potassium diphenylphosphide,
reaction with *p*-bromotoluene 226*f*
Potassium diphenylphosphinite ... 71
Potassium ethyl xanthate 92
Potassium iodide 34
Potassium *O,O*-diethyl thiophosphite 71
Potassium *O,O*-diethylthiophosphite 92
Potassium thiocyanate 92
Preparation of nucleophile, in
reactions 284
Pretermination step 20, 216
Product isolation 286
Propagation cycle 187
Propagation steps 8
electron transfer 220
Propene 11
Protic solvents 127
Pyrazines 35
Pyridazines 35
Pyridine derivatives 38
Pyridine family, halogen
mobility 125, 126*t*
Pyrimidines 35
Pyrrolyl anion 63

Q

Quantum yield determinations
in ammonia 127
2-Quinolyl radical, reaction rate
with nucleophiles 195*t*
2-Quinolyl radicals, coupling rates 194

R

Radical anion
decomposition 188–92

Radical anion—*Continued*
dissociation 8
fragmentation 216–24
fragmentation to reactive
intermediates 205–16
miscellaneous fragmentation ..222–24
Radical anion fragmentation, vs.
free electron density 190*t*
Radical anion intermediates, molecular orbital considerations ... 16
Radical anion stability, solvent
effect 191
Radical character, evidence 227
Radical nucleophilic substitution
mechanism 5
Rate constants 178
Reaction with disubstituted
substrates 26–29
Reaction rates, effect of spin
densities 190
Reactions in liquid ammonia 283–87
apparatus 283*f*
Reactions stimulated by
electrodes 176–85
Reactions stimulated by solvated
electrons 163–70
Reactive intermediates 205
formation by hydrogen atom
abstraction 224–25
Redox-catalyzed processes,
efficiency 184
Reduction 239
Repulsion 143
Ring closure reactions 29, 30, 41–47
benzo(b)furan synthesis 112
substrates with one leaving
group 108–13

S

Scrambling of aryl rings 214*f*
Second-order perturbation 144
Selenium nucleophiles 92–95
Sigma-complexes 4
Sigma molecular orbitals 145
Silicon nucleophiles 50
Sodium anthracene 105
Sodium azide, photostimulated
reaction with bromobenzene . 63
Sodium borohydride, photoreduction 278–80
Sodium dithionite 65, 92
Soft nucleophiles 193
Solvated electron reactions 163–70
mixing 168
Solvated electron stimulated
reactions 102–6

Solvent, electron ejection from .198–205
Solvents
  protic 127
  requirements 126
Spin densities 190
Spontaneous electron transfer 163
Spontaneous reactions 34–36
  initiation 161–63
  proposed mechanism 161
Substituted benzenes and acetone
  enolate ion 103t
meta-Substituted fluorobenzene,
  reduction in ammonia 102
Substitution radical nucleophilic,
  unimolecular 5, 6
Substitution reactions 8
Substrate, homolytic bond
  dissociation 170
Substrate reactivity, phenyl
  selenide ion 124
Substrates 101–42
  comparative reactivity in
    ammonia 123–26
  with one leaving group 101–13
  with two leaving groups 113–17
    halogen mobility 114
  relative reactivity 123
Sulfide ions 92
Sulfur nucleophiles 80–92
Superoxide ion as nucleophile ..276, 278

T

Tellurium nucleophiles 95–99
Termination, by solvated electrons 239
Termination step 7f, 20, 239–47
  carbon–cyanide bond
    fragmentation 48
  electrochemically induced
    reaction 240
  fragmentation 220
  hydrogen abstraction 126
  independent of the initiation
    step 241–46
  possibilities 241
  that depend on the initiation ..239–41
1,2,3,4-Tetrahydrocarbazole,
  preparation 109
Tetrahydrofuran 130

N,N,N′,N′-Tetramethylphos-
  phonamide 71
Tetraphenylhydrazine 229
Tetraphenylsilane 102
THF—See Tetrahydrofuran
Thiolate ions, photostimulated
  reaction with haloaromatic
  compounds 219t
Tin nucleophiles 51–56
Toluene radical anion,
  decomposition 155
Tolyldiphenylphosphines,
  formation 102
Tributylstannyllithium
  reaction with bromomesitylene . 55
  reaction with halobenzenes 55
  reaction with p-halotoluenes ... 56t
  reaction with haloarenes 51
  reaction with halotoluenes 52t
Trimethylsilyl anion 50, 130
Trimethylstannylsodium, reaction
  with halobenzenes 52t
Triorganosilyl anions, reaction
  with haloarenes 50
Triorganostannyl ions, reaction
  with halobenzenes 51
Triphenylbismutine 106
Triphenylgermyllithium, reaction
  with haloarenes 51
Triphenylphosphine, cyclic
  voltammetry 170
Triphenylsulfonium ion, solvated
  electron effects 105
Triplet excitation energies 171
Tungsten bulb 17
Tunneling, electron advance 169

U

Unactivated aryl halides 2
Unsaturated ketones, cyclic,
  reduction 50
Unsymmetrical dialkyl ketones,
  photostimulated reactions ...23–26

V

Vinylic system reactions 273–75
Vinylmercury halides, coupling ... 272

Copy Editor and Indexer: Deborah Corson
Production Editor: Anne G. Bigler
Jacket Artist: Kathleen Schaner

Typesetting: Circle Graphics, Washington, D.C.,
and Service Composition Company, Baltimore, MD
Printing: Maple Press Company, York, PA